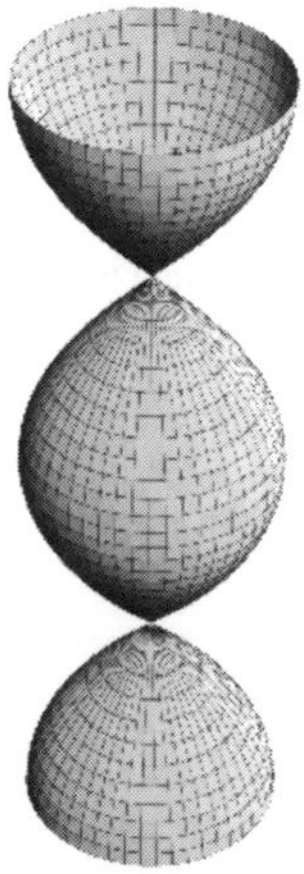

Topology in Ordered Phases

Proceedings of the 1st International Symposium on TOP 2005

Sapporo, Japan 7 – 10 March 2005

With CD-ROM

Editors

Satoshi Tanda
Hokkaido University, Japan

Toyoki Matsuyama
Nara University of Education, Japan

Migaku Oda
Hokkaido University, Japan

Yasuhiro Asano
Hokkaido University, Japan

Kousuke Yakubo
Hokkaido University, Japan

World Scientific

NEW JERSEY · LONDON · SINGAPORE · BEIJING · SHANGHAI · HONG KONG · TAIPEI · CHENNAI

Published by

World Scientific Publishing Co. Pte. Ltd.
5 Toh Tuck Link, Singapore 596224
USA office: 27 Warren Street, Suite 401-402, Hackensack, NJ 07601
UK office: 57 Shelton Street, Covent Garden, London WC2H 9HE

British Library Cataloguing-in-Publication Data
A catalogue record for this book is available from the British Library.

TOPOLOGY IN ORDERED PHASES
(With CD-ROM)
Proceedings of the 1st International Symposium on TOP2005

ISBN 981-270-006-4

Printed by Mainland Press Pte Ltd

PREFACE

This issue contains Proceedings of the 1st International Symposium on Topology in Ordered Phases (TOP2005). The symposium was held from 5 to 7 March, 2005 at Sapporo Grand Hotel, Japan. It was sponsored by "The 21st century center of the excellence program at Hokkaido University, Topological Science and Technology".

TOP2005 was open to experiments and theories having connection with Topology, including wide scientific fields such as materials science, superconductivity, charge density waves, superfluidity, optics, and field theory. The structure of TOP2005 was designed to stimulate exchange of ideas and international cooperation through a timely discussion of recent results among scientists with different research background.

A total of 64 papers were presented at the symposium, including 19 invited talks. The number of participants was 102. The proceedings contain 59 papers out of those presentations. We would like to thank all reviewers for their careful reading of submitted papers. It is our hope that the proceedings will be useful for many researchers in topological science and technology.

Finally we would like to thank all participants for their fruitful and exciting discussion throughout the symposium.

December 2005

S. Tanda

Satoshi Tanda (Chairman, Editor of the proceedings)

T. Matsuyama

Toyoki Matsuyama (Co-chairman, Editor of the proceedings)

CONTENTS

V. Topology in Quantum Phenomena

TOP 2005 Symposium

Sponsor

Hokkaido University, The 21st Century COE Program

International Advisory Committee

(Alphabetical order)

M. V. Berry	(Univ. Bristol, UK)
A. Cleland	(UC Santa Barbara, USA)
R. Jackiw	(Massachusetts Inst. Tech., USA)
K. Maki	(Univ. Southern California, USA)
J. Mesot	(ETH Zurich, Switzerland)
D. Osheroff	(Stanford Univ. USA)
V. A. Osipov	(Bogoliubov Lab., Russia)
M. Paalanen	(Helsinki Univ. Tech., Finland)
A. Tonomura	(Hitachi Ltd., Japan)
G. Volovik	(Helsinki Univ. Tech., Finland)

Organizing Committee

Chairperson:

S. Tanda	(Hokkaido Univ., Japan)

Vice-Chairperson:

T. Matsuyama	(Nara Univ. Education, Japan)

M. Oda	(Hokkaido Univ., Japan)
K. Yakubo	(Hokkaido Univ., Japan)
K. Nemoto	(Hokkaido Univ., Japan)
N. Nishiguchi	(Hokkaido Univ., Japan)
Y. Asano	(Hokkaido Univ., Japan)
K. Inagaki	(Hokkaido Univ., Japan)
H. Amitsuka	(Hokkaido Univ., Japan)

Invited Speakers
(Alphabetical order)

M. Berry	(Univ. Bristol, UK)
P. Hakonen	(Helsinki Univ. Tech., Finland)
M. Hayashi	(Tohoku Univ., Japan)
R. Jackiw	(Massachusetts Inst. Tech., USA)
H. Kuratsuji	(Ritsumeikan Univ., Japan)
K. Machida	(Okayama Univ., Japan)
A. Maeda	(Univ. Tokyo, Japan)
K. Maki	(Univ. Southern California, USA)
P. Monceau	(CNRS, France)
J. Mesot	(ETH Zurich, Switzerland)
D. Mihailovic	(Jozef Stefan Inst., Slovenia)
Y. Nogami	(Okayama Univ., Japan)
M. Nakahara	(Kinki Univ., Japan)
M. Padgett	(Glasgow Univ., UK)
K. Shirahama	(Keio Univ., Japan)
S. Tanimura	(Osaka City Univ., Japan)
G. Tatara	(Osaka Univ., Japan)
Z. Tesanovic	(Johns Hopkins Univ., USA)
A. Tonomura	(Hitachi Ltd, Japan)

Number of Presentations: 64
Number of Invited Talks: 19
Number of Participants: 102
Number of participating countries: 7

TOP 2005
The 1st International Symposium on Topological Science and Technology
TOPOLOGY IN ORDERED PHASES

I

Topology as Universal Concept

OPTICAL VORTICULTURE

M. V. BERRY
H H Wills Physics Laboratory, Bristol University,
Tyndall Avenue, Brisrol BS8 1TK, UK

Lines of topological singularity in the phase and polarization of light are being intensively studied now,[1] motivated in part by a theoretical paper published thirty years ago.[2] However, the subject has a very long prehistory, that is not well known.

In puzzling over Grimaldi's observations of edge diffraction in the 1660s, Isaac Newton narrowly missed discovering phase singularities in light. The true discovery of phase singularities was made by William Whewell[3] in 1833, not in light but in the pattern of ocean tides. The first polarization singularity was observed (but not understood) by Arago in 1817, in the pattern of polarization of the blue sky. A different polarization singularity was predicted by Hamilton in the 1830s, in the optics of transparent biaxial crystals (this was also the first 'conical intersection' in physics).

After reviewing this history, the general structure of the singularities, as we understand them today, will be presented. Phase singularities have several aspects:[4,5] as vortices, around which the current (lines of the Poynting vector) circulates; as lines on which the phase of the light wave is undefined; as nodal lines, where the light intensity is zero; and as dislocations,[2] where the wavefronts possess singularities closely analogous to the edge and screw dislocations of crystal physics. Polarization singularities are lines[5,6] of two types: C lines, where the polarization is purely circular, and L lines, where the polarization is purely linear.

Then, three modern applications of optical singularities will be described. The first[7] is the pattern of optical vortices behind a spiral phase plate, which is a device, commonly used to study phase singularities, that introduces a phase step into a light beam. The intricate dance of the vortices as the height of the step is varied (especially complicated near half-integer multiples of 2π) is a surprising illustration of how vortices behave in practice. Experiment confirms the theory.[8]

The second application is to knotted and linked vortex lines. A mathematical construction[8,10] leads to solutions of the wave equation whose vortices have the topology of any chosen knot on a torus. The knots are described by two integers m, n (if m and n have a common factor N, the 'knot' consists of N linked loops). The construction can be implemented experimentally.[11] Vortex knots and links also exist in quantum waves.[12]

The third application is a prediction of quantum effects near the phase singularities of classical light. This is motivated by a philosophical aspect[13,14] of singularities in physics. They have a dual role: as the most important predictions from any physical theory, and also as a signal that the theory is breaking down. In light, the phase singularities are threads of darkness, offering a window through which can be seen the faint fluctuations of the quantum vacuum;[15] the radius of this 'quantum core' can be calculated. Analogous cores exist in sound waves.

Related articles are contained in the CD-ROM ("M_V_Berry" folder). Extracts from the readme file: "*Welcome to the Bristol vorticulture CD_ROM On this disk are most of the 86 papers, articles and PhD theses on the subject of wave dislocations (phase singularities, optical vortices) and polarization singularities published between 1974 (with Nye & Berry's seminal 'Dislocations in wave trains'[v1]) and January 2005, by authors working in the Physics Department, University of Bristol, UK.* "

References

1. M. V. Berry *et al.*, J. Optics A **6**, (Editorial introduction to special issue) (2004).
2. J. F. Nye and M. V. Berry, Proc. Roy. Soc. Lond. **A336**, 165 (1974).
3. W. Whewell, Phil. Trans. Roy. Soc. Lond. **123**, 147 (1833).
4. M. V. Berry, in SPIE **3487**, 1 (1998).
5. J. F. Nye, *Natural focusing and fine structure of light: Caustics and wave dislocations*, Institute of Physics Publishing, Bristol (1999).
6. J. F. Nye and J. V. Hajnal, Proc. Roy. Soc. Lond. **A409**, 21 (1987).
7. M. V. Berry, J. Optics. A **6**, 259 (2004).
8. J. Leach *et al.*, New Journal of Physics **6**, 71 (2004).
9. M. V. Berry and M. R. Dennis, Proc. Roy. Soc. Lond. **457**, 2251 (2001).
10. M. V. Berry and M. R. Dennis, J. Phys. A **34**, 8877 (2001).
11. J. Leach *et al.*, Nature **432**, 165 (2004).
12. M. V. Berry, Found. Phys. **31**, 659 (2001).
13. M. V. Berry, in *Proc. 9th Int. Cong. Logic, Method., and Phil. of Sci.*, edited by D. Prawitz, B. Skyrms, and D. Westerståhl (1994), pp. 597.
14. M. V. Berry, Physics Today, May, 10 (2002).
15. M. V. Berry and M. R. Dennis, J. Optics A **6**, S178 (2004).

ON UNIVERSALITY OF MATHEMATICAL STRUCTURE IN NATURE: TOPOLOGY

TOYOKI MATSUYAMA

Department of Physics, Nara University of Education,
Takabatake-cho, Nara 630-8528, JAPAN
E-mail: matsuyat@nara-ed.ac.jp

An introductory talk on a purpose of the project "Topological Science and Technology" is given so as specialists in various fields can share a common perception.

1. Introduction

This symposium is organized by the project "Topological science and technology. The scope of the project is very wide. Physics, technology, engineering, biology, medical research, information science and so on. I try to explain what is the purpose of our project in talking about a conceptual or spiritual aspect of this project but not about the technical details.

First I will talk about scientific methods to seek for universality in variety of nature. Secondly a universality of the topology as a kind of logic. Some examples which have been discovered already will be explained. Finally I will remark about the future of our project.

2. Universality in variety

In observing many phenomena in nature, we find the marvelous variety apparently. It is a hope of our human being to understand the essentials of nature. Then we have taken two strategies for the aim as shown in Fig. 1. One is to decompose a material into elements and seek for a universal law in each element. The typical area of science is the particle physics. The final goal is the theory of everything. I call this way as the science of elements.

The another way is to study mathematical structures in each phenomenon. We can find a mathematical logic in the universal structures. Please imagine a way that God created our universe. He must be to design by using some mathematical modules which are very excellent. God must

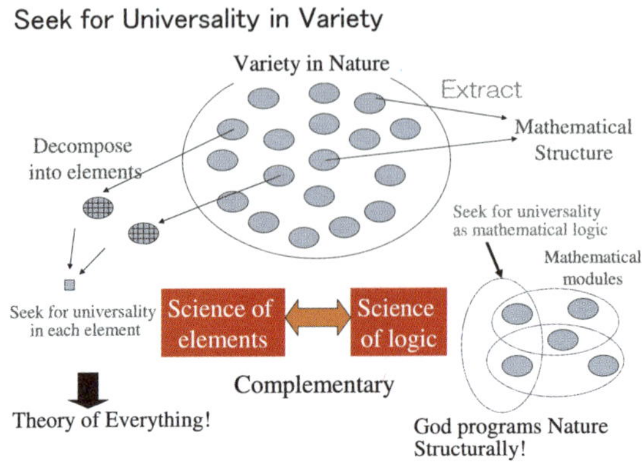

Figure 1. Two categories of scientific methods

be the most excellent mathematician programming nature structurally. I call the second as science of logics.

Both methods are complementary, of course. I think that the concept of topology is a strong candidate of the universality in the nature. Its universality is very high. The project aims the new paradigm shift in science by the concept of topology.

As you know, every matters are composed of atoms. Inside the atom, we find a core, around which the electron moves with electromagnetic interaction. The core is composed of neutrons and protons by a strong interaction. Further these particles are composed of quarks. Nowadays the fundamental elements confirmed experimentally are "electron" and "quark". In addition, it is known that four types of interaction exist in the universe. They were bifurcated in the evolution of the universe after its birth. This is the typical example of the science of elements. (Fig. 2)

In Fig. 3, the various subjects of researches, physics, mathematics, chemistry, and also social science are included. We have discovered that there exist same mathematical structures in the particle physics, the condensed matter physics which have the important topological meanings. This fact suggests that we can find an interesting physics by seeking the topology. This kind of relation appears in the luminessence phenomena, population changing of Deer in closed islands and so on. Some subjects which seem to be completely different at first sight, might be governed by the same mathematical logic.

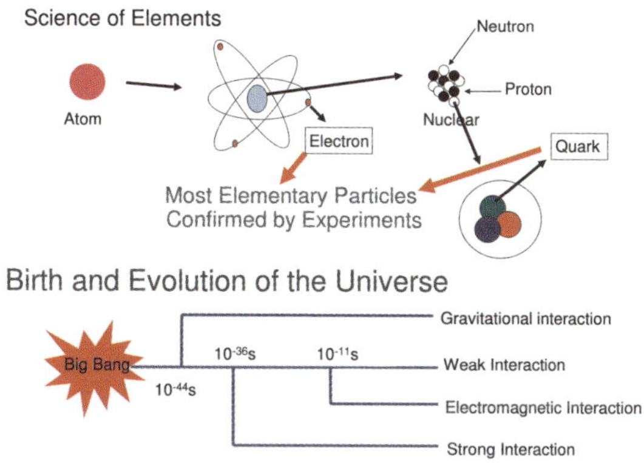

Figure 2. Science of elements

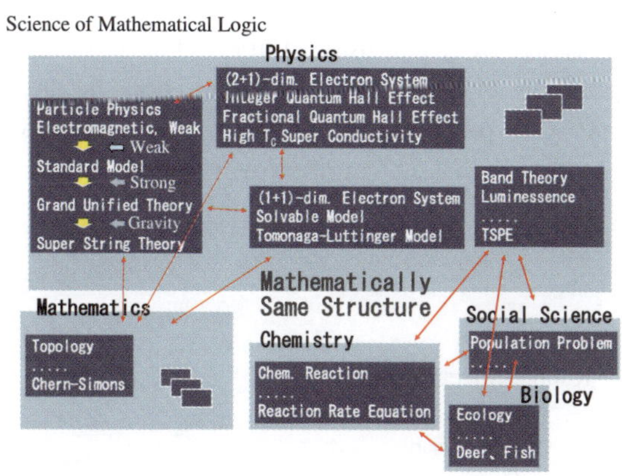

Figure 3. Science of mathematical logic

3. Universality of Topology

The birth place of the concept of topology is at Königsberg. Euler considered how to cross the bridges without a repetition. Modern topology was started by Poincare. Differential and combinatoric topologies were borne. They developed to algebraic topology. The relation is shown by Fig. 4. The most wide class of topology is a general topology, that is, rubber sheet geometry.

In the rubber sheet geometry (Fig. 5), two objects which become the same shape by a continuous deformation, is considered to be equivalent.

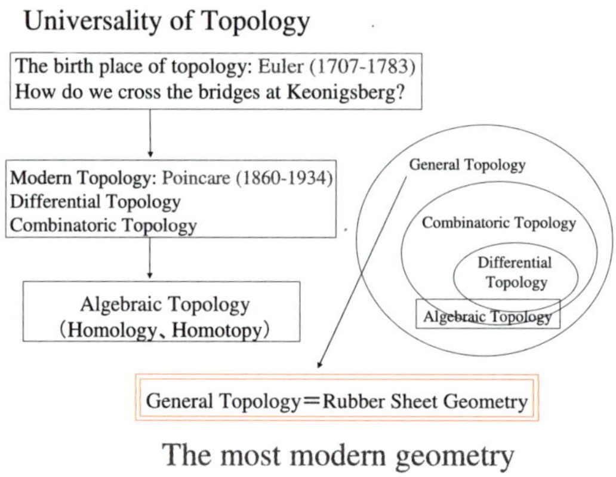

Figure 4. Universality of topology

The shape is classified by a winding number. The paths on plane without any hole are same one. If the plane has a hole, then the paths are classified by the times that each path winds the hole. It is called the winding number or the topological invariant[1] and is invariant without cutting the path (Fig. 5).

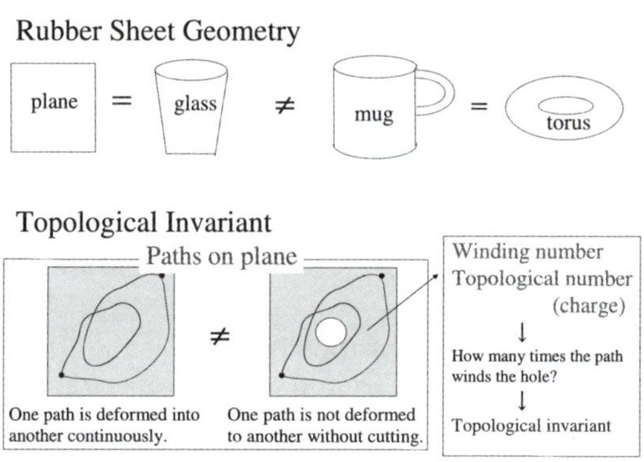

Figure 5. Rubber Sheet Geometry and topological Invariant

4. Topology in Nature

Now we turn to the examples of topology in nature. The first example is the space-time and geometry. (Fig. 6) Physics were started by the Newton mechanics. It is the world of the Euclid geometry. In the previous century, the concept of space-time was evolved by Einstein. After that, the development is just follow the development of geometries, as Minkowski, Rieman and so on[2,3,4]. The concept is developing still now of course.

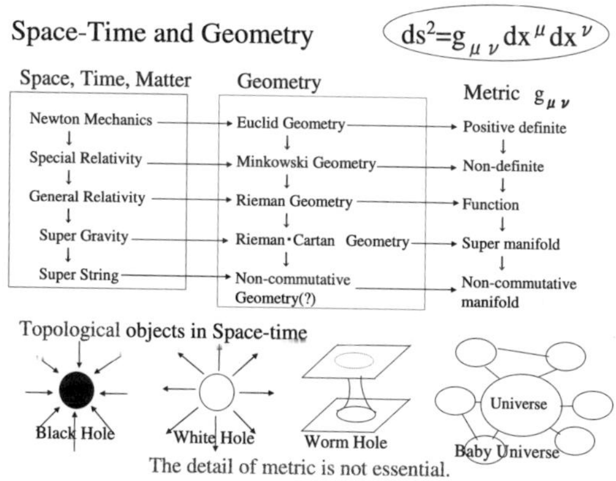

Figure 6. Space-Time and Geometry

The next example is the topological soliton. (Fig. 7) A magnetism appears by ordering microscopic magnets to one direction. In addition, it may appear a nontrivial order that the configuration of vectors is circular. It is a topological soliton called as the vortex. Another example is a topological soliton appeared in space-time, which is called as the instanton. Both solitons produce important physics[5].

The next example is the quantum anomaly[6] appeared in quantum field theories. Nature have several symmetries. The symmetry tells us that there exists a conservation law. Let ' s assume that there are two symmetries. It means that two conservation laws hold. In nature, we need the prescription of the quantization to get correct answer. There is the case that two symmetries cannot hold at same time under the quantization. If one conservation law holds, another is broken. This is the quantum anomaly.

The most famous example is π^0-decay to two γ in the field of the particle physics. (Fig. 8) It is named as the Adler-Bell-Jackiw Anomaly. Another

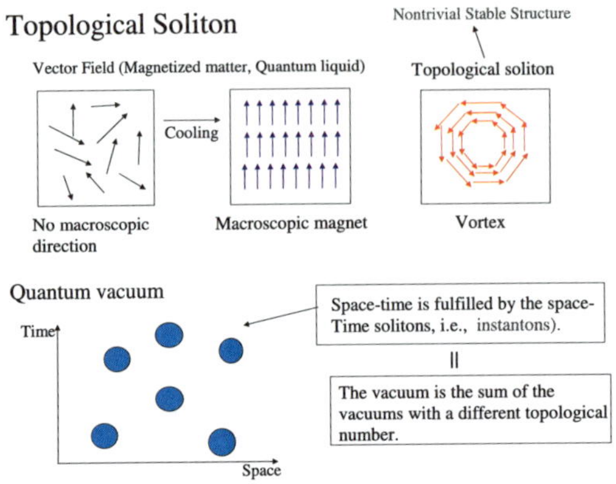

Figure 7. Topological soliton

example is the parity anomaly on a plane. This anomaly has an application in the quantum Hall effect. (Fig. 8)

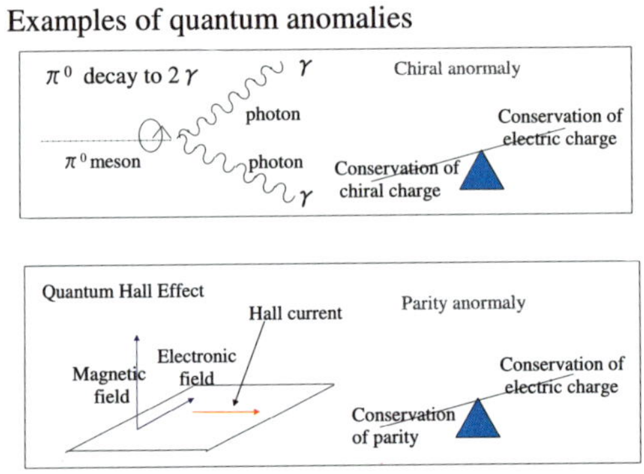

Figure 8. Examples of quantum anomalies

Anomaly has a deep topological origin. It can be seen in the path integral formulation of quantization. (Fig. 9) But it is a little bit technical so that I will skip it here.

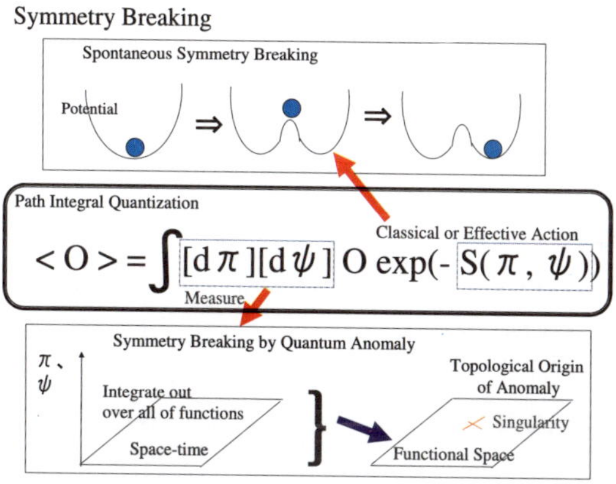

Figure 9. Symmetry breaking

5. Quantum field theory and condensed matter physics

The quantum filed theory is developed mainly in the field of particle physics. The theory makes possible to treat many particle system so that the theory was a powerful tool also in condensed matter physics. Further the theory gives us a picture that an interaction is exchanges of particles. The fundamental interactions are described by the exchange of special kind of particles called as gauge particles. The particles obey the gauge principle which has a topological origin. The theory satisfying the principle is the gauge theory.

About 15 years ago, a new collaboration between the quantum field theory and the condensed matter physics started[7,8]. It was summarized in Fig. 10. (The figure is revised as including the parity-preserving QED_3, following the suggestion by Prof. Tesanovic.)

What's happened in the collaboration of the quantum field theory with the condensed matter physics. It is shown some typical example in Fig. 11. In the evaluation of any physical quantity, it is very hard to get an exact result. One way is to use a perturbation technique in the weak or strong coupling region. We can calculate a physical quantity order by order.

However we may take a topological strategies in a special situation which is not always. If we can proof that the quantity is written by the topological invariant, we can get exact answer. This is one of powers of the topology.

Topological Approach

Quantum Field Theory	Condensed Matter Physics
(1+1)-dimensions	
Chiral Anomaly	Josephson Junction Incommensurate CDW
Topological Soliton	Polymer (Fractional Charge)
Berry Phase	Optical Fiber
(2+1)-dimensions	
Parity Anomaly	Integer Quantum Hall Effect Farady Rotation
Chern-Simons Term	Fractional Statistics (Anyon) Fractional Quantum Hall Effect
(Parity-violating or parity-preserving QED_3)	High T_C Superconductivity He_3 Vortex
(3+1)-dimensions	
Gauge Field General Relativity	Aharanov- Bohm Effect Dislocation of Crystal

Figure 10. Quantum field theory and condensed matter physics

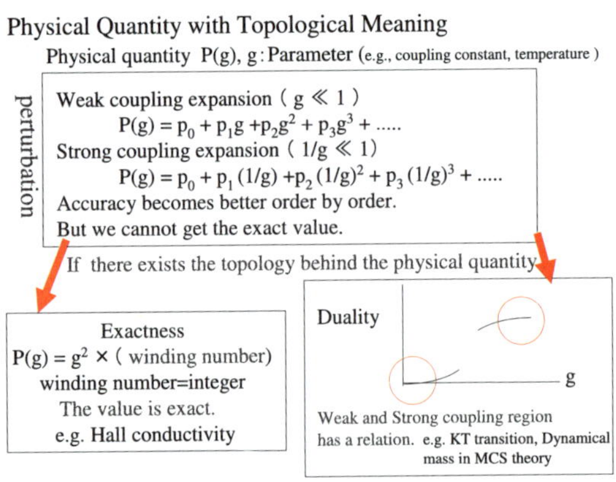

Figure 11. Powers of topology

6. Summary and perspective

I showed you some example of topology in nature, mainly in physics. Especially in the area of quantum field theories, many novel topological objects as black hole, space-time topology, topological soliton and instanton, anormaly, and so on were dis-covered. As one of unified points of view, the topologically non-trivial structures can exist in three different kinds of spaces, i.e., configuration space, energy-momentum space (k-space) and phase space. The places where non-trivial topologies exist can be seen in Fig. 12. Topology spreads through the condensed matter physics. It was recognized that the non-trivial phenomena in the low dimensional electron system with a strong correlation, i.e., the quantum Hall effect or the high T_C superconductivity, have a strong connection with the topology. Now the topology starts to affect on the technology[9].

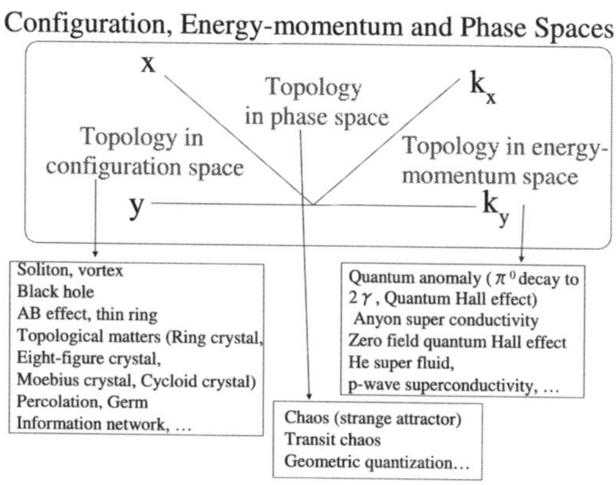

Figure 12. Where is the topologies?

Of course, the topologies hide other place as the more abstract internal space or the functional space of the path integral meager. I hope that we can discover a new place in future. It will extend the topological world in nature.

I told about the science of elements and the science of logic at the begining. What is the linking of the two complementary methods. In the science of elements, the most important structure is the hierarch. It means the existence of the stability and transition. In the science of logic, the stability will be the topological invariance and the transition will mean the topology changing. A key element in nature might be the tree structure. (Fig. 13)

It was just 100 years ago that Einstein started the evolution in physics which entirely changed the concept of space, time and matter after the Newton's evolution. His general relativity gives us a unified picture on space-time and matter as shown in Fig. 14. The guiding principle in his work, the general relativity, is realized by the Rieman geometry mathematically. After that, we also were able to understand the fundamental interactions geometrically by the gauge principle. The power of topological consideration in various aspects of Nature is apparent.

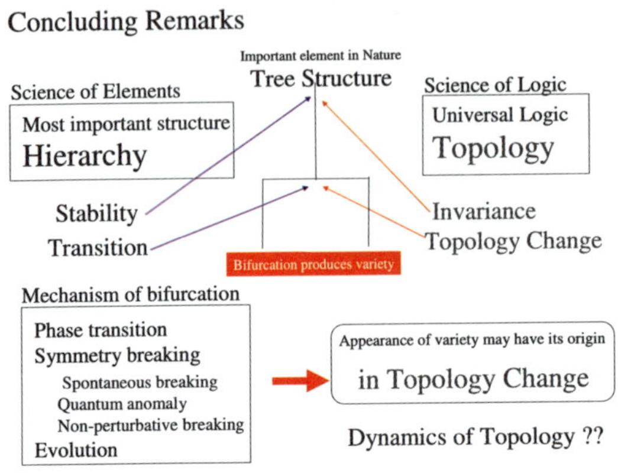

Figure 13. Tree structure

Recent developments in a research of complex and network systems suggest that the topology might be a key ingredient in the life science and the information science. Now the topology which is the most modern geometry might be the universal logic in studying Nature which includes life and information as Fig. 14.

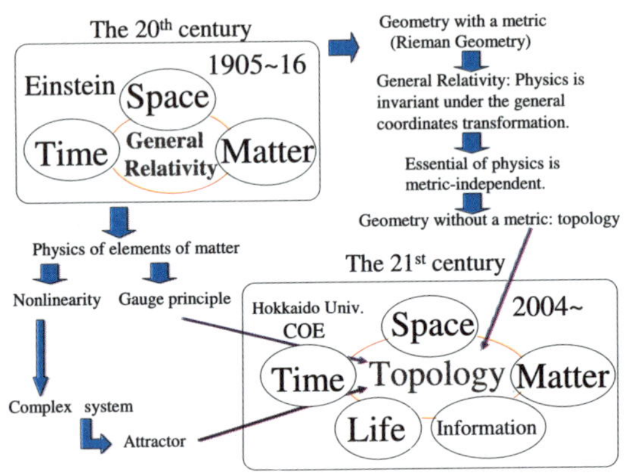

Figure 14. Toward a new paradaim in the 21st century

Acknowledgments

This work has been partially supported by the 21st COE program on "Topological Science and Technology" from the Ministry of Education, Culture, Sport, Science and Technology of Japan.

References

1. S. Chern, "Complex Manifolds without Potential Theory", Springer Verlag (1979)
2. S. Weinberg, "Gravitation and Cosmology", Wiley (New York, 1972).
3. C.W. Misner, K.S. Thorne and J.A. Wheeler, "Gravitation, Freeman (San Francisco, 1973).
4. S.W. Hawking and G.F.R. Ellis, "The Large Scale Structure of Space-Time", Cambridge Univ. Press. (Cambridge, 1973).
5. S. Coleman, "Aspect of Symmetry", Cambridge Univ. Press. (Cambridge, 1985).
6. S.B. Treiman, R. Jackiw, B. Zumino and E. Witten, "Current Algebra and Anomalies", World Scientific (Singapore, 1985).
7. E. Fradkin, "Field Theories of Condensed Matter Systems", Addison-Wesley (1991).
8. F. Wilczek, "Fractional Statistics and Anyon Superconductivity", World Scientific (Singapore, 1990).
9. S. Tanda, T. Tsuneta, Y. Okajima, K. Inagaki, K. Yamaya and N. Hatakenaka, *Nature* **417**, 397 (2002).

TOPOLOGY IN PHYSICS*

R. JACKIW†

Center for Theoretical Physics
MIT, Cambridge, MA 02139-4307
E-mail: jackiw@lns.mit.edu

The phenomenon of quantum number fractionalization is explained. The relevance of non-trivial phonon field topology is emphasized.

1. Introduction

Discussions of the spatial forms of physical materials use in a natural way geometrical and topological concepts. It is to be expected that arrangements of matter should form patterns that are described by pre-existing mathematical structures drawn from geometry and topology. But theoretical physicists also deal with abstract entities, which do not have an actual material presence. Still geometrical and topological considerations are relevant to these ephemeral theoretical constructs. I have in mind fields, both classical and quantum, which enter into our theories of fundamental processes. These fields $\phi(x)$ provide a mapping from a "base" space or space-time on which they are defined into the field "target" manifold on which they range. The base and target spaces, as well as the mapping, may possess some non-trivial topological features, which affect the fixed time description and the temporal evolution of the fields, thereby influencing the physical reality that these fields describe. Quantum fields of a quantum field theory are operator valued distributions whose relevant topological properties are obscure. Nevertheless, topological features of the corresponding classical fields are important in the quantum theory for a variety of reasons: (i) Quantized fields can undergo local (space-time dependent) transformations (gauge transformations, coordinate diffeomorphisms) that involve classical

*TOP2005 symposium, sapporo, japan, march 2005.
†This work is supported in part by funds provided by the U.S. Department of Energy (D.O.E.) under cooperative research agreement DE-FC02-94ER40818.

functions whose topological properties determine the allowed quantum field theoretic structures. (ii) One formulation of quantum field theory uses a functional integral over classical fields, and classical topological features become relevant. (iii) Semi-classical (WKB) approximations to the quantum theory rely on classical dynamics, and again classical topology plays a role in the analysis.

Topological effects in quantum electrodynamics were first appreciated by Dirac in his study of the quantum mechanics for (hypothetical) magnetic monopoles. This analysis leads directly to contemporary analysis of Yang-Mill theory – the contemporary generalization of Maxwell's electrodynamics – and has yielded several significant results: the discovery of the θ-vacuum angle; the recognition that c-number parameters in the theory may require quantization for topological reasons (like Dirac's monopole strength); the realization that the chiral anomaly equation is just the local version of the celebrated Atiyah-Singer index theorem.

Here I shall not describe the Yang-Mills investigations; they are too technical and too specialized for this general audience. Rather I shall show you how a topological effect in a condensed matter situation leads to charge fractionalization. This phenomenon has a physical realization in 1-dimensional (lineal) polymers, like polyacetylene, and in 2-dimensional (planar) systems, like the Hall effect.

The polyacetylene story is especially appealing, because it can be told in several ways: in pictorial terms which only involves counting, or in the first quantized formalism for quantum mechanical equations, or in the second quantized formalism of a quantum field theory [1].

2. The Polyacetylene Story (Counting Argument)

Polyacetylene is a material consisting of parallel chains of carbon atoms, with electrons moving primarily along the chains, while hopping between chains is strongly suppressed. Consequently, the system is effectively 1-dimensional. The distance between carbon atoms is about 1Å.

If the atoms are considered to be completely stationary, *i.e.* rigidly attached to their equilibrium lattice sites, electron hopping along the chain is a structureless phenomenon. However, the atoms can oscillate around their rigid lattice positions for a variety of reasons, like zero-point motion, thermal excitation, etc. It might be thought that these effects merely give rise to a slight fuzzing of the undistorted-lattice situation.

In fact this is not correct; something more dramatic takes place. Rather

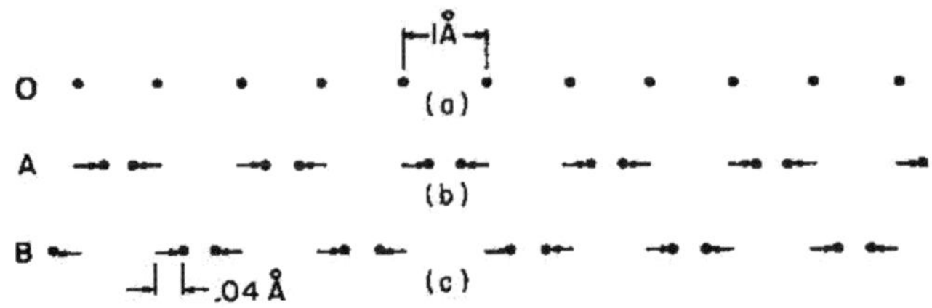

Figure 1. (a) The rigid lattice of polyacetylene; (O) the carbon atoms are equally spaced 1 Å apart. (b), (c) The effect of Peierls' instability is to shift the carbon atoms .04Åto the right (A) or to the left (B), thus giving rise to a double degeneracy.

than oscillating about the rigid-lattice sites, the atoms first shift a distance of about .04 Å and then proceed to oscillate around the new, slightly distorted location. That this should happen was predicted by Peierls, and is called the Peierls instability. Due to reflection symmetry, there is no difference between a shift to the right or a shift to the left; the material chooses one or the other, thus breaking spontaneously the reflection symmetry, and giving rise to doubly degenerate vacua, called A and B.

If the displacement is described by a field ϕ which depends on the position x along the lattice, the so-called phonon field, then Peierls' instability, as well as detailed dynamical calculations indicate that the energy density $V(\phi)$, as a function of constant ϕ, has a double-well shape. The symmetric point $\phi = 0$ is unstable; the system in its ground state must choose one of the two equivalent ground states $\phi = \pm \mid \phi_0 \mid = \pm .04$Å. In the ground states, the phonon field has uniform values, independent of x.

By now it is widely appreciated that whenever the ground state is degenerate there frequently exist additional stable states of the system, for which the phonon field is non-constant. Rather, as a function of x, it interpolates, when x passes from negative to positive infinity, between the allowed ground states. These are the famous solitons, or kinks. For polyacetylene they correspond to domain walls which separate regions with vacuum A from those with vacuum B, and vice versa. One represents the chemical bonding pattern by a double bond connecting atoms that are closer together, and the single bond connecting those that are further apart.

Consider now a polyacetylene sample in the A vacuum, but with two solitons along the chain. Let us count the number of links in the sample without solitons and compare with number of links where two solitons are present. It suffices to examine the two chains only in the region where they differ, *i.e.* between the two solitons. Vacuum A exhibits 5 links, while the

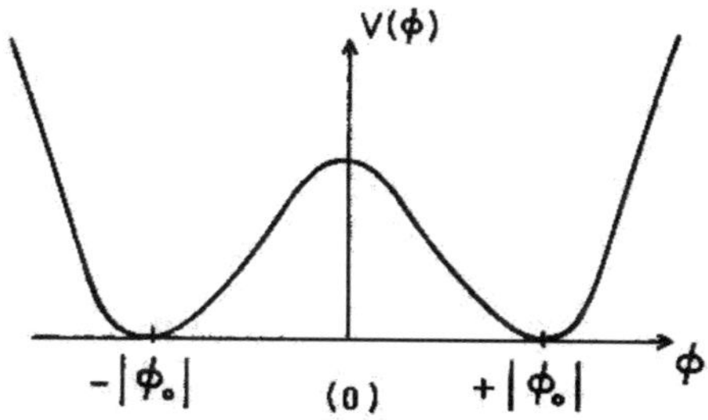

Figure 2. Energy density $V(\phi)$, as a function of a constant phonon field ϕ. The symmetric stationary point, $\phi = 0$, is unstable. Stable vacua are at $\phi = +|\phi_0|$, (A) and $\phi =$ -$|\phi_0|$, (B).

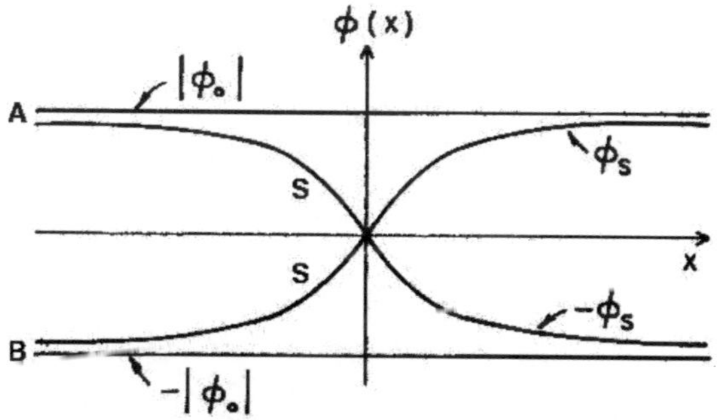

Figure 3. The two constant fields, $\pm \mid \phi_0 \mid$, correspond to the two vacua (A and B). The two kink fields, $\pm\phi_s$, interpolate between the vacua and represent domain walls.

addition of two solitons decreases the number of links to 4. The two soliton state exhibits a deficit of one link. If now we imagine separating the two solitons a great distance, so that they act independently of one another, then each soliton carries a deficit of half a link, and the quantum numbers of the link, for example the charge, are split between the two states. This is the essence of fermion fractionization.

It should be emphasized that we are not here describing the familiar situation of an electron moving around a two-center molecule, spending "half" the time with one nucleus and "half" with the other. Then one might say that the electron is split in half, on the average; however fluctuations in any quantity are large. But in our soliton example, the fractionization is without fluctuations; in the limit of infinite separation one achieves an eigenstate with fractional eigenvalues.

We must however remember that the link in fact corresponds to two states: an electron with spin up and another with spin down. This doubling

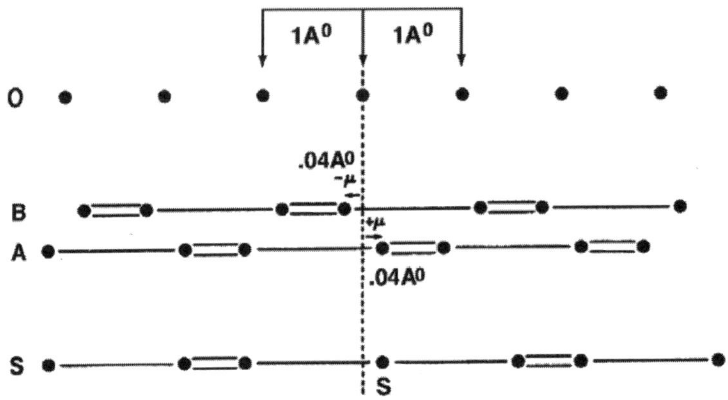

Figure 4. Polyacetylene states. The equally spaced configuration (O) possesses a left-right symmetry, which however is energetically unstable. Rather in the ground states the carbon atoms shift a distance μ to the left or right, breaking the symmetry and producing two degenerate vacua (A, B). A soliton (S) is a defect in the alteration pattern; it provides a domain wall between configurations (A) and (B).

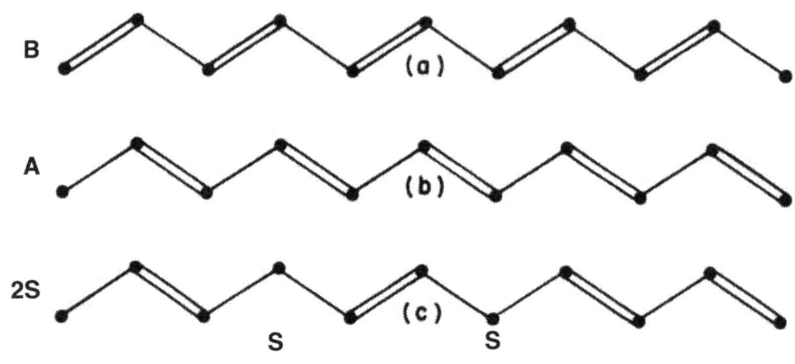

Figure 5. (a), (b) Pattern of chemical bonds in vacua A and B. (c) Two solitons inserted into vacuum A.

obscures the dramatic charge $\frac{1}{2}$ effect, since everything must be multiplied by 2 to account for the two states. So in polyacetylene, a soliton carries a charge deficit of one unit of electric charge. Nevertheless charge fractionization leaves a spur: the soliton state has net charge, but no net spin, since all of the electron spins are paired. If an additional electron is inserted into the sample, the charge deficit is extinguished, and one obtains a neutral state, but now there is a net spin. These spin-charge assignments (charged

– without spin, neutral – with spin) are unexpected, but in fact have been observed, and provide experimental verification for the soliton picture and fractionalization in polyacetylene.

Notice that in this simple counting argument no mention is made of topology. This feature emerges only when an analytic treatment is given. I now turn to this.

3. The Polyacetylene Story (Quantum Mechanics)

I shall now provide a calculation which shows how charge 1/2 arises in the quantum mechanics of fermions in interaction with solitons. The fermion dynamics are governed by an one-dimensional Dirac Hamiltonian, $H(\phi)$, which also depends on a background phonon field ϕ, with which the fermions intact. The Dirac Hamiltonian arises not because the electrons are relativistic. Rather it emerges in a certain well-formulated approximation to the microscopic theory, which yields a quantal equation that is a 2x2 matrix equation, like a Dirac equation. In the vacuum sector, ϕ takes on a constant value ϕ_0, appropriate to the vacuum. When a soliton is present, ϕ becomes the appropriate, static soliton profile ϕ_s. We need not be any more specific. We need not insist on any explicit soliton profile. All that we require is that the topology [*i.e.* the large distance behavior] of the soliton profile be non-trivial.

In the present lineal case the relevant topology is that infinity corresponds to two points, the end points of the line, and the phonon field in the soliton sector behaves differently at the points at infinity.

To analyze the system we need the eigenmodes, both in the vacuum and soliton sectors.

$$H(\phi_0)\psi_E^v = E\psi_E^v \tag{1}$$

$$H(\phi_s)\psi_E^s = E\psi_E^s \tag{2}$$

The Dirac equation is like a matrix-valued "square root" of the wave equation. Because a square root is involved, there will be in general negative energy solutions and positive energy solutions. The negative energy solutions correspond to the states in the valence band; the positive energy ones, to the conduction band. In the ground state, all the negative energy levels are filled, and the ground state charge is the integral over all space of the charge density $\rho(x)$, which in turn is constructed from all the negative energy wave functions.

$$\rho(x) = \int_{-\infty}^{0} dE\, \rho_E\,(x),\;\; \rho_E(x) = \psi_E^*\,(x)\, \psi_E\,(x) \tag{3}$$

Of course integrating (3) over x will produce an infinity; to renormalize we measure all charges relative to the ground state in the vacuum sector. Thus the soliton charge is

$$Q = \int dx \int_{-\infty}^{0} dE\, \{\rho_E^s\,(x) - \rho_E^v\,(x)\}. \tag{4}$$

Eq. (4) may be completely evaluated without explicitly specifying the soliton profile, nor actually solving for the negative energy modes, provided H possesses a further property. We assume that there exists a conjugation symmetry which takes positive energy solutions of (1) and (2) into negative energy solutions. (This is true for polyacetylene.) That is, we assume that there exists a unitary 2x2 matrix M, such that

$$M\psi_E = \psi_{-E}. \tag{5}$$

An immediate consequence, crucial to the rest of the argument, is that the charge density at E is an even function of E.

$$\rho_E(x) = \rho_{-E}(x) \tag{6}$$

Whenever one solves a conjugation symmetric Dirac equation, with a topologically interesting background field, like a soliton, there always are, in addition to the positive and negative energy solutions related to each other by conjugation, self-conjugate, normalizable zero-energy solutions. That this is indeed true can be seen by explicit calculation. However, the occurrence of the zero mode is also predicted by very general mathematical theorems about differential equations. These so-called “index theorems” count the zero eigenvalues, and insure that the number is non-vanishing whenever the topology of the background is non-trivial. We shall assume that there is just one zero mode, described by the normalized wave function ψ_0.

To evaluate the charge Q in (4), we first recall that the wave functions are complete, both in the soliton sector and in the vacuum sector.

$$\int_{-\infty}^{\infty} dE\, \psi_E^*\,(x)\, \psi_E(y) = \delta(x-y) \tag{7}$$

As a consequence, it follows that

$$\int_{-\infty}^{\infty} dE\,[\rho_E^s\,(x)\;-\rho_E^v\,(x)] = 0. \tag{8a}$$

In the above completeness integral over all energies, we record separately the negative energy contributions, the positive energy contributions, and for the soliton, the zero-energy contribution. Since the positive energy charge density is equal to the negative one, by virtue of (6), we conclude that (8a) may be equivalently written as an integral over negative E.

$$\int_{-\infty}^{0} dE\,[2\rho_E^s\,(x) - 2\rho_E^v\,(x)] + \psi_0^*\,(x)\,\psi_0\,(x) = 0 \tag{8b}$$

Rearranging terms give

$$Q = \int dx \int_{-\infty}^{0} dE[\rho_E^s(x) - \rho_0^v(x)] = -\frac{1}{2}\int dx\psi_0(x)\psi_0(x) = -\frac{1}{2}. \tag{9}$$

This is the final result: the soliton's charge is $-\frac{1}{2}$; a fact that follows from completeness (7) and conjugation symmetry (6). It is seen in (9) that the zero-energy mode is essential to the conclusion. The existence of the zero mode in the conjugation symmetric case is assured by the non-trivial topology of the background field. The result is otherwise completely general.

4. The Polyacetylene Story (Quantum Field Theory)

The quantum mechanical derivation that I just presented does not address the question of whether the fractional half-integer charge is merely an uninteresting expectation value or whether it is an eigenvalue. To settle this, we need a quantum field theory approach, that is we need to second quantize the field. For this, we expand Ψ, which now is an anti-commuting quantum field operator, in eigenmodes of our Dirac equation in the soliton sector as

$$\begin{aligned}\Psi &= \sum^{E}(b_E\,\psi_E^s + d_E^\dagger\,\psi_{-E}^s) + a\psi_0\\ \Psi^\dagger &= \sum^{E}(b_E^\dagger\,\psi_E^{s*} + d_E\,\psi_{-E}^{s*}) + a^\dagger\psi_0.\end{aligned} \tag{10}$$

The important point is that while the finite energy modes $\psi_{\pm E}^s$ enter with annihilation particle (conduction band) operators b_E and creation anti-particle (valence band) operators $d_E^\dagger$, the zero mode does not have a partner and is present in the sum simply with the operator a. The zero energy state is therefore doubly degenerate. It can be empty $\mid - >$, or filled $\mid + >$, and the $a, a^\dagger$ operators are realized as

$$a \mid + >=\mid - >, \ a^\dagger \mid + >= 0, \ a \mid - >= 0, \ a^\dagger \mid + >=\mid + > . \tag{11}$$

The charge operator $Q = \int dx \psi^\dagger \psi$ must be properly defined to avoid infinities. This is done, according to Schwinger's prescription in the vacuum sector, by replacing the formal expression by

$$Q = \frac{1}{2} \int dx \, (\psi^\dagger \psi - \psi \psi^\dagger). \tag{12}$$

We adopt the same regularization prescription for the soliton sector and insert our expansion (10) into (12). We find with the help of the orthonormality of wave functions

$$\begin{aligned} Q &= \frac{1}{2} \sum_E (b_E^\dagger \, b_E + d_E \, d_E^\dagger - b_E \, b_E^\dagger - d_E^\dagger \, d_E) + \frac{1}{2}(a^\dagger a - a a^\dagger) \\ &= \sum_E (b_E^\dagger \, b_E - d_E^\dagger \, d_E) + a^\dagger a - \frac{1}{2}. \end{aligned} \tag{13}$$

Therefore the eigenvalues for Q are

$$Q \mid - >= -\frac{1}{2} |- >, \ \ Q \mid + >= \frac{1}{2} |+ > ! \tag{14}$$

5. Conclusion

This then concludes my polyacetylene story, which has experimental realization and confirmation. And the remarkable effect arises from the non-trivial topology of the phonon field in the soliton sector.

Many other topological effects have been found in the field theoretic descriptions of condensed matter and particle physics. Yet we must notice that mostly these arise in phenomenological descriptions, not in the fundamental theory. In condensed matter the fundamental equation is the many-body Schrödinger equation with Coulomb interactions. This does not show any interesting topological structure. Only when it is replaced by effective, phenomenological equations do topological considerations become relevant for the effective description. Fundamental (condensed matter) Nature is simple!

Similarly in particle physics, our phenomenological, effective theories, like the Skyrme model, enjoy a rich topological structure. Moreover, even the Yang-Mills theory of our fundamental "standard particle physics model" supports non-trivial topological structure, which leads to the QCD vacuum angle. In view of my previous observation, can we take this as indirect evidence that thisYang-Mills based theory also is a phenomenological, effective description and at a more fundamental level – yet to be discovered – we shall find a simpler description that does not have any elaborate mathematical structure. Perhaps in this final theory Nature will be described by simple counting rules – like my first polyacetylene story. Surely this will not be the behemoth of string theory.

This work is supported in part by funds provided by the U.S. Department of Energy (D.O.E.) under cooperative research agreement DE-FC02-94ER40818.

References

1. This research was performed in collaboration with C. Rebbi, and independently by W.P. Su, J.R. Schrieffer and A. Heeger. For a summary see R. Jackiw and J.R. Schrieffer "Solitons with Fermion number 1/2 in Condensed Matter and Relativistic Field Theories" *Nucl. Phys.* **B190**, 253 (1981).

ISOHOLONOMIC PROBLEM AND HOLONOMIC QUANTUM COMPUTATION

SHOGO TANIMURA

Graduate School of Engineering, Osaka City University, Osaka 558-8585, Japan
E-mail: tanimura@mech.eng.osaka-cu.ac.jp

Geometric phases accompanying adiabatic processes in quantum systems can be utilized as unitary gates for quantum computation. Optimization of control of the adiabatic process naturally leads to the isoholonomic problem. The isoholonomic problem in a homogeneous fiber bundle is formulated and solved completely.

1. Introduction

The isoholonomic problem was proposed in 1991 by a mathematician, Montgomery[1]. The isoholonomic problem is a generalization of the isoperimetric problem, which requests finding a loop in a plane that surrounds the largest area with a fixed perimeter. On the other hand, the isoholonomic problem requests finding the shortest loop in a manifold that realizes a specified holonomy. This kind of problem naturally arose in studies of the Berry phase[2–4] and the Wilczek-Zee holonomy[5], which appear in a state of a controlled quantum system when the control parameter is adiabatically changed and returned to the initial value. Experimenters tried to design efficient experiments for producing these kinds of holonomy. Montgomery formulated the isoholonomic problem in terms of differential geometry and gauge theory. Although he gave partial answers, construction of a concrete solution has remained an open problem.

Recently, in particular after the discovery of factorization algorithm by Shor[6] in 1994, quantum computation grows into an active research area. Many people have proposed various algorithms of quantum computation and various methods for their physical implementation. Zanardi, Rasetti[7] and Pachos[8] proposed utilizing the Wilczek-Zee holonomy for implementing unitary gates and they named the method holonomic quantum computation. Since holonomy has its origin in geometry, it dose not depend on detail of dynamics and hence it does not require fine temporal tuning of control parameters.

It should be noted, however, that holonomic quantum computation requires two seemingly contradicting conditions. The first one is the adiabaticity condition. To suppress undesirable transition between different energy levels we need to change the control parameter quasi-stationarily. Hence a safer control demands longer execution time to satisfy adiabaticity. The second one is the decoherence problem. When a quantum system is exposed to interaction with environment for a long time, the system loses coherence and a unitary operator fails to describe time-evolution of the system. Hence a safer control demands shorter execution time to avoid decoherence. To satisfy these two contradicting conditions we need to make the loop in the control parameter manifold as short as possible while keeping the specified holonomy. Thus, we are naturally led to the isoholonomic problem.

We would like to emphasize that a quantum computer is actually not a digital computer but an analog computer in its nature. Hence, the geometric and topological approaches are useful for building and optimizing quantum computers.

This paper is based on collaboration with D. Hayashi and M. Nakahara[9]. We are further developing our studies on optimal and precise control of quantum computers with Y. Kondo, K. Hata and J.J. Vartiainen[10–12]. We thank Akio Hosoya, Tohru Morimoto and Richard Montgomery for their kind interest in our work.

2. Wilczek-Zee holonomy

A state vector $\psi(t) \in \mathbb{C}^N$ evolves according to the Schrödinger equation

$$i\hbar \frac{d}{dt}\psi(t) = H(t)\psi(t). \tag{1}$$

The Hamiltonian admits a spectral decomposition $H(t) = \sum_{l=1}^{L} \varepsilon_l(t) P_l(t)$ with projection operators $P_l(t)$. Therefore, the set of energy eigenvalues $(\varepsilon_1, \ldots, \varepsilon_L)$ and orthogonal projectors $(P_1, \ldots, P_L)$ constitutes a complete set of control parameters of the system. Now we concentrate on the eigenspace associated with the lowest energy ε_1. We write $P_1(t)$ as $P(t)$ for simplicity. Suppose that the degree of degeneracy $k = \operatorname{tr} P(t)$ is constant. For each t, we have the eigenvectors such that

$$H(t)v_\alpha(t) = \varepsilon_1(t)v_\alpha(t), \qquad (\alpha = 1, \ldots, k). \tag{2}$$

We assume that they are normalized as $v_\alpha^\dagger(t)v_\beta(t) = \delta_{\alpha\beta}$. Then

$$V(t) = \Big(v_1(t), \ldots, v_k(t)\Big) \tag{3}$$

forms an $N \times k$ matrix satisfying $V^\dagger(t)V(t) = I_k$ and $V(t)V^\dagger(t) = P(t)$. Here I_k is the k-dimensional unit matrix. The adiabatic theorem guarantees that the state remains the eigenstate associated with the eigenvalue $\varepsilon_1(t)$ of the instantaneous Hamiltonian $H(t)$ if the initial state was an eigenstate with $\varepsilon_1(0)$. Therefore the state vector is a linear combination

$$\psi(t) = \sum_{\alpha=1}^{k} \phi_\alpha(t)\, v_\alpha(t) = V(t)\phi(t). \tag{4}$$

The vector $\phi = {}^t(\phi_1, \ldots, \phi_k) \in \mathbb{C}^k$ is called a reduced state vector. By substituting it into the Schrödinger equation (1) we get

$$\frac{d\phi}{dt} + V^\dagger \frac{dV}{dt}\phi(t) = -\frac{i}{\hbar}\varepsilon_1(t)\phi(t). \tag{5}$$

Its solution is formally written as

$$\phi(t) = \exp\Big(-\frac{i}{\hbar}\int_0^t \varepsilon_1(s)ds\Big)\, \mathbb{T}\exp\Big(-\int_0^t V^\dagger \frac{dV}{ds}ds\Big)\,\phi(0), \tag{6}$$

where $\mathbb{T}$ stands the time-ordered product. Then $\psi(t) = V(t)\phi(t)$ becomes

$$\psi(t) = e^{-\frac{i}{\hbar}\int_0^t \varepsilon_1(s)ds}\, V(t)\, \mathbb{T}\, e^{-\int V^\dagger dV}\, V^\dagger(0)\psi(0). \tag{7}$$

In particular, when the control parameter comes back to the initial point as $P(T) = P(0)$, the state vector $\psi(T)$ also comes back in the same eigenspace as $\psi(0) = V(0)\phi(0)$. The Wilczek-Zee holonomy $\Gamma \in U(k)$ is defined via

$$\psi(T) = e^{-\frac{i}{\hbar}\int_0^T \varepsilon_1(s)ds}\, V(0)\Gamma\,\phi(0) \tag{8}$$

and is given explicitly as

$$\Gamma = V(0)^\dagger\, V(T)\, \mathbb{T}\, e^{-\int V^\dagger dV}. \tag{9}$$

If the condition $V^\dagger \frac{dV}{dt} = 0$ is satisfied, the curve $V(t)$ is called a horizontal lift of the curve $P(t)$. Then the holonomy Eq.(9) is reduced to $\Gamma = V^\dagger(0)V(T)$.

3. Formulation of the problem

The isoholonomic problem is formulated in terms of the homogeneous fiber bundle $(S_{N,k}(\mathbb{C}),\, G_{N,k}(\mathbb{C}), \pi, U(k))$. The Stiefel manifold $S_{N,k}(\mathbb{C})$ is the set of orthonormal k-frames; a k-frame V spans the degenerate energy eigenspace in $\mathbb{C}^N$;

$$S_{N,k}(\mathbb{C}) = \{V \in M(N,k;\mathbb{C}) \,|\, V^\dagger V = I_k\}, \tag{10}$$

where $M(N,k;\mathbb{C})$ is the set of $N \times k$ complex matrices. An element of the unitary group $h \in U(k)$ acts on $V \in S_{N,k}(\mathbb{C})$ from the right as $(V,h) \mapsto Vh$ by means of a matrix product. The Grassmann manifold $G_{N,k}(\mathbb{C})$ is defined as the set of projection matrices to k-dimensional subspaces in $\mathbb{C}^N$,

$$G_{N,k}(\mathbb{C}) \;= \{P \in M(N,N;\mathbb{C}) \,|\, P^2 = P,\ P^\dagger = P,\ \mathrm{tr} P = k\}. \tag{11}$$

The projection map $\pi : S_{N,k}(\mathbb{C}) \to G_{N,k}(\mathbb{C})$ is defined as $\pi : V \mapsto P := VV^\dagger$. Then it can be proved that the Stiefel manifold $S_{N,k}(\mathbb{C})$ becomes a principal bundle over $G_{N,k}(\mathbb{C})$ with the structure group $U(k)$. The canonical connection form on $S_{N,k}(\mathbb{C})$ is defined as a one-form

$$A = V^\dagger dV, \tag{12}$$

which takes its value in the Lie algebra $\mathfrak{u}(k)$. The holonomy associated with this connection is called the Berry phase in case of $k = 1$ and the Wilczek-Zee holonomy in general. We define Riemannian metrices, $\|dV\|^2 = \mathrm{tr}\,(dV^\dagger dV)$ for the Stiefel manifold and $\|dP\|^2 = \mathrm{tr}\,(dPdP)$ for the Grassmann manifold. For any curve $P(t)$ in $G_{N,k}(\mathbb{C})$, there is a curve $V(t)$ in $S_{N,k}(\mathbb{C})$ such that $\pi(V(t)) = P(t)$. If the curve $V(t)$ satisfies

$$V^\dagger \frac{dV}{dt} = 0, \tag{13}$$

it is called a horizontal lift of the curve $P(t)$. When the curve $P(t)$ is a closed loop, such that

$$V(T)V^\dagger(T) = V(0)V^\dagger(0), \tag{14}$$

the holonomy associated with the loop is defined as $V(T) = V(0)\Gamma$ and is given as

$$\Gamma = V^\dagger(0)V(T) \in U(k). \tag{15}$$

We formulate the isoholonomic problem as a variational problem. The length of the horizontal curve $V(t)$ is evaluated by the functional

$$S[V,\Omega] = \int_0^T \left\{ \mathrm{tr}\left(\frac{dV^\dagger}{dt}\frac{dV}{dt}\right) - \mathrm{tr}\left(\Omega\, V^\dagger \frac{dV}{dt}\right)\right\} dt, \tag{16}$$

where $\Omega(t) \in \mathfrak{u}(k)$ is a Lagrange multiplier to impose the horizontal condition (13) on the curve $V(t)$. Thus the isoholonomic problem is stated as follows; find a horizontal curve $V(t)$ that attains an extremal value of the functional Eq.(16) and satisfies the boundary conditions Eqs.(14) and (15).

4. Derivation and solution of the Euler-Lagrange equation

We derive the Euler-Lagrange equation associated the functional S and solve it explicitly. A variation of the curve $V(t)$ is defined by an arbitrary smooth function $\eta(t) \in \mathfrak{u}(N)$ such that $\eta(0) = \eta(T) = 0$ and an infinitesimal parameter $\epsilon \in \mathbb{R}$ as

$$V_\epsilon(t) = (1 + \epsilon\eta(t))V(t). \tag{17}$$

By substituting $V_\epsilon(t)$ into Eq.(16) and differentiating with respect to ϵ, the extremal condition yields

$$0 = \left.\frac{dS}{d\epsilon}\right|_{\epsilon=0} = \int_0^T \mathrm{tr}\left\{\dot{\eta}(V\dot{V}^\dagger - \dot{V}V^\dagger - V\Omega V^\dagger)\right\} dt. \tag{18}$$

Thus we obtain the Euler-Lagrange equation

$$\frac{d}{dt}(\dot{V}V^\dagger - V\dot{V}^\dagger + V\Omega V^\dagger) = 0. \tag{19}$$

The extremal condition with respect to $\Omega(t)$ reproduces the horizontal equation $V^\dagger\dot{V} = 0$.

Next, we solve Eqs.(13) and (19). Equation (19) is integrated to yield

$$\dot{V}V^\dagger - V\dot{V}^\dagger + V\Omega V^\dagger = \mathrm{const} = X \in \mathfrak{u}(N). \tag{20}$$

Conjugation of the horizontal condition Eq.(13) yields $\dot{V}^\dagger V = 0$. Then, by multiplying V on Eq.(20) from the right we obtain

$$\dot{V} + V\Omega = XV. \tag{21}$$

By multiplying $V^\dagger$ on Eq.(21) from the left we obtain

$$\Omega = V^\dagger XV. \tag{22}$$

We can show $\dot{\Omega} = 0$ by a straightforward calculation. Hence, $\Omega(t)$ is actually a constant matrix. The solution of Eqs.(21) and (22) is

$$V(t) = e^{tX}\, V_0\, e^{-t\Omega}, \qquad \Omega = V_0^\dagger X V_0. \tag{23}$$

We call this solution the horizontal extremal curve. Then Eq.(20) becomes

$$(XV - V\Omega)V^\dagger - V(-V^\dagger X + \Omega V^\dagger) + V\Omega V^\dagger = X,$$

which is arranged as

$$X - (VV^\dagger X + XVV^\dagger - VV^\dagger XVV^\dagger) = 0. \tag{24}$$

Here we used Eq.(22). We may take, without loss of generality,

$$V_0 = \begin{pmatrix} I_k \\ 0 \end{pmatrix} \in S_{N,k}(\mathbb{C}) \tag{25}$$

as the initial point. We can parametrize $X \in \mathfrak{u}(N)$, which satisfies Eq.(22), as

$$X = \begin{pmatrix} \Omega & W \\ -W^\dagger & Z \end{pmatrix} \tag{26}$$

with $W \in M(k, N-k; \mathbb{C})$ and $Z \in \mathfrak{u}(N-k)$. Then the constraint equation (24) implies that $Z = 0$. Finally, we obtained a complete set of solution (23) of the horizontal extremal equation (13) and (19).

5. Solution to the boundary value problem

The remaining problem is to find the controller matrices Ω and W that satisfy the closed loop condition

$$V(T)V^\dagger(T) = e^{TX} V_0 V_0^\dagger e^{-TX} = V_0 V_0^\dagger \tag{27}$$

and the holonomy condition

$$V_0^\dagger\, V(T) = V_0^\dagger\, e^{TX}\, V_0\, e^{-T\Omega} = U_{\text{gate}} \tag{28}$$

for a requested unitary gate $U_{\text{gate}} \in U(k)$. Montgomery[1] presented this boundary value problem as an open problem. Here we give a prescription to construct a controller matrix X that produces the specified unitary gate U_{gate}. It turns out that the working space should have a dimension $N \geq 2k$ to apply our method. In the following we assume that $N = 2k$. The time interval is normalized as $T = 1$.

Our method consists of three steps. In the first step, we diagonalize a given unitary matrix $U_{\text{gate}} \in U(k)$ as

$$R^\dagger U_{\text{gate}} R = U_{\text{diag}} = \text{diag}(e^{i\gamma_1}, \ldots, e^{i\gamma_k}) \qquad (0 \leq \gamma_j < 2\pi) \tag{29}$$

with $R \in U(k)$. The small circle is a circle in a two-sphere $\mathbb{C}P^1 \subset G_{N,k}(\mathbb{C})$ that surrounds a solid angle which is equal to twice of the Berry phase. In the second step, combining k small circles we construct $k \times k$ matrices

$$\Omega_{\text{diag}} = \text{diag}(i\omega_1, \ldots, i\omega_k), \qquad W_{\text{diag}} = \text{diag}(i\tau_1, \ldots, i\tau_k), \tag{30}$$

with $\omega_j = 2(\pi - \gamma_j)$ and $\tau_j = e^{i\phi_j}\sqrt{\pi^2 - (\pi - \gamma_j)^2}$. We combine them into a $2k \times 2k$ matrix

$$X_{\text{diag}} = \begin{pmatrix} \Omega_{\text{diag}} & W_{\text{diag}} \\ -W_{\text{diag}}^{\dagger} & 0 \end{pmatrix}.$$

In the third step, we construct the controller X as

$$X = \begin{pmatrix} R & 0 \\ 0 & I_k \end{pmatrix} \begin{pmatrix} \Omega_{\text{diag}} & W_{\text{diag}} \\ -W_{\text{diag}}^{\dagger} & 0 \end{pmatrix} \begin{pmatrix} R^{\dagger} & 0 \\ 0 & I_k \end{pmatrix} = \begin{pmatrix} R\Omega_{\text{diag}}R^{\dagger} & RW_{\text{diag}} \\ -W_{\text{diag}}^{\dagger}R^{\dagger} & 0 \end{pmatrix}. \tag{31}$$

In the paper[9] we calculated explicitly controllers of various unitary gates; the controlled NOT gate, the discrete Fourier transformation gate and so on.

6. Conclusion

We formulated and solved the isoholonomic problem in the homogeneous fiber bundle. The problem was reduced to a boundary value problem of the horizontal extremal equation. We determined the control parameters that satisfy the boundary conditions. This result is applicable for producing arbitrary unitary gates.

References

1. R. Montgomery, *Commun. Math. Phys.* **128**, 565 (1991).
2. B. Simon, *Phys. Rev. Lett.* **51**, 2167 (1983).
3. M. V. Berry, *Proc. Roy. Soc. Lond.* **A392**, 45 (1984).
4. H. Kuratsuji and S. Iida, *Prog. Theor. Phys.* **74**, 439 (1985).
5. F. Wilczek and A. Zee, *Phys. Rev. Lett.* **52**, 2111 (1984).
6. P. W. Shor, *Proc. 35nd Annual Symposium on Foundations of Computer Science* (IEEE Computer Society Press) 124 (1994).
7. P. Zanardi and M. Rasetti, *Phys. Lett.* **A264**, 94 (1999); quant-ph/9904011.
8. J. Pachos, P. Zanardi and M. Rasetti, *Phys. Rev.* **A61**, 010305(R) (1999); quant-ph/9907103.
9. S. Tanimura, M. Nakahara and D. Hayashi *J. Math. Phys.* **46**, 022101 (2005); quant-ph/0406038.
10. M. Nakahara, Y. Kondo, K. Hata and S. Tanimura, *Phys. Rev.* **A70**, 052319 (2004); quant-ph/0405050.
11. M. Nakahara, J. J. Vartiainen, Y. Kondo, S. Tanimura and K. Hata; quant-ph/0411153.
12. Y. Kondo, M. Nakahara, K. Hata and S. Tanimura; quant-ph/0503067.

II

Topological Crystals

TOPOLOGICAL CRYSTALS OF $NbSe_3$

SATOSHI TANDA[1], TAKU TSUNETA[2], TAKESHI TOSHIMA[1], TORU MATSUURA[1], AND MASAKATSU TSUBOTA[1]

[1] *Department of Applied Physics, Hokkaido University, Sapporo, Hokkaido 060-8628, Japan*

[2] *Low Temperature Laboratory, Helsinki University of Technology, Otakaari 3A, Espoo, Finland*

We report the discovery of a Möbius crystal of $NbSe_3$, conventionally grown as ribbons and whiskers. We also reveal their formation mechanisms of which two crucial components are the spherical selenium (Se) droplet, which a $NbSe_3$ fiber wraps around due to surface tension, and the monoclinic ($P2_1/m$) crystal symmetry inherent in $NbSe_3$, which induces a twist in the strip when bent. Our crystals provide a non-fictitious Möbius world governed by a non-trivial real-space topology.

1. Introduction

The Möbius strip, which can be made by simply twisting an ordinary strip by 180 degrees and then joining the two ends, provides an exotic one-sided world, and has inspired many people ranging from artists, like M. C. Escher who put several ants on its surface in his paintings, to scientists who put Cooper pairs [1], electrons [2] and Ising spins[3,4] instead of the ants on its surface. Besides fictitious and theoretical worlds, one may wonder if such the one-sided world can be realized in a crystal? Crystal rigidity at first sight seems to prevent either bending or twisting. How can a crystal grow in the shape of Möbius strip? Here we report the discovery of a Möbius crystal of $NbSe_3$, conventionally grown as ribbons and whiskers. We also reveal their formation mechanisms of which two crucial components are the spherical selenium (Se) droplet, which a $NbSe_3$ fiber wraps around due to surface tension, and the monoclinic ($P2_1/m$) crystal symmetry inherent in $NbSe_3$, which induces a twist in the strip when bent. Our crystals provide a non-fictitious Möbius world governed by a non-trivial real-space topology.

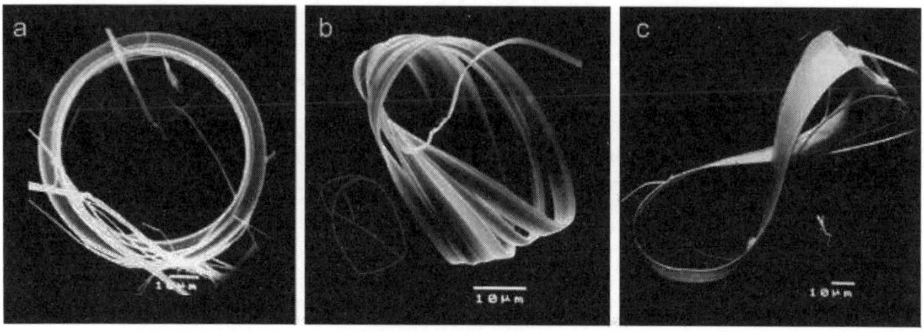

Figure 1. SEM images of the three types of $NbSe_3$ topological materials classified by their twists: (a), the ring (0π-twisted), (b)the Möbius strip (π-twisted), and (c) the figure-8 strip (2π-twisted). The scale is indicated on each image. The growth conditions of these materials are same.

2. Experimental

$NbSe_3$ crystal is a typical low-dimensional inorganic conductor with monoclinic symmetry and displays phase transitions at 52 K and 145 K into a charge-density-wave (CDW) ground state as a consequence of electron-phonon interactions; a periodic charge density modulation accompanied by a periodic lattice distortion. The $NbSe_3$ crystal has been synthesized with the chemical vapor transport method and their shapes are usually fibro-crystalline ribbons and whiskers.

Here we discovered a Möbius strip of single crystals of $NbSe_3$ under the following growth conditions. Starting materials were composed of a mixture of Se and Nb powder. The purity of all materials used in this work was 99.99%. The mixture was soaked at 740 °C for a few hours to a few days in an evacuated (less than 10^{-6} torr) quartz tube and then quenched to room temperature. A crucial difference in growth condition compared to the conventional method is the use of a furnace with a large temperature gradient, leading to significant nonequilibrium state inside the quartz tube, in which Se molecules are circulating through vapor, mist, and liquid (droplet) phases, analogous to the water circulation on the earth.

Figure 1 shows scanning electron microscope photographs of (a) the ring, (b) the Möbius, and (c) figure-8 $NbSe_3$ crystals. The ring diameters and widths are typically 100 μm and 1 μm width, respectively. Typical Möbius crystals have ~50 μm diameters and widths less than 1 μm. The figure-8 strip crystals with a double twist have ~ 200μm circumferences and 1 μm widths. We confirmed by both X-ray diffraction patterns and

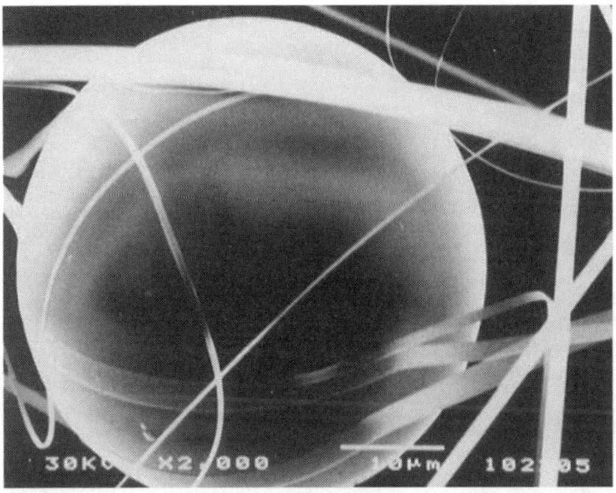

Figure 2. A SEM picture of $NbSe_3$ fibers (the white streaks in the picture) circulating a solidified drop of selenium (the sphere). The diameter of the drop is about 50 μm, which is comparable to that of the topological materials. As the picture shows, thin $NbSe_3$ fibers that touch a Se droplet are bent due to surface tension. This encirclement leads to the formation of the rings.

transmission electron diffraction that the crystallinity of the ring and the Möbius was equivalent to that of the conventional ribbon crystal: single-phase and monoclinic ($P2_1/m$) with the lattice constants a = 10.01 Å, b = 3.48 Å, c = 15.63 Å, and $\beta = 108.5°$. It is convenient to introduce a twist number n to specify the obtained topological materials. Our crystals are thereby labeled as the ring ($n = 0$), Möbius ($n = 1$), and figure-8 ($n = 2$), forming a subgroup of $n\pi$-twisted-loop crystals. By using this terminology, carbon nanotubes [5] can also be categorized to $n = 0$ loop.

We propose the following mechanism of ring (0π twisted) formation: The ribbon-shaped $NbSe_3$ crystals grown in the viscous Se droplet are bent due to Se surface tension, such that a growing crystal can then eat its own tail. Consequently, the crystals form perfectly seamless rings. Figure 2 shows the SEM picture of a circulating $NbSe_3$ fiber on the Se droplets during growth and strongly supports the proposed mechanism. An alternative story is also possible in the presence of a liquid selenium droplet on the wall of the quartz tube. After Se evaporates from the drop, Se chains remain at the edge of the drop [6], forming a ring. Vaporized $NbSe_3$ molecules crystallize from the Se ring which acts as the nucleation centre (Fig. 3). In either case, the Se droplets are required for $NbSe_3$ ring formation.

Growth of the Möbius strip seems difficult compared with that of the ring because of twisting. Note that bending of a bar or beam is accom-

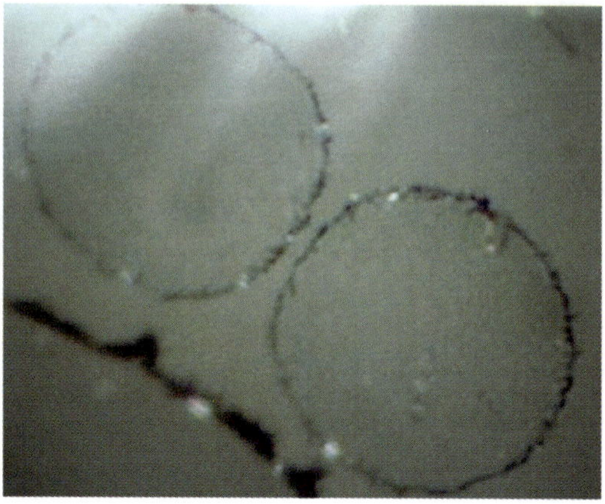

Figure 3. Annularly aggregated $NbSe_3$ crystallites appeared after evaporation of a Se droplet. The diameter of these annuluses is roughly 200 μm. This is another way of crystallites, which drift with the circulation of selenium inside a closed qualtz tube, to be formed in the shape of a ring.

Figure 4. A twisted $NbSe_3$ ribbon on a Se drop. A twist always appears on bending a beam that is elastically anisotropic. This kind of process is relevant to the formation of a Möbius strip.

panied with twisting in spite of the crystal symmetry unless the cross section of the bar is a perfect circle[7]. Crystal symmetry also promotes the bending-twisting conversion: low symmetry crystals, such as monoclinic and triclinic, transform bending to twisting through off-diagonal matrix elements of the compliance tensor. For example, the element s_{35} combines

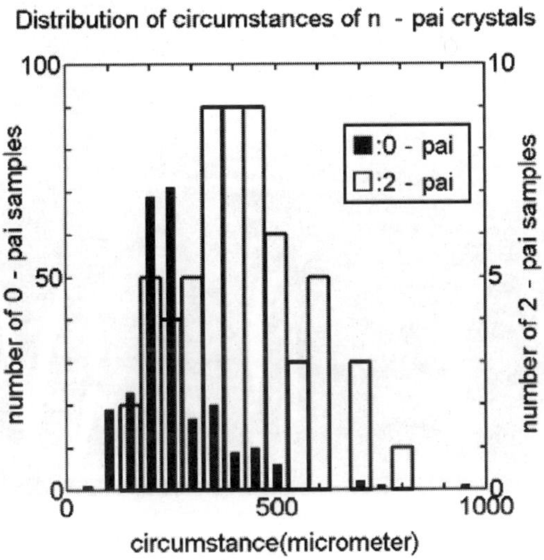

Figure 5. The distribution histogram of the circumference of the three types of material: orange, the ring-shaped crystals (0π), light blue, the Möbius crystals (π), and dark blue, the figure-8 crystals (2π). The samples are taken from batches with the same growth condition.

the bending around x_1 axis with the twisting around x_3 axis[8]. Figure 4 shows a clear evidence of the twisting of $NbSe_3$ on the Se droplets during encircling growth. In addition, the droplet rotations that we often observed in experiments might help to produce the twist.

The figure-8 crystals arise as a result of either the double encircling or double twisting (2π). According to the famous White theorem in the topology fields[9], a double encircling loop is topologically equivalent to the (2π) twisted loop, so-called isotope discussed in a ring DNA supercoil system [10]. Figure 5 shows the distribution of the circumference of these topological crystals. The circumference of the figure-8 crystal is about two times as large as the other two types on average. From this we concluded that the double-encircling mechanism is preferable to the twisting; $NbSe_3$ fibers encircle a Se droplet twice before eating its own tail.

Figure 6 summarizes the processes of the $n\pi$ crystals deduced from the SEM pictures. It turns out that Se droplet is necessary for the encircling process and crystal symmetry is the key for inducing a twist.

CDW is a manifestation of a quantum effect that occurs on a macroscopic scale as a result of coherent superposition of a large number of micro-

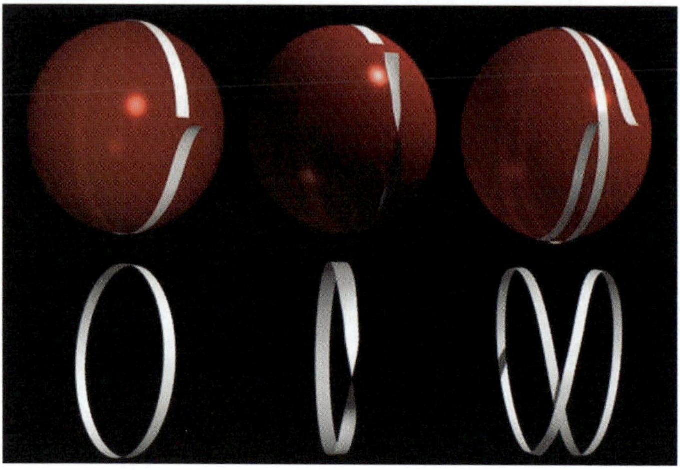

Figure 6. The schematic illustration describing the dominant formation mechanisms of each three class of our topological crystals. The red spheres and white ribbons represent droplets of selenium and ribbon-shaped crystals of $NbSe_3$, respectively. (a) Rings(0π-twisted): A $NbSe_3$ ribbon is spooled to a Se droplet by surface tension, and then its both ends bond to each other to form a ring. (b) Möbius crystals (π-twisted): The spooling of a ribbon can also produce a twist, which is essential for the formation of a Möbius crystal, due to its anisotropic elastic properties. (c) figure-8 crystals (2π-twisted): The loop in this picture, formed by encircling the droplet twice, has no twist. However, it can transform into another loop that have a twist of 2π (see the actual crystal in figure 1(c)). These two types of loop, the one of double encircling and the one with 2π twist, are in a same topological class. Although figure-8 crystals can be made by a spooling process that involves twisting in a similar manner to Möbius crystals, our observation suggests that the formation process described above predominates.

scopic degrees of freedom. Do the ring, Möbius, and figure-8 crystals exhibit a CDW transition like the conventional $NbSe_3$ ribbon and fiber crystals? The following three measurements were performed: (1) The satellites in the electron diffraction patterns show CDW formation with the CDW wave vector $Q_1 = 0.24 \pm 0.01$ (Fig. 7). (2) Anomalies due to CDW phase transitions in the temperature dependence of the resistivity were observed at 141-144 K and 52-54 K, which are less than those of the conventional ribbon-shaped crystals. (3) Nonlinear conduction due to CDW sliding was also observed. The threshold field is similar to that for the conventional ribbon-shaped $NbSe_3$ crystals [11,12,13]. These results comprise convincing evidence for an annular (topological) and Möbius strip CDW formation. The formation of CDW indicates that the samples are good crystals and relatively free from the expected disorder originating from bending of the

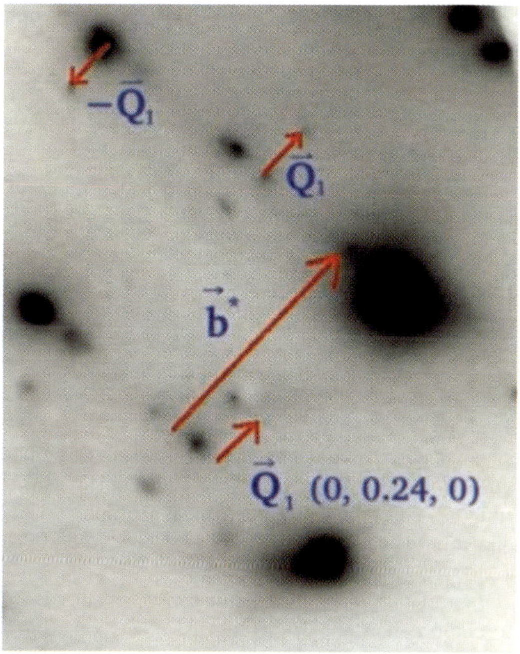

Figure 7. The electron diffraction pattern of a $NbSe_3$ ring crystal taken at 135K. It shows the satellite due to CDW formation with periodic lattice distortion. The CDW wave vector is $\vec{Q_1} = 0.24 \pm 0.01\vec{b}^*$, which agrees with that of ribbon shaped $NbSe_3$ crystals.

crystal axis. The transport phenomena including interference effect of these materials are now being studied in detail.

By investigating the formation mechanism of topological crystals, we have developed a new growth technique for topological crystals by using a spherical-droplet as a spool. This spherical-droplet spool technique might be applicable to a wide class of materials and it may be possible to grow the crystals of an arbitrary size by controlling the nonequilibrium conditions in the furnace, i.e., the size of the droplets. This technique provides a powerful way for studying the almost unexplored area of topological effects in condensed matter. For instance, the minimum diameter of our ring samples was 300 nm which can be regarded as mesoscopic. Such a sample thereby enables us to investigate Aharonov-Bohm effect as a topological effect in CDW and/or superconducting states. Our newly discovered crystals will open a new area for exploring the topological effects in quantum mechanics, like Berry's phase[14,15], in addition to the potential for constructing new devices.

The authors are grateful to K. Inagaki, K. Yamaya, Y. Okajima, N. Hatakenaka, T. Sambongi, T. Matsuyama, M. Hayashi, G. E. Volovik, P. Hakonen, M. Paalanen, M. Nishida, K. Kagawa, and M. Jack for useful discussions. We also thank H. Kawamoto, M. Shiobara, Y. Sakai, K. Ikeda, K. Asada, Y. Nogami, K. Ikeda and S. Yasuzuka for experimental support. We also would like to thank S. Mori for the contribution of the illustrations. This reserch was supported by the Japan Society for the Promotion of Science, the Ministry of Education, Japan.

References

1. Hayashi, M. and Ebisawa, H., Little-Parks Oscillation of Superconducting Möbius Strip., *J. Phys. Soc. Jpn.* **70**, 3495–3498 (2001); Hayashi, M., Ebisawa, H., and Kuboki, K., Superconductivity on a Möbius strip: Numerical studies of order parameter and quasiparticles., *Phys. Rev. B* **72** 024505 (2005).
2. Mila, F., Stafford, C., and Capponi, S., Persistent currents in a Möbius ladder: A test of interchain coherence of interacting electrons., *Phys. Rev. B* **57** 1457–1460 (1998).
3. Ito, H. and Sakaguchi, T., 2D Ising spin system on the Möbius strip., *Phys. Lett.* **A 160**, 424–428 (1991).
4. Kaneda, K. and Okabe, Y., Finite-Size Scaling for the Ising Model on the Möbius Strip and the Klein Bottle., *Phys. Rev. Lett.* **86**, 2134–2137 (2001).
5. Iijima, S., Helical microtubules of graphitic carbon., *Nature* **354**, 56–58 (1991).
6. Deegan, R. D., Bakajin, O., Dupont, T. F., Huber, G., Nagel, S. R., Witten, T. A., Capillary flow as the cause of ring stains from dried liquid drops., *Nature* **389** 827–829 (1997).
7. Landau, L. D. and Lifshitz, E. M., Theory of Elasticity., (Pergamon Press, Oxford, 1959).
8. Hearmon, R. F. S., An introduction to applied anisotropic elasticity., (Oxford Univ. Press, London, 1961).
9. White, J. H., Self-linking and Gauss integral in higher dimensions., *Amer. J. Math.*, **91**, 693–728 (1969).
10. Vologodskii, A.V., Anshelevich, V.V., Lukashin, A.V., Frank-Kamenetskii, M.D. Statistical mechanics of supercoils and the torsional stiffness of the DNA double helix., *Nature* **280** 294–298 (1979).
11. Tsutsumi, K., Takagaki, T., Yamamoto, M., Shiozaki, Y., Ido, M., Sambongi, T., Yamaya, K., and Abe, Y., Direct Electron-Diffraction Evidence of Charge-Density-Wave Formation in $NbSe_3$, *Phys. Rev. Lett* **39** 1675–1676 (1977).
12. For reviews on CDWs, see , edited by Monceau P., Electronic Properties of

Inorganic Quasi-One-Dimensional Compounds. , (Reidel, Dordrecht, 1985).
13. Grüner, G., Density Waves in Solids., (Addison-Wesley, Reading, 1994).
14. Berry, M. V., Quantal phase factors accompanying adiabatic changes., *Proc. R. Soc. Lond Ser. A* **392**, 45–57 (1984).
15. Ando, T., Nakanishi, T., and Saito, R., Berry's Phase and Absence of Back Scattering in Carbon Nanotubes., *J. Phys. Soc. Jpn.* **67**, 2857–2862 (1998).

SUPERCONDUCTING STATES ON A MÖBIUS STRIP

M. HAYASHI, T. SUZUKI AND H. EBISAWA

Graduate School of Information Sciences, Tohoku University, 6-3-09 Aramaki Aoba-ku, Sendai 980-8579, Japan and CREST-JST

K. KUBOKI

Department of Physics, Kobe University, Kobe 657-5801, Japan

The superconducting states on a Möbius strip are studied based on Ginzburg-Landau theory and Bogoliubov-de Gennes theory. It is shown that, in a Möbius strip made of an anisotropic superconductor, the Little-Parks oscillation, which occurs when an magnetic flux is threading a superconducting ring, is significantly modified. Especially, when the flux is close to a half-odd-integer times the flux quantum, a new type of states appears, which we call the "nodal state". In these states the superconducting gap has a node in the middle of the strip along the circumference. We discuss the stability and the electronic properties of these states in two-dimensional case, where the thickness of the strip is negligible. A possible extension of this analysis to the thicker strips is also addressed.

1. Introduction

The realization of crystals with unusual shapes, e.g., ring, cylinder, eight-figure, Möbius strip etc., by Tanda and coworkers[1,2,3] has stimulated renewed interest in the effects of the system geometry on the physical properties. Especially, the synthesis of Möbius strip made of transition metal calcogenides ($NbSe_3$, TaS_3 etc.) opens new possibility to examine the physical properties of superconductivity or charge-density-wave in topologically nontrivial spaces.

Recently, Hayashi and Ebisawa[4] have studied s-wave superconducting (SC) states on a Möbius strip based on the Ginzburg-Landau (GL) theory and found that the Little-Parks oscillation, which is characteristic to the ring-shaped superconductor, is modified for the Möbius strip and a new state, which does not appear for ordinary rings, shows up when the number of the magnetic flux quanta threading the ring is close to a half-odd-integer. Vodolazo and Peeters[12] have studied the eight-figure SC ring and have predicted intriguing behaviors caused by its topological form. Yakubo, Avishai and Cohen[5] have studied the spectral properties of the metallic Möbius strip with impurities and clarified statistical characteristics of the fluctuation of the persistent current as a function of the magnetic flux threading the ring. The persistent current in a more simplified version of the Möbius strip has also been studied by Mila, Stafford and Caponi[6]. Wakabayashi and Harigaya[7] have studied the Möbius strip made of a nanographite ribbon, and the effects of Möbius geometry on the edge localized states, which is peculiar to the graphite ribbon, have been clarified. A study from a more fundamental point of view can be found in the paper by Kaneda and Okabe[8] where the Ising model on Möbius strip and its domain wall structures are studied.

In this paper, we report our studies on the physical properties of a SC Möbius strip[4,9,10]. Since $NbSe_3$ can be SC on doping or under hydrostatic pressure, this system is now experimentally realizable. With actual system in mind, we consider a system consisting of an array of one-dimensional SC chains, as depicted in Fig. 1. We assume that the chains are weakly coupled by inter-chain hopping and the strip can be treated as an anisotropic superconductor.

In this paper we first present the studies based on GL equation in Sec. 2 and then those based on microscopic Bogoliubov-de Gennes (BdG) equation in Sec. 3.

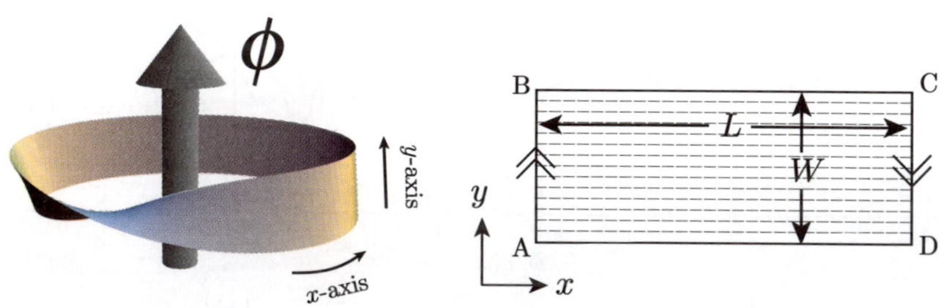

(a) (b)

Figure 1. (a) Structure of the Möbius strip. The bold arrow shows the direction of the magnetic flux. The setting of x- and y-axis is also indicated. (b) Developed figure. The broken lines represent the direction of the SC chains comprising the ring. In this figure, the segment A-B and C-D are identified with the orientation indicated in the figure.

2. Ginzburg-Landau Theory

We consider the strip as shown in Fig. 1. The width, circumference and inter-chain spacing are denoted by W, L and a, respectively. The GL free energy of the system can be written as

$$F = \sum_{i=-K+1}^{K} \int_0^L \mathrm{d}x \left[\frac{1}{2m^*} \left| \left(\frac{\hbar}{i}\partial_x + \frac{e^*}{c}A_x \right) \psi_i \right|^2 + \alpha \left|\psi_i\right|^2 + \frac{\beta}{2}\left|\psi_i\right|^4 \right]$$
$$+ \sum_{i=-K+1}^{K-1} \int_0^L \mathrm{d}x\, v \left|\psi_{i+1} - \psi_i\right|^2 . \qquad (1)$$

Here the x-coordinate $(0 < x \leq L)$ is taken along the azimuthal direction of the ring (see Fig. 1). $\psi_i(x)$ and A_x are the order parameter of the i-th chain and the x-component of the vector potential, respectively. The number of the chains is assumed to be even $(= 2K)$ for simplicity. v is a parameter of the interchain Josephson coupling. The vector potential is taken to be a constant $A_x = \phi/L$, where ϕ is the magnetic flux enclosed by the ring. We assume that the magnetic flux on the strip is negligible. α and β are constants, where $\alpha = \alpha_0(t-1)$ with t begin T/T_c (T_c is the SC transition temperature in the bulk). ¿From an approximate calculation based on GL theory [4], one can obtain the phase diagram of the SC Möbius strip as shown in Fig. 2.

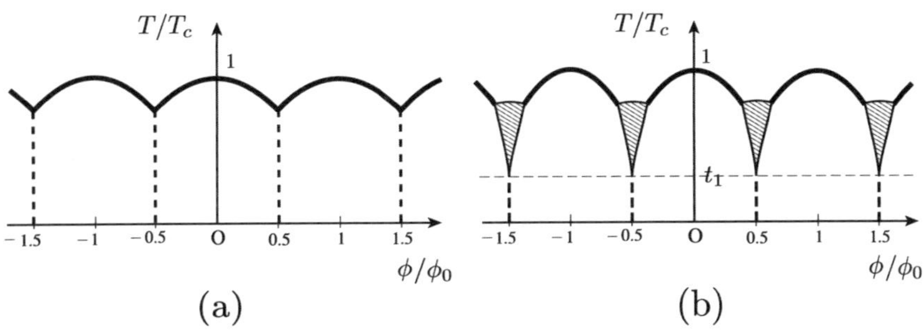

Figure 2. The phase diagram of a SC Möbius strip based on GL free energy. (a) for the case of $\frac{\pi}{2\sqrt{3}} r_{\parallel} < r_{\perp}$ and (b) $r_{\perp} < \frac{\pi}{2\sqrt{3}} r_{\parallel}$ (see text for details).

Here we find two important parameters which determines the SC behaviors of the Möbius strip. They are defined by $r_{\perp} = \xi_{\perp}(0)/W$ and $r_{\parallel} = \xi_{\parallel}(0)/L$ where $\xi_{\parallel}(0)^2 \equiv \hbar^2/(2m^*\alpha_0)$ is the coherence length parallel to the chains and $\xi_{\perp}(0)^2 \equiv a^2 v/\alpha_0$ is that perpendicular to the chains obtained by applying continuum approximation to Eq. (1). When $\frac{\pi}{2\sqrt{3}}r_{\parallel} < r_{\perp}$, the magnetic phase diagram of the system is essentially same as that for an ordinary SC ring. In this case, the critical temperature shows well-known Little-Parks oscillation. However, when $r_{\perp} < \frac{\pi}{2\sqrt{3}}r_{\parallel}$, a series of new states appears when the flux is close to a half-odd-integer times the flux quantum, as shown in Fig. 2 (b) by hatched regions. The minimum temperature of the stability regions of these states is given by $t_1 = 1 - \left(\frac{3\pi^2}{4\sqrt{2}}\frac{r_{\parallel}^2}{r_{\perp}}\right)^2$. The energy gap in these states has a real-space line-node at the center the strip along the circumference. Thus we call these states the nodal states. Order parameter configuration in the nodal state ($\phi \sim \phi_0/2$) is given in Fig. 3.

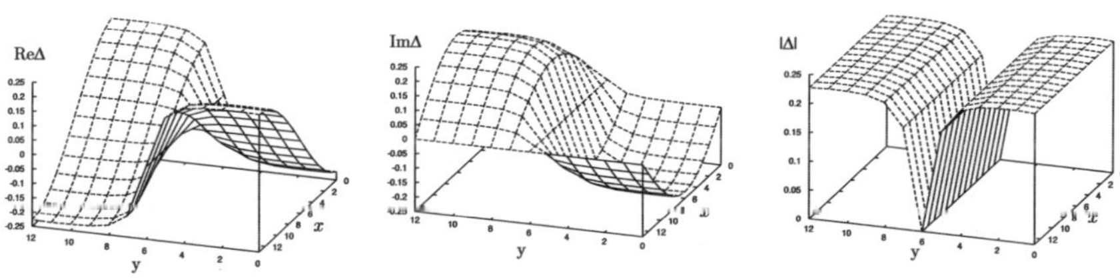

(a) (b) (c)

Figure 3. Order parameter configuration in the nodal state at $\phi \sim \phi_0/2$. (a) real part, (b) imaginary part and (c) amplitude are shown.

The results above are obtained based on an approximate analytical calculation, in which we have assumed a possible configuration of the order parameter and compared the free energy of these states with that of other possible states. Therefore we cannot avoid the arbitrariness of the order parameter choice. To overcome this difficulty we performed a numerical minimization of the free energy, Eq. (1). This kind of procedure has been used for the analysis of various mesoscopic superconductors[11].

We have discretized the Eq. (1) and found as many metastable states as possible using numerical minimization method. The obtained free energy for three different temperatures are given in Fig. 4. The calculation has been done with $L = W = 10a$, $\xi_{\parallel}(0) = 1.5a$ and $\xi_{\perp}(0) = 1.2a$. One can see the existence of the nodal state near $\phi = \phi_0/2$. It is a true equilibrium state at $t = 0.78$, although it becomes metastable at $t = 0.5$ and below. This behavior is qualitatively in good agreement with the previous result, Fig. 2.

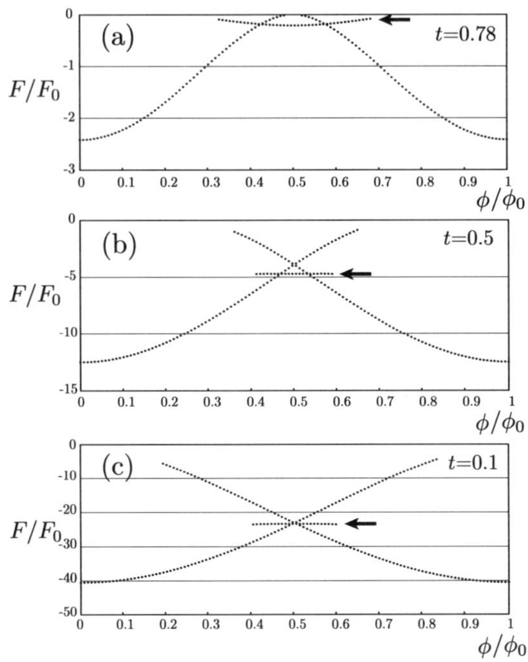

Figure 4. Free energy of metastable states in SC Möbius strip for (a) $t = 0.78$, (b) $t = 0.5$ and (c) $t = 0.1$. Arrows indicates the branch corresponding to the nodal state.

3. Bogoliubov-de Gennes Theory

The electronic properties of the nodal states are also of interest from both theoretical and experimental aspects. Especially, we can expect low energy bound states of quasiparticles in the node. The quasiparticle energy levels are obtained by solving BdG equation[13]. We employ the tight-binding-model for electrons and the Hamiltonian is given by

$$
\begin{aligned}
H = & -t_x \sum_{j\sigma} (e^{i\phi_x} c^\dagger_{j+\hat{x},\sigma} c_{j,\sigma} + e^{-i\phi_x} c^\dagger_{j,\sigma} c_{j+\hat{x},\sigma}) \\
& -t_y \sum_{j\sigma} (c^\dagger_{j+\hat{y},\sigma} c_{j,\sigma} + c^\dagger_{j,\sigma} c_{j+\hat{y},\sigma}) \\
& -V \sum_{j} n_{j\uparrow} n_{j\downarrow} - \mu \sum_{j\sigma} c^\dagger_{j,\sigma} c_{j,\sigma}
\end{aligned}
\tag{2}
$$

where $c_{j\sigma}$ is the annihilation operator of electron at site j with spin $\sigma(=\uparrow, \downarrow)$, and $V(> 0)$ and μ are the attractive interaction and the chemical potential, respectively. Here $j = (j_x, j_y)$ $(1 \leq j_x \leq N_x, 1 \leq j_y \leq N_y)$ numbers the sites, where N_x and N_y are the numbers of sites along the x- and y-direction, respectively, and $\hat{x} = (1,0)$, $\hat{y} = (0,1)$. $n_{j\sigma} = c^{\dagger}_{j\sigma} c_{j\sigma}$ is the electron number operator. The transfer integrals for x- and y-directions are denoted by t_x and t_y, respectively, and the Peierls phase $\phi_x = (\pi/N_x)(\phi/\phi_0)$ represents the effect of the AB flux ϕ threading the Möbius strip. The interaction term is decoupled within a mean-field approximation as

$$n_{j\uparrow} n_{j\downarrow} \to \Delta_j c^{\dagger}_{j\downarrow} c^{\dagger}_{j\uparrow} + \Delta^*_j c_{j\uparrow} c_{j\downarrow} - |\Delta_j|^2 \tag{3}$$

with $\Delta_j \equiv \langle c_{j\uparrow} c_{j\downarrow} \rangle$ being the SC order parameter. By tuning the system parameters we have succeeded in obtaining the nodal state for the case of $\Phi = \phi_0/2$ and calculated the local density of states (LDOS). The LDOS is calculated from

$$N(j,E) = -\frac{1}{\pi} \mathrm{Im} \sum_n \frac{u^*_{jn} u_{jn}}{E - E_n + i\Gamma}. \tag{4}$$

where Γ is the broadening of the single energy level, introduced to simulate the actual experiment, and j and n are the site number and the index of the states, respectively. u_{jn} is the n-th wave function evaluated at the site j. As one can see from the Fig. 5, the indication of the bound states can be seen in inner chains, although well-developed gap is observed in outer chains. The details of the bound states (such as circumference dependence) is discussed elsewhere[10].

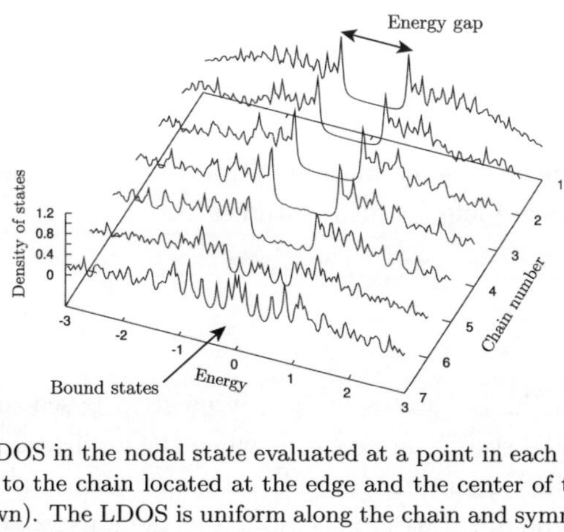

Figure 5. The LDOS in the nodal state evaluated at a point in each chain. The chain 1 and 7 correspond to the chain located at the edge and the center of the strip (only half of the strip is shown). The LDOS is uniform along the chain and symmetric with respect to the central chain "7". Indication of the bound states can be seen in inner chains.

4. Three-Dimensionality

Above mentioned results are for two-dimensional strips, where the thickness is negligible. This, however, is not the case of experimentally realized Möbius strips. Therefore we discuss briefly in this section what happens if the thickness is not negligible and show that the node becomes a embedded vortex line if the strip is thick enough. Here, it is useful to generalize the notion of the "Möbius geometry"in the way previously suggested by Volovik[14]. Here we consider $\pi/2$- and π-Möbius geometry (Fig. 6 (b) and (c), respectively), which is obtained by twisting a bar with a square cross section by $\pi/2$ or π before making it a ring by closing the both ends. In closing, the points A and A' (B and B') are overlapped in Fig. 6. The π-Möbius geometry is equivalent to the ordinary Möbius strip obtained when the strip is very thick.

Now we put a singly quantized vortex line embedded at the center of the bar as shown by bold arrows in Fig. 6. The phase shift around the vortex line induces an additional phase shift along the bar when we twist it. In case of $\pi/2$-Möbius, we obtain $\pi/2$ additional phase shift between the points, A and A', in Fig. 6 (b). Then, if we put 1/4 flux quantum to the $\pi/2$-Möbius ring, which generates $-\pi/2$ Peierls phase along the circumference, the total phase shift between A and A' vanishes, which means that the total current flowing in the ring is also vanishing and this state can be energetically stable, if the energy cost due to the vortex line is relatively small. In case of π-Möbius, we see in the same way that single vortex state can be stable when the number of the flux quantum is 1/2, which corresponds to the nodal state discussed for Möbius strips in the preceding sections.

¿From these considerations, we expect that the generalized Möbius superconductors show much more variety of states, although actual phase diagram may depends on the material parameters. More detailed analysis of these problems are left for future studies.

5. Summary

We have studied the SC states on a Möbius strip based on GL and BdG theory. The "nodal state"can be stable when the applied flux is a half-odd-integer times a flux quantum depending on the anisotropy of the system. The existence of bound states in the node was demonstrated by calculating the local density of states. Some possible extension of the nodal states to thicker Möbius strip and generalized Möbius geometry was presented.

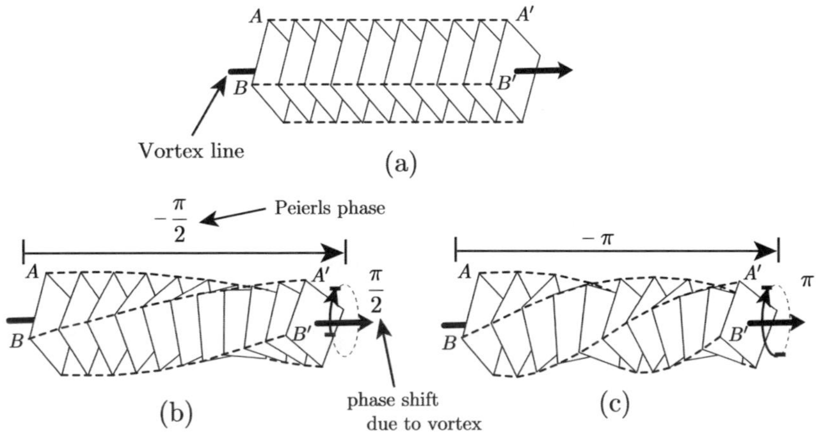

Figure 6. (a) Untwisted, (b) $\pi/2$- and (c) π-twisted bar with a square cross section. The latter two give generalized Möbius geometries.

References

1. S. Tanda et al., Nature **417** 397 (2002).
2. S. Tanda et al., Physica B **284-288** 1657 (2000).
3. Y. Okajima et al., Physica B **284-288** 1659 (2000).
4. M. Hayashi and H. Ebisawa, J. Phys. Soc. Jpn. **70** 3495 (2002).
5. K. Yakubo, Y. Avishai and D. Cohen, Phys. Rev. B **67** 125319 (2003).
6. F. Mila, C. Stafford and S. Capponi, Phys. Rev. B **57** 1457 (1998).
7. K.Wakabayashi and K. Harigaya, J. Phys. Soc. Jpn. **72** 998 (2003).
8. K. Kaneda and Y. Okabe, Phys. Rev. Lett. **86** 2134 (2001).
9. M. Hayashi, H. Ebisawa and K. Kuboki, cond-mat/0502149.
10. T. Suzuki, M. Hayashi, H. Ebisawa and K. Kuboki, in preparation.
11. B. J. Baelus, F. M. Peeters and V. A. Schweigert, Phys. Rev. B **61**, 9734 (2000).
12. D. Y. Vodolazov and F. M. Peeters, Physica C **400**, 165 (2004).
13. P. G. de Gennes, *Superconductivity of Metals and Alloys* (W. A. Benjamin Inc., 1966, New York).
14. G. E. Volovik, *The Universe in Helium Droplet* (Oxford Science Publishing, 2003, New York).

STRUCTURE ANALYSES OF TOPOLOGICAL CRYSTALS USING SYNCHROTRON RADIATION

Y. NOGAMI,[1,2] T. TSUNETA, [3,4] K. YAMAMOTO,[1] N. IKEDA,[5] T. ITO,[1] N. IRIE,[1] AND S. TANDA[3]

[1] *Division of Frontier and Fundamental Sciences, Graduate School of Natural Science and Technology, Okayama University, Okayama 700-8530, Japan.*
[2] *CREST, Japan Science and Technology Agency, Saitama 332-0012, Japan.*
[3] *Department of Applied Physics, Hokkaido University, Sapporo 060-8628, Japan.*
[4] *Low Temperature Laboratory, Helsinki University of Technology, Finland.*
[5] *Japan Synchrotron Radiation Research Institute, SPring-8, Hyogo 679-5198, Japan.*

Structure analyses of topological crystals were done using intensive synchrotron radiation from SPring-8. Firstly, directions of the crystal axes were determined using a highly sensitive oscillation camera under vacuum. The atomic arrangement of topological crystals was then determined by the newly developed X-ray camera under vacuum. Small but systematic shrinkage of the a axis with thickness was observed in ring crystals.

1. Introduction

Recently, S. Tanda *et al.* found that the instability against bending and twisting deformation of whisker crystals of quasi one-dimensional conductor NbSe3. Growing along the one-dimensional b axis, $NbSe_3$ micro-whisker was spooled by a Se droplet and bent by the surface tension of the droplet. Utilizing this effect, Tanda *et al.* produced ring (cylinder, tube), Möbius strip and figure-of-eight strip crystals, as shown in Fig. 1. Since both ends of a crystal are joined, these crystals belong to entirely different topological classes from that of ordinary crystals, and are therefore named topological crystals[1]. Surprisingly, there is no seam in their appearance, and the direction of microcrystallinity seems to change smoothly(see Fig. 1).

Strictly speaking, these SEM pictures do not reflect the microscopic atomic arrangement of topological crystals. From a crystallographic point of view, the mechanical stress caused by joining crystal ends may cause a structural difference between the conventional whisker and topological

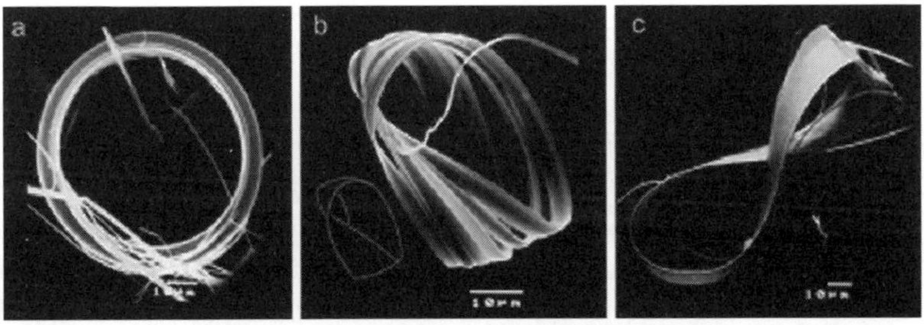

Figure 1. Topological crystals of $NbSe_3$[1].

crystals, so we chose to conduct structure analyses of topological crystals using X-ray.

Furthermore, structure analyses of topological crystals are important for understanding their electronic properties. A conventional whisker crystal of $NbSe_3$ has a quasi one-dimensional structure with the space group of monoclinic $P2_1/m$ and quasi one-dimensional Fermi surfaces leading to multiple charge-density-wave (CDW) transitions at 141 K and 58 K. A small decrement in the CDW critical temperature of topological crystals[2] indicates a possible change of structure. Furthermore, if there is little lattice imperfection, topological crystals with no end will be ideal for investigating the basic quantum mechanical problem of topological interference in the CDW quantum state.

Thus, there is an urgent need to determine the structure of topological crystals in order to understand their properties more deeply. In this paper, we present structural identification of topological crystals(rings, a cut ring, figure-of-eight strips) of $NbSe_3$.

2. Experimental

Due to the smallness of the sample crystal size (1000–10000 μm^3), we used intensive synchrotron radiation at beam line BL02B1 of SPring-8. The beam was monochromatized at $\lambda = 0.99184$ Å to suppress the fluorescent X-ray of selenium SeKα. To suppress the air scattering of X-ray we used a highly sensitive oscillation camera, named the low temperature vacuum camera (LTVAC), in BL02B1.

3. Results and discussion

Figure 2 shows oscillation photographs (rotation angle is 2 deg.) of a thicker ring crystal observed at room temperature. These diffraction patterns reflect the direction of the diffraction plane, and hence that of reciprocal lattice vectors. Many diffraction points in Fig. 2(a) resemble those in the oscillation diffraction pattern from a single crystal. However, short arcs can be seen especially in the right part of Fig. 2(a), which indicate the presence of continuous rotation of the diffraction plane in the ring crystal.

Furthermore, the diffraction feature in Fig. 2(b) consisting of many arcs is similar to the Debye ring pattern from powder crystals. These arcs are a 2θ constant diffraction pattern and the ellipsoidal (= not circular) appearance of the arcs is due to the deformation caused by use of the curved imaging plate (IP). This figure shows that reciprocal lattice vectors (= diffraction planes) are bent smoothly reflecting the ring crystal shape and form the Bragg ring. In terms of crystallography, when the Bragg ring intersects with the Ewald sphere almost orthogonally, the diffraction feature consists of spots (Fig. 2(a)). On the other hand, Bragg rings are located nearly on the surface of the Ewald sphere, and the diffraction feature consists of long arcs (Fig. 2(b)). In this sense, these figures reflect the ring topology of the unit cell vectors.

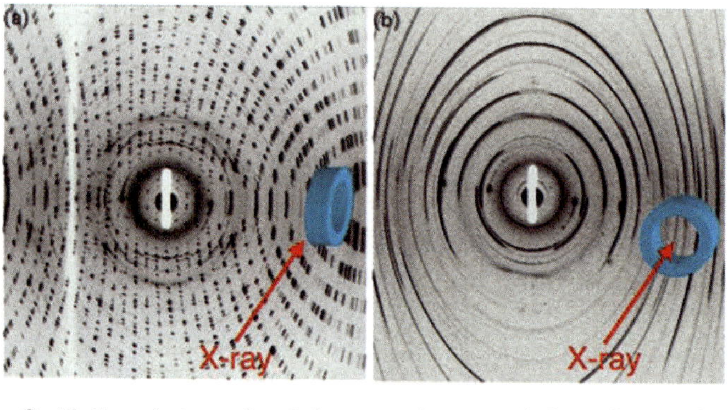

Figure 2. Oscillation photographs of the same ring crystal. Inset figures show relative crystal orientation to the beam.

Figure 3 is an oscillation photograph (rotation angle is 2 deg.) of a figure-of-eight crystal observed by LTVAC at room temperature. One can see two sets of strong nearly-parallel rows (see the right part of the figure). These rows are assemblies of short arcs, which indicate a small rotation angle of the diffraction planes. Thereby *two* sets of strong nearly-parallel rows come from two semi-straight parts with different directions around the crossover point of 8. On the other hand, one can see many thinner long arcs, around the center of Fig. 3. Long arcs mean a large rotation angle of the diffraction plane. Hence, this part should come from the hairpin curves of 8.

Figure 3. Oscillation photograph of figure-of-eight crystal.

Next, we conducted Rietvelt structure analysis[3] of the topological crystal. It is difficult to apply the single-crystal structure analysis method to topological crystals due to rotation of the diffraction plane reflecting sample shape, so we developed a new X-ray camera and averaged the orientation of the diffraction plane by using a two-axis sample rotator. Details of the developed X-ray camera together with the new analysis method have been presented elsewhere[4]. Figure 4 shows the result of structure analysis of a thin ring with thickness of 7.6 μm. R factors are quite small, showing that this analysis was successful. Note that we used only one topological crystal and did not destroy the sample to obtain powder-like diffraction features.

Comparing the results of similar Rietvelt structure analyses among the topological crystals, we noticed a systematic decrease of interchain direction a length especially in ring crystals as a function of the thickness, as shown in Fig. 5.

One possible explanation for this decrement in a length is the effect of elastic energy along the one-dimensional axis b. As the thickness t in a ring crystal increases, the difference in circumference lengths between inner and outer parts increases. Accordingly, the b length along the circumference direction (see the right figure) increases in the outer part but decreases in

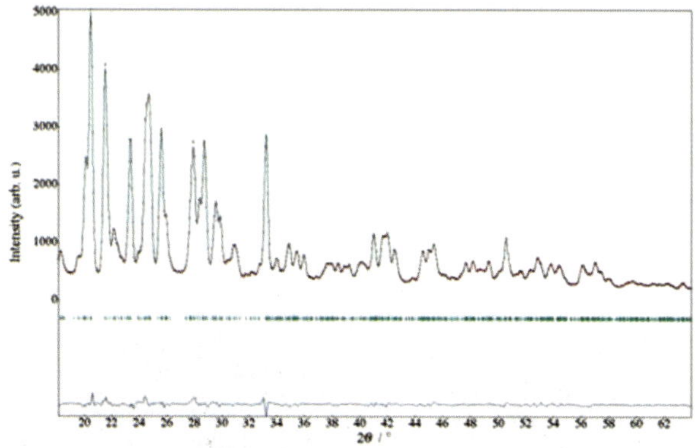

Figure 4. Rietvelt analysis of thin ring with 7.6 μm thickness. $R_{wp} = 2.38$, $S = 0.68$, $R_I = 0.72$.

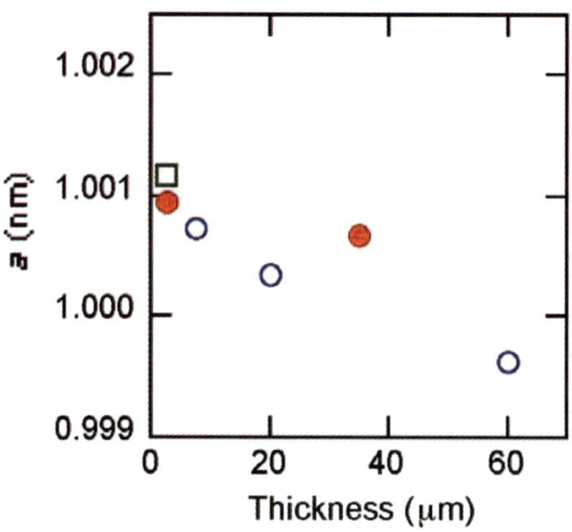

Figure 5. a length as a function of the thickness. Open circles denote the results of ring crystals, closed circles those of figure-of-eight crystals, open square that of the cut ring crystal.

the inner part as shown in Fig. 6. Since the compressibility of the one-dimensional axis b is small, the above deformation causes the loss of much elastic energy. To prevent this energy loss, the thickness t (the origin of the deformation along the b direction) tends to decrease. This means a decrement in the a length along the radial direction (see the right figure of Fig. 6). In the cut ring, the a length does not decrease, possibly owing to the relaxation by cutting.

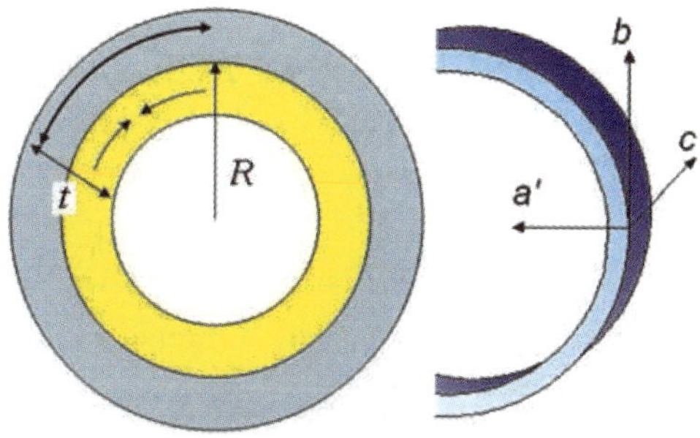

Figure 6. Schematic presentation of deformation in a ring crystal. The b length along the circumference direction increases in the outer part but decreases in the inner part. The right figure denotes the direction of the axes.

Another explanation is the self-pressure effect. The lattice parameter tends to shrink in the inner part of the ring. Note that the inner diameter is nearly zero in the thicker ring measured (thickness 20 μm and 60 μm).

Acknowledgments

This work was partially supported by the 21st century COE program on "Topological Science and Technology" and by a grant-in-aid for the scientific research of priority areas "Novel Function of Molecular Conductors under Extreme Conditions" from the Ministry of Education, Culture, Sport, Science and Technology of Japan. We acknowledge the support provided by Okayama University's priority research program on "Novel Quantum Effects and Phenomena in Materials with Structural Hierarchy-Integrated Approach to the Reorganization of Material Structure Science".

References

1. S. Tanda *et al.*, *Nature* **417**, 397 (2002).
2. Y. Okajima *et al.*, *Physica* **B284-288**, 1659 (2000).
3. F. Izumi and T. Ikeda, *Mater. Sci. Forum* **321-324**, 198 (2000).
4. K. Yamamoto *et al.*, *in this book*.

TRANSPORT MEASUREMENT FOR TOPOLOGICAL CHARGE DENSITY WAVES

T. MATSUURA, K. INAGAKI, S. TANDA

Department of Applied Physics, Hokkaido University,
Kita 13, Nishi 8,
Sapporo 060-8628, Japan
E-mail: toru@eng.hokudai.ac.jp

T. TSUNETA

Low Temperature Laboratory, Helsinki University of Technology,
Otakaari 3A, Espoo, Finland

Y. OKAJIMA

Asahikawa National College of Technology,
Harukouidai 2-2-1-6, Asahikawa, Japan

We have investigated transport properties of charge density wave (CDW) rings. The realization of CDW rings by synthesizing of niobium triselenide ring crystals provides a new system for investigation of topological effects in macroscopic quantum state. DC nonlinear conductivity measurement and the AC conductivity measurement are useful methods to investigate CDW dynamics. To investigate the topological effects, we cut the ring samples and measured AC conductivity again. We found anomalies of conductivity in DC and AC measurements. These anomalies could not be explained a simple parallel circuit model. These results suggest that the topology of CDW rings was reflected in CDW dynamics.

1. Introduction

Form is an essential concept in science. Topology is mathematical concept treating continuity of form. In recent years, topologically non-trivial crystals of charge density wave materials have been discovered.[1] The crystal forms are topologically identified as rings, the Möbius strips, and figures of 8, because each of them has a topological invariant. We call them topological crystals. The invariant is associated with boundary conditions, degree of freedom, curvatures and symmetry of a physical system. For example, the curvatures are finite values at each point of the topological crystals.

Lattice translation operators are unable to be defined. We have usually considered only simple condition like plain and infinite space in condensed matter physics. The topological invariant of crystal would request to change the base of condensed matter physics essentially.

The invariant would affect not only the crystal structure but also electrical property especially the CDW. The CDW is one of macroscopic quantum states of quasi one- or two-dimensional electron systems. Due to electron-phonon coupling, electron density waves develop with periodic lattice displacement.[2] The CDW has a long phase-phase correlation length of approximately 1∼10 μm. As the CDW is incommensurate to the lattice period, the CDW carries charge without dissipation. This is a possible mechanism of superconductivity proposed by Fröhlich before BCS theory.[3] However, the Fröhlich's superconductivity has not been found yet, because the CDWs are pinned by disorders in crystals. The disorders are impurities, dislocations, and boundaries such as edges of the crystals. Since the periodic potential is broken at each electrode, the electrodes also act as pinning centers. The number of impurities and dislocations could be decreased but the edges of the crystals seem impossible to be eliminated. However, since we now have succeeded in synthesizing of the ring crystals of CDW materials, the endless CDW system has been realized.

There would be two ways how the CDW is affected by the topological invariant of crystal. The one way would be caused by local curvatures. Since the topological crystals are bent and twist, the crystal structures are locally modulated. The modulation affects to the CDW through the electron-phonon interaction. Moreover, the CDW would be directly affected by the curvature because the CDW is one of the electron crystals. The other way would be caused by the periodic boundary conditions. The lattice has a loop structure. Therefore, the loop current is able to flow in the system without the electrodes. It is an available candidate for realization of the Fröhlich's superconductivity. On the other hand, observation of interference effects of the CDW phase associated with Berry's phase is expected[4]. It is important to investigate conductivity of the topological crystals smaller than the CDW coherence length.

2. Experimental

We measured DC nonlinear conduction of the CDW ring with 6 electrodes. The IV characteristic has showed 4 discontinuities.[5] The simple parallel circuit model has explained 2 discontinuities. However, other 2 discontinuities

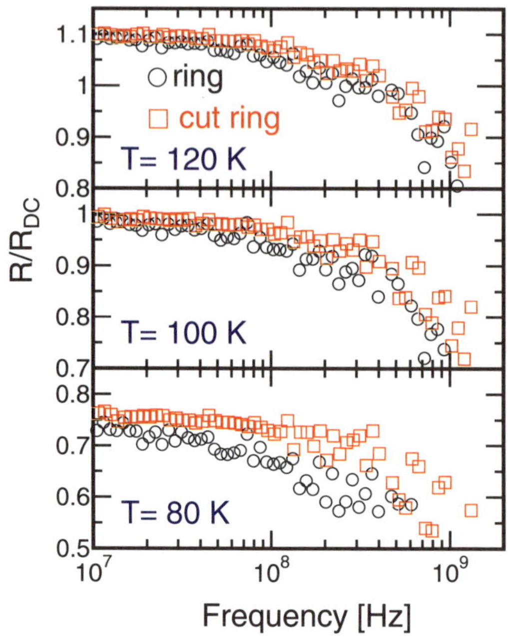

Figure 1. Comparison of frequency dependence of normalized resistively of ring and cut ring below T_{C1}=145 K. The CDW motion contributes decreasing of resistively in radio frequency range. The difference of the normalized conductivity increases in low temperature.

have not been explained yet. We consider that they are caused by an interference effect of CDW between two arms. The problem of the existence of the electrodes has remained. The electrons condensed to CDW state are conversed to normal electrons at the electrodes when the CDW contributes as DC current of nonlinear conduction. The CDW phase coherence would be broken at the electrodes. AC conductivity measurement is another useful method to investigate CDW dynamics.[6] In radio frequency range, the CDW oscillates around the pinning center and contributes AC current. The CDW dynamics is considered as the ridged body dynamics. The CDW does not flow over the pinning potential when the amplitude of AC electrical field is small. The conversion at the electrodes does not occurred. It is expected that the pinning frequency or effective mass be increased by phase-phase correlation around each two electrode. We have measured AC conductivity of the ring crystal by two electrodes measurement in frequency range of 10 kHz to 1.3 GHz. Frequency dependence of CDW conductivity of the ring crystal shows that the pinning is stronger than that of normal CDW. To confirm whether the effect is caused by the topology, we cut the ring using focused ion beam milling and measure conductivity of the cut ring again. By the cutting, the number of current path is decreased. Then the conductivity has become small. We have compared frequency dependence of normalized conductivity of the ring before and after cutting. We have found that the CDW motion of the cut ring is suppressed below T_{C1}=145 K rather than that of ring (Figure 1). We consider that the ring CDW has been reflected the information of the topological invariant of crystal.

It is important to investigate the conductivity without electrodes. By applying AC magnetic field, the loop current can be induced. It is difficult to measure the loop current because the ring is very small. We now plan measurement for the loop current using a sensitive mechanical magnetometer; that is a very light cantilever.[7] The ring should be attached the end of the cantilever. Then the cantilever is put in DC magnetic field. The oscillation of the cantilever produces a rotation of the ring against the magnetic field. Since flux penetrating the ring varies in time, AC current is induced in the ring. The conductivity of the ring crystal decreases the quality factor of the cantilever. This idea has been developed from discussions with Prof. Cleland and Prof. Awschalom of University of California at Santa-Barbara.

This research has been supported by 21COE program on Topological Science and Technology from the Ministry of Education, Culture, Sport, Science and Technology of Japan.

References

1. S. Tanda, T. Tsuneta, Y. Okajima, K. Inagaki, K. Yamaya, and N. Hatakenaka, *Nature* **417**, 397 (2002).
2. G. Grüner, *Rev. Mod. Phys.* **60**, 1129 (1988).
3. H. Fröhlich, *Proc. Roy. Soc. Lond.* **A232**, 296 (1954).
4. M. V. Berry, *Proc. R. Soc. Lond.* **A392**, 45 (1984).
5. Y. Okajima, H. Kawamoto, M. Shiobara, K. Matsuda, S. Tanda and K. Yamaya *Physica B*, **284**, 1659 (2000)
6. G. Grüner, L. C. Tipple, J. Sanny, W. G. Clark, *Phys. Rev. Lett.* **45**, 936 (1980); D. Reagor, S. Sridhar, G. Grüner, *Phys. Rev. B* **34**, 2212 (1986)
7. A. N .Cleland and M. L. Roukes, *Nature* **392**, 160 (1998); J. G. E. Harris, R. Knobel, K. D. Maranowski, A. C. Gossard, N. Samarth, and D. D. Awschalom, *Phys. Lev. Lett.* **86**, 4644 (2001).

THEORETICAL STUDY ON LITTLE-PARKS OSCILLATION IN NANOSCALE SUPERCONDUCTING RING

T. SUZUKI, M. HAYASHI AND H. EBISAWA

Graduate School of Information Sciences, Tohoku University, 6-3-09 Aramaki Aoba-ku, Sendai 980-8579, Japan and CREST-JST

We study magnetic response of superconducting order parameter of one-dimensional ring. We solve Bogoliubov–de Gennes equation numerically without losing self-consistency, and obtain energy level of quasiparticle. It is found that magnetic oscillation has half period of quantum flux $\Phi_e \equiv hc/|e|$, however, the behavior is unexpected— absolute value of pair potential increases when magnetic flux approaches $\frac{1}{4}\Phi_e$.

1. Introduction

In experimental studies on mesoscopic superconductivity, peculiar magnetic phenomena have been reported [1,2], and not been solved theoretically until now. Understanding magnetism of nanoscale system is required for quantum device in the future. Several theorists believe that difference in coherence between normal-conducting and superconducting electrons may cause novel magnetism, and it is known that orbital magnetism of quantum dot shows fluctuating magnetization determined by magnetic-field dependence of electronic energy. Hence, for superconducting system, we expect that some bound states of quasiparticle may contribute to characteristic response in AB oscillation, Little-Parks oscillation, AAS oscillation, and so on.

Magnetism of superconducting system has been studied widely, based on Bogoliubov-de Gennes theory [3], or using various tight-binding models [4]. We study Little-Parks oscillation of superconducting ring using Hubbard model, and discuss the magnetic response owing to one-dimensionality.

2. Model, calculation, and results

We study superconducting ring (Fig. 1) threading magnetic flux Φ, by applying Bogoliubov-de Gennes (BdG) theory to one dimensional Hubbard model with negative "U" (attractive interaction).

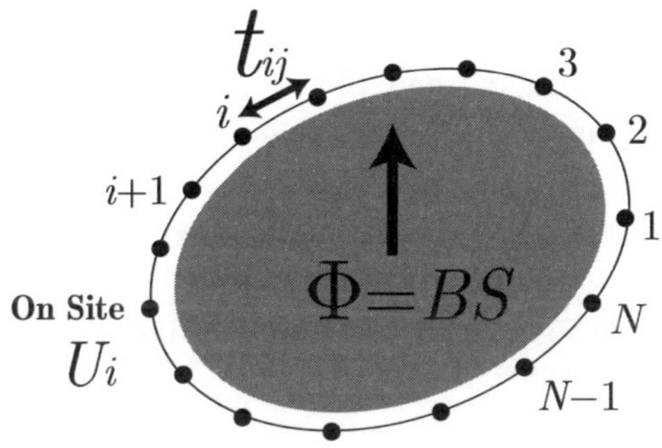

Figure 1. Model of superconducting ring.

Magnetic effect is involved in Peierls phase. Hamiltonian of system is

$$\mathcal{H} = -\sum_{i=1}^{N}\sum_{j=1}^{N}\sum_{\sigma=\uparrow\downarrow} t_{ij}(\Phi)\, c_{i\sigma}^{\dagger} c_{j\sigma} - \mu \sum_{i=1}^{N}\sum_{\sigma=\uparrow\downarrow} c_{i\sigma}^{\dagger} c_{i\sigma} + \sum_{i=1}^{N}\left(\Delta_i\, c_{i\uparrow}^{\dagger} c_{i\downarrow}^{\dagger} + \Delta_i^{*}\, c_{i\downarrow} c_{i\uparrow}\right), \quad (1)$$

where $t_{ij}(\Phi)$ represents nearest-neighbor hopping $t_{i,i\mp 1} = t \times \exp\left(\pm 2\pi \mathrm{i}\, \Phi/(N\Phi_{\mathrm{e}})\right)$, N is number of site, $\Phi_{\mathrm{e}} \equiv hc/|e|$ is quantum flux, and μ is chemical potential.

We solve BdG equation

$$\begin{pmatrix} \widehat{H} & \widehat{\Delta} \\ \widehat{\Delta}^{*} & -\widehat{H}^{*} \end{pmatrix} \begin{pmatrix} u(\vec{r}_i) \\ v(\vec{r}_i) \end{pmatrix} = E \begin{pmatrix} u(\vec{r}_i) \\ v(\vec{r}_i) \end{pmatrix} \quad (2)$$

numerically without losing self-consistency of pair potential $\Delta_i \equiv U\langle c_{i\downarrow} c_{i\uparrow}\rangle$ for each site. In numerical calculation, Eq. (2) is rewritten in matrix form $\widehat{\mathsf{H}} W = EW$, where $\widehat{\mathsf{H}}$ is $2N \times 2N$ matrix and $W_{2i-1} = u(\vec{r}_i)$, $W_{2i} = v(\vec{r}_i)$. The Hamiltonian matrix is determined by

$$\begin{aligned} H_{2i-1,2j-1} &= -t_{ij} - \delta_{ij}\, \mu, \quad H_{2i-1,2i} = \Delta_i \\ H_{2i,2j} &= +t_{ij}^{*} + \delta_{ij}\, \mu, \quad H_{2i,2i-1} = \Delta_i^{*} \end{aligned}$$

for $i, j = 1, 2, \cdots, N$. Calculated pair potential

$$\Delta_i = -U \sum_{\substack{n=1 \\ (E_n \geq 0)}}^{N} u_n(\vec{r}_i) v_n(\vec{r}_i) \tanh\left(\frac{E_n}{k_B T}\right)$$

is independent of site index, and $|\Delta| = |\Delta_i|$ shows magnetic oscillation with half period of quantum flux $\Phi_e \equiv hc/|e|$, namely, Little-Parks (LP) oscillation.

From our numerical calculation (Fig. 2) at zero chemical potential (half-filled in cosine band of tight-binding model), we find that the LP oscillation has unexpected behavior ; absolute value of pair potential increases as magnetic flux approaches $\frac{1}{4}\Phi_e$.

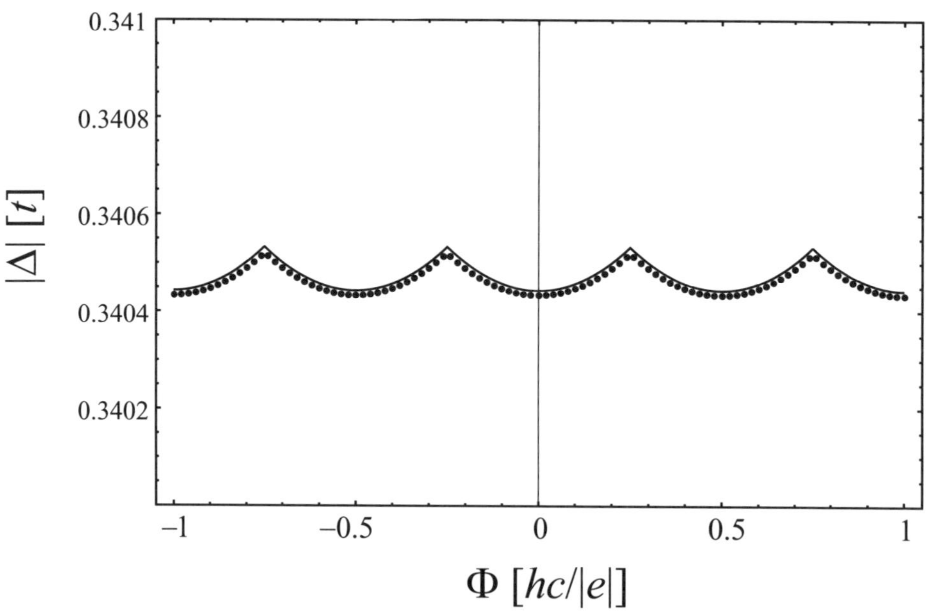

Figure 2. Little-Parks oscillation at $\mu = 0$, $|U| = 2t$, $N = 100$, and $T = 0$ (zero temperature). Numerical result by solving BdG equation (matrix diagonalization) is plotted by solid circle •. Analytical result (Eq. (4), $\omega_c = 0.245t$) is shown by solid line.

We also obtain gap equation of this model analytically as

$$-\frac{1}{U} = \frac{1}{N} \sum_{k=1}^{N} \frac{\tanh\left(\frac{\beta}{2} E_k^+\right) + \tanh\left(\frac{\beta}{2} E_k^-\right)}{4E_k}, \tag{3}$$

where $\beta \equiv (k_B T)^{-1}$, T is temperature,

$$E_k \equiv \sqrt{\xi_k^2 + |\Delta|^2}, \qquad E_k^\pm \equiv E_k \pm \eta_k,$$

$$\xi_k \equiv -2t \cos x_k \cos y - \mu, \quad \eta_k \equiv +2t \sin x_k \sin y,$$

$$x_k \equiv \frac{2\pi}{N}\left(k + \tfrac{1}{2}n(\phi)\right), \quad y \equiv \frac{2\pi}{N}\left(\phi - \tfrac{1}{2}n(\phi)\right),$$

and $\phi \equiv \Phi/\Phi_e$. Integer $n(\phi)$ is winding number regarding argument of pair potential $\Delta_j = |\Delta| \exp(2\pi i\, nj/N)$, $j = 1, 2, \cdots, N$. Particularly, at zero temperature, we can solve Eq. (3) approximately. The analytic solution is

$$|\Delta| \approx \frac{F(\mu + \omega_c)F(\mu - \omega_c)}{2\tau} \exp\left(-\pi\sqrt{4\tau^2 - \mu^2}/|U|\right), \tag{4}$$

where

$$F(\varepsilon) \equiv \sqrt{(2\tau - \mu)(2\tau + \varepsilon)} + \sqrt{(2\tau + \mu)(2\tau - \varepsilon)},$$

$$\tau \equiv t \times \cos\left[\frac{2\pi}{N}\left(\phi - \tfrac{1}{2}n(\phi)\right)\right],$$

and ω_c is cut-off energy. Approximation in Eq. (4) is based on assumption

$$|\xi_{k_F \pm 1}| \ll |\Delta| \ll 2\tau - \mu,$$

where integer k_F is determined by the nearest energy level to chemical potential μ. Pair potential calculated analytically agrees with that calculated numerically. Figure 2 shows the agreement.

3. Discussion

By using Eq. (4), it is ensured that such opposite magnetic response is possible when chemical potential is away from band edges. We find that, when chemical potential is close to band edges (beyond validity of analytical results), pair potential shows normal LP oscillation without the opposite response. We will discuss the magnetic response in terms of density of states in one dimension, elsewhere.

References

1. F. B. Muller-Allinger, A. C. Mota, *et al.*, *Phys. Rev.* **B62**, R6120 (2000).
2. H. Vloeberghs,, V. V. Moshchalkov, C. V. Haesendonch, R. Jonchheere, Y. Bruynseraede, *Phys. Rev. Lett.* **69**, 1268 (1992).
3. J. Cayssol, T. Kontos, G. Montambaux, *Phys. Rev.* **B67**, 184508-1 (2003).
4. A. Ghosal, M. Randeria, N. Trivedi, *Phys. Rev.* **B65**, 014501 (2001).

FRUSTRATED CDW SATES IN TOPOLOGICAL CRYSTALS

K. KUBOKI, T. AIMI AND Y. MATSUDA
Department of Physics, Kobe University,
Kobe 657-8501, Japan
E-mail: kuboki@kobe-u.ac.jp

M. HAYASHI
Graduate School of Information Science, Tohoku University,
Sendai 980-8579, Japan

We study theoretically possible coexistence of CDW and superconducting (SC) orders in a ring-shaped crystal of $NbSe_3$. Since the transfer integrals of inner and outer chains may differ due to the bending of the crystal, the electronic states may be different from chain to chain. This may lead to a coexisting state of CDW and superconductivity, and we examine this possibility by applying the Bogoliubov de Gennes method to a simple tight-binding model.

1. Introduction

Recently single crystals of $NbSe_3$ with unusual shape, e.g., ring, cylinder and Möbius strip have been synthesized by Tanda and his group[1,2]. The properties of ordered states in these systems, such as superconducting (SC) and charge density wave (CDW) states, can be very different from those in ordinary crystals. Then these topologically nontrivial systems (topological crystals) have attracted much attention[3–8]. Hayashi and Ebisawa studied s-wave SC states on a Möbius strip based on the Ginzburg-Landau (GL) theory, and found that the Little-Parks oscillation, which is characteristic to the ring-shaped superconductor, is modified for the Moebius strip. A new state, which does not appear for ordinary rings, shows up when the number of the magnetic flux quanta threading the ring is close to a half odd integer times the flux quantum $\phi_0 = hc/(2e)$. (h,c and e are the Planck constant, the speed of light and the electron charge, respectively). This SC state has a real-space node in the middle of the strip, and is called a "nodal state". Although the analysis in Ref.3 is a phenomenological one, more microscopic calculations based on the GL and Bogoliubov de

Gennes (BdG) equations showed that the nodal state can actually appear[4]. The appearance of the nodal state can be understood as a consequence of unusual boundary conditions required by the sample geometry. The persistent current in Moebius strip have been investigated by Mila *et al.*[5] and Yakubo *et al.*[6]. Wakabayashi and Harigaya studied the Moebius strip made of nanographite ribbon[7], and Ising model on a Moebius strip was studied by Kaneda and Okabe[8].

In this article we examine the electronic states in crystals with ring geometry. $NbSe_3$ is a quasi one-dimensional conductor, and has a CDW ground state. In a ring-shaped sample the one-dimensional conducting chains (along b axis) at the outer (inner) side are stretched (shrunk), and so the electronic states of both sides may be different because of the difference of the transfer integrals and intersite interactions. $NbSe_3$ is known to be superconducting under pressure[9], and we investigate the possibility of having SC states in ring-shaped samples, because shrinkage of the inner chains might have similar effect as that of the pressure on the electronic states.

2. Model and BdG Equations

In a ring-shaped $NbSe_3$ crystal, the circumferential direction of a loop is the b-axis (the best conductive axis)[2]. We treat a two-dimensional plane containing circumferential and radial directions of the ring, and assume that the system is uniform along the third direction. We denote the former (latter) direction as x (y), and the impose periodic (open) boundary condition. Namely we consider a two-dimensional system consisting of weakly coupled chains. The Hamiltonian of our model is given as

$$\begin{aligned} H = &- \textstyle\sum_{i\sigma} t_x(i_y)(c^\dagger_{i\sigma} c_{i+\hat{x},\sigma} + h.c.) - \mu \sum_{i\sigma} c^\dagger_{i\sigma} c_{i\sigma} \\ &+ U \textstyle\sum_i n_{i\uparrow} n_{i\downarrow} + V_1 \sum_i n_i n_{i+\hat{x}} + V_2 \sum_i n_i n_{i+2\hat{x}} \\ &- t_y \textstyle\sum_{i\sigma} (c^\dagger_{i,\sigma} c_{i+\hat{y},\sigma} + h.c.) \end{aligned} \tag{1}$$

where $n_i \equiv n_{i\uparrow} + n_{i\downarrow}$, and μ, U, V_1 and V_2 are the chemical potential, attractive on-site interaction, nearest- and next-nearest-neighbor repulsive interactions, respectively. $t_x(i_y)$ is the intrachain transfer integral for i_y-th chain ($1 \leq i_y \leq N_y$), and it decreases as i_y increases. (Namely $i_y = 1$ and $i_y = N_y$ correspond to innermost and outermost chain, respectively.) t_y is the inter-chain transfer integral, and the system size along the x axis is N_x.

By treating H using a mean-field approximation

$$U n_{i\uparrow} n_{i\downarrow} \to U \langle n_{i\uparrow} \rangle n_{i\downarrow} + U \langle n_{i\downarrow} \rangle n_{i\uparrow} - U \langle n_{i\uparrow} \rangle \langle n_{i\downarrow} \rangle$$

$$+ U \Delta_i c_{i\uparrow}^\dagger c_{i\downarrow}^\dagger + U \Delta_i^* c_{i\downarrow} c_{i\uparrow} - U |\Delta|^2,$$

$$V_{1(2)} n_i n_{i+(2)x} \to V_{1(2)} \langle n_i \rangle n_{i+(2)x} + V_{1(2)} n_i \langle n_{i+(2)x} \rangle - V_{1(2)} \langle n_i \rangle \langle n_{i+(2)x} \rangle, \tag{2}$$

with $\Delta_i \equiv \langle c_{i\uparrow} c_{i\downarrow} \rangle$ being the SC order parameter (OP), we obtain the following mean-field Hamiltonan

$$H_{MFA} = \sum_i \sum_j (c_{i\uparrow}^\dagger, c_{i\downarrow}) \hat{h}_{ij} \begin{pmatrix} c_{j\uparrow} \\ c_{j\downarrow}^\dagger \end{pmatrix} + E_0, \tag{3}$$

where E_0 is a constant term. The procedure of the self-consistent numerical calculation is the following. We substitute an initial set of the SCOP and the electron density, $\langle n_{j\sigma} \rangle$, in the matrix elements of $\hat{h}_{ij}$, and diagonalize it by solving the BdG equation. Then we recalculate the OP's using eigenvalues and eigenfunctions of $\hat{h}_{ij}$, and the iteration is performed until the values of OP's are converged.

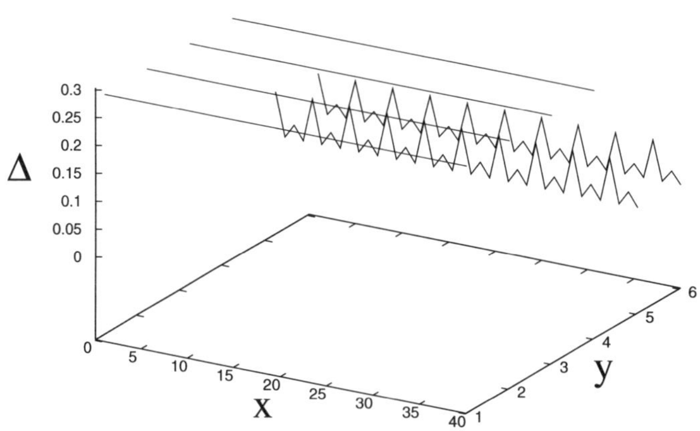

Figure 1. Superconducting order parameters in a system with $N_x = 40$, $N_y = 6$.

3. Results

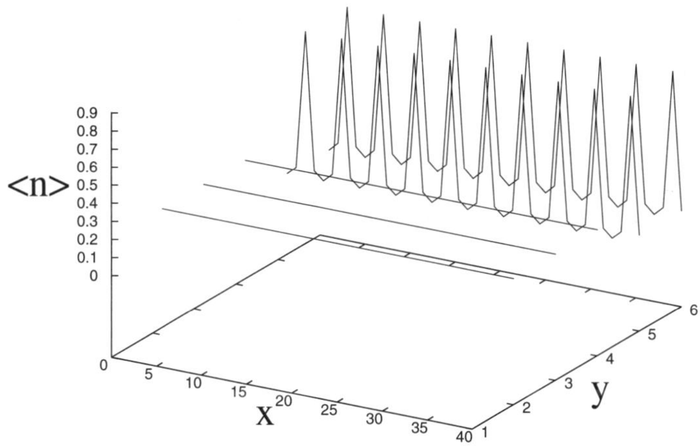

Figure 2. Electron density in a system with $N_x = 40$, $N_y = 6$.

In this section we show the results of self-consistent numerical calculations at zero temperature. For one dimensional system (i.e., $t_y = 0$) both the CDW and SC states can appear for the same values of the parameters μ, U, V_1 and V_2, if the transfer integral t_x is different. This is due to the fact that the electron density can be different when t_x is varied. Then we couple chains with different t_x ($1.00 \leq t_x \leq 1.01$) with finite t_y to simulate the system of ring-shaped crystals. In Fig.1 the spatial variation of the SCOP is shown for $U = -3$, $V_1 = 0.56$, $V_2 = 0.2V_1$, $\mu = -1.52$ and $t_y = 10^{-3}$. The electron density is shown in Fig.2 for the same values of the parameters. It is seen that the SC state appear for the chains with larger t_x (i.e., in inner chains) , while the coexisting state occurs for chains with smaller t_x (i.e., outer chains).

4. Summary

We have studied the behaviors of the superconducting (SC) and CDW states in a model describing ring-shaped $NbSe_3$ crystals. It is found that coexistence of SC and CDW orders may be possible if the transfer integrals of inner and outer chains are different.

Acknowledgments

This work was financially supported by the Sumitomo Foundation.

References

1. S. Tanda et al., *Nature* **417**, 397(2002).
2. T. Tsuneta et al., *Physica* **B329-333**, 1544 (2003).
3. M. Hayashi and H. Ebisawa, *J. Phys. Soc. Jpn.*, **70**, 3495 (2001).
4. M. Hayashi, H. Ebisawa, and K. Kuboki, *Phys. Rev.* **B72**, 024505 (2005).
5. F. Mila, C. Stafford, and S. Capponi, *Phys. Rev.* **B57**, 1457 (1998).
6. K. Yakubo, Y. Avishai, and D. Cohen. *Phys. Rev.* **B57**, 125319 (2003).
7. K. Wakabayashi and K. Harigaya, *J. Phys. Soc.Jpn.*, **72**, 998 (2003).
8. K. Kaneda and Y. Okabe, *Phys. Rev. Lett.*, **86**, 2134 (2001).
9. A. Briggs et al., *J. Phys..* **C13**, 2117 (1980).

LAW OF GROWTH IN TOPOLOGICAL CRYSTAL

M. TSUBOTA, S. TANDA, K. INAGAKI, T. TOSHIMA AND T. MATSUURA

Department of Applied Physics,
Hokkaido University,
Sapporo 060-8628, Japan
E-mail: tsubota@eng.hokudai.ac.jp

We studied growth of topological crystal. We investigated reaction time dependence of the number of produced ring crystals of TaS_3. As a result, we found the number of ring crystals decreased after it reached a peak. We proposed a model to explain the result. It is necessaty to consider the number of droplet by chalcogen and decomposition rate. A result of numerical simulation using the model reproduced the experimental result qualitatively. We discussed possible mechanisms about the decomposition process. Effects of the solid to vapor phase transition, the solid to liquid phase transition and mechanical force were considered. In those processes, we thought the solid to liquid phase transition was most plausible from binary phase diagrams. Our result will lead to an optimum condition of mass production of the topological crystals.

1. Introduction

Topological crystals have been researched since the first ring crystal of $NbSe_3$ was discovered in 2000[1]. In addition to $NbSe_3$, topological crystals of $TaSe_3$[2] and TaS_3[3] were discovered. A change of topology could affect physical properties. For example, carbon nanotube changed metallic or semiconductive by relation to chiral vector[4]. One would expect to observe a phenomenon which had not been discovered in the non-topological (ribbon and whisker) crystals.

The topological crystals had been synthesized by chemical vapor transportation method in an oversaturated condition of chalcogen. A growth model had been proposed as a $NbSe_3$ whisker encircles a selenium droplet by the surface tension and joining its two ends[5]. However, details of the growth mechanism were not perfectly understood. We believed that a universal mechanism of growth of the topological crystals should appear from comparison of different materials. Therefore we produced TaS_3 topological crystals and found the rules of growth by comparison between $NbSe_3$ and

TaS_3 topological crystals.

2. Experiments and results

The samples were prepared by the chemical vapor transport method. We soaked a mixture Ta and S powder at 530 °C in an evacuated quartz tube for a period varying from a few hours to several hours. Quartz tube was 1.7 cm in internal diameter and 20 cm in length, while evacuated to the order of 10^{-6} Torr. The furnace which was used experiment had a large temperature gradient. Several hours later, we took out a quartz tube from furnace and quench low temperature side in liquid nitrogen. Figure 1 shows a scanning electron microscope (SEM) photograph of the synthesized TaS_3 topological crystal. The diameter is 5 μm. The width is 0.2 μm.

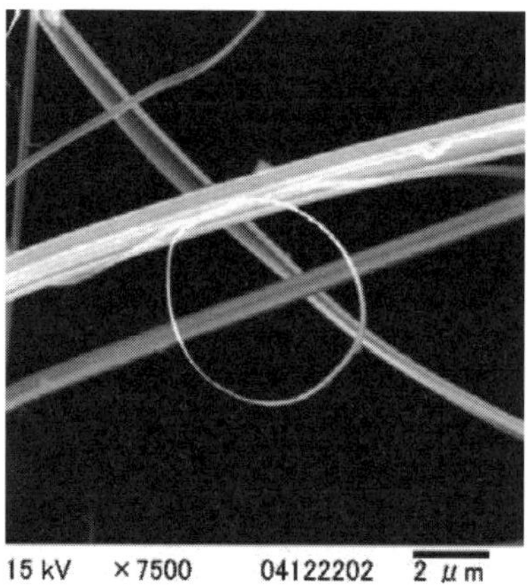

Figure 1. Scanning electron microscope image of a TaS_3 topological crystal. The diameter is 5 μm and the width is 0.2 μm.

We investigated reaction time dependence of the number of produced ring crystals. As a result, the number of ring crystals decreased after it reached a peak at five hours (Fig. 2). Solid circles represented the number of ring crystals at different reaction time. This behavior was interesting because if the crystals were produced as time passes, the number of ring crystals would increase monotonically until the materials ran out. This was similar to $NbSe_3$[6] and possibly universal to the formation of the topological crystals.

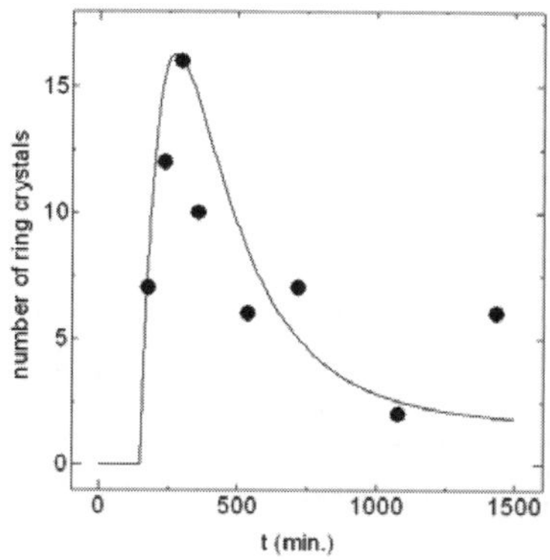

Figure 2. The number of TaS_3 ring crystals as a function of reaction time (solid circles). Solid line is numerical simulation using (1)-(4) of reaction time dependence. We suppose ring crystal is not produced for several minutes at first because it is necessary that whisker grow up to a point. Together the number of ring crystals decreases after it reaches a peak.

3. Discussion

It was necessary to consider the number of droplet and decomposition rate. TaS_3 topological crystal was produced because the whisker encircled a sulfur droplet. The quantity of droplet was directly proportional to the remain of material. So we proposed the following formulae:

$$N = n + n_r + n_w \tag{1}$$

$$\frac{dN}{dt} = \frac{dn}{dt} + \frac{dn_r}{dt} + \frac{dn_w}{dt} = 0 \tag{2}$$

$$\frac{dn_r}{dt} = \alpha n \times kn - \beta_r n_r \tag{3}$$

$$\frac{dn_w}{dt} = \alpha n - \beta_w n_w \tag{4}$$

Here N is the total quantity in a quartz tube, n is the quantity of matter which have not synthesized. n_r and n_w is the quantity of ring crystals and whiskers, respectively, kn is the quantity of droplet, and α, β_r, and β_w is coefficients of growth, and decomposition of rings and whiskers, respectively. Figure 2 shows the result of numerical simulation using (1)-(4). The result reproduced qualitatively the observed dependency shown

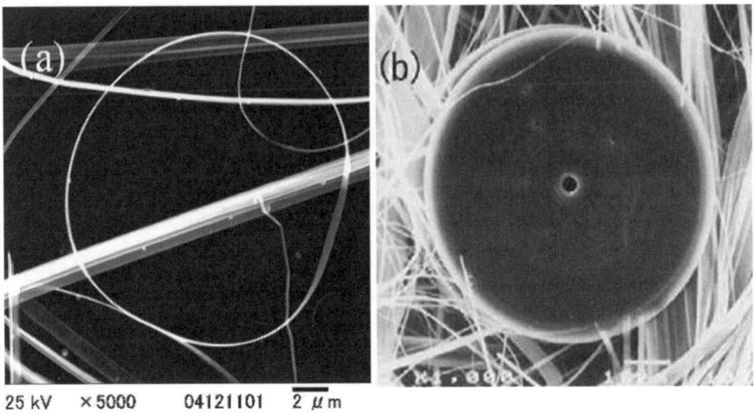

Figure 3. Two ring crystals are produced by a difference of material. (a) TaS_3 ring crystal. The diameter is 20 μm and the thickness is 0.5 μm. (b) $NbSe_3$ ring crystal[6]. The diameter is 65 μm and the thickness is 30 μm.

in Fig. 2. The value of solid line in Fig. 2 were chosen by trial and error. N is 10^4, k is 6×10^{-7}, α is 10^{-2}, β_r is 8×10^{-3} and β_w is 2×10^{-3}.

We discussed possible mechanism for the decomposition process. If topological crystals were decomposed by mechanical force, whiskers would be also decomposed. However, the number of whiskers increased monotonically as time passes. Next, we considered phase transition. Topological crystals had been synthesized by vapor deposition[5]. In the case of this, the crystals almost never became vapor. So we could explain the crystal growth, but could not explain that the number of ring crystals decreased. Therefore we considered solid and liquid equilibrium. If solid phase and liquid phase were equilibrium, the solid to liquid and liquid to solid phase transition would happen. In case a part of ring crystal transformed the solid to liquid phase, it would be whisker to break the link. We thought this was most likely to decompose. The number of TaS_3 ring crystals was fewer compared with $NbSe_3$ ones[6]. We considered this was for the reason the ratio of microparticles in sulfur droplet was fewer than selenium droplet[7]. In other words, decomposition rate might vary by each droplet.

We could explain a growth mechanism of ring crystals from the model. Fig. 3 (a) shows a ring crystal of TaS_3 and Fig. 3 (b) shows a thick ring crystal of $NbSe_3$. All ring crystals of TaS_3 were thin, though the several ring crystals of $NbSe_3$ were thick[6]. For instance, the thickness of TaS_3 ring crystals, however thick, was about 1 μm. In the case of $NbSe_3$ ring crystal, the thickness was 30 μm. Possibly TaS_3 ring crystal was decomposed before it become thick. A size of ring crystal might be influenced by a difference of droplet.

In conclusion, we found the law of growth in topological crystals. Our result will lead to an optimum condition of mass production of the topological crystals.

This work is partly supported by Grant-in-Aid for the 21st Century COE program " Topological Science and Technology ".

References

1. S. Tanda, H. Kawamoto, M. Shiobara, Y. Sakai, S. Yasuzuka and Y. Okajima, *Physica* **B284**, 1657 (2000).
2. T. Matsuura, S. Tanda, K. Asada, Y. Sakai, T. Tsuneta, K. Inagaki and K. Yamaya, *Physica* **B329**, 1550 (2003).
3. H. Okawa, private communication.
4. N. Hamada, S. Sawada and A. Oshiyama, *Phys. Rev. Lett.* **68**, 1579 (1992).
5. S. Tanda, T. Tsuneta, Y. Okajima, K. Inagaki, K. Yamaya and N. Hatakenaka, *Nature* **417**, 397 (2002).
6. T. Tsuneta and S. Tanda, *Journal of Crystal Growth* **264**, 223 (2004).
7. H. Okamoto, *Phase Diagrams for Binary Alloys*, Published by ASM INTERNATIONAL (2000).

SYNTHESIS AND ELECTRIC PROPERTIES OF NbS_3: POSSIBILITY OF ROOM TEMPERATURE CHARGE DENSITY WAVE DEVICES

H. NOBUKANE, K. INAGAKI, S. TANDA

Department of Applied Physics, Hokkaido University,
Kita 13, Nishi 8,
Sapporo 060-8628, Japan
E-mail: nobukane@eng.hokudai.ac.jp

M. NISHIDA

Graduate School of Advanced Science of Matter, Hiroshima University,
Kagamiyama,
Higasi-hiroshima, 739-8530, Japan,

We have investigated sliding of a charge density wave (CDW) on one-dimensional conductor NbS_3 at room temperature. We confirmed existence of the CDW by transmission electron diffraction. The Peierls vector was determined as $q = (0, 1/4b^*, 1/4c^*)$ from satellites. Nonlinear conductor with threshold field of 13.3 V/cm was exhibited in NbS_3 crystals. Moreover, we made prototype of the CDW field effect transistor. The result is that source-to-drain voltage was modulated by applied gate voltage, as the larger constant dc current around a threshold value E_T flowed.

1. Introduction

Recently quantum devices instead of semiconductors have widely been studied as new devices. A Josephson junction (JJ) and a single electron tunnelling transistor (SET) are examples. Since a high-temperature superconductivity was discovered in 1986 [1], many researchers have demonstrated that JJ devices can be operated above the liquid nitrogen temperature (77 K) [2,3]. There is no superconductor that can be operated at room temperature so far because the highest transition temperature is about 150 K now.

A CDW is one of the macroscopic quantum states in which the electronic charge density in a quasi-one-dimensional conductor is modulated, as

$$\rho(x) = \rho\cos(2k_F x + \phi(x)).$$

where k_F is the Fermi wavevector. The Peierls gap opens up over Fermi surface below the transition temperature. The CDW can transport current when an applied field exceeds a threshold value E_T [4]. The CDW and the JJ are dual in that the roles of charge and flux are interchanged, as well as those of current and voltage. J.H. Miller Jr. discussed the principle of a soliton tunnelling transistor (STT) for the Macroscopic Quantum Tunnelling accompanied by quantum pair creation of soliton domain wall. They also performed an experimental study on the STT in a single crystal of $NbSe_3$ at 35 K [5]. They claimed that the principle was confirmed by their experiment. However, their experiment has some issues: because $NbSe_3$ leaves a small portion of Fermi surface below Peierls transition temperature [4], uncondensed electrons screen the charges induced by the gate so that the effectiveness of the gate should be reduced.

Here we remark a quasi-one-dimensional material NbS_3, which undergoes the transition at 340 K and has no uncondensed electrons below the Peierls transition temperature [6]. Therefore, we believe that NbS_3 would be a candidate of the CDW device running at room temperature. In this study, we synthesized NbS_3 single crystals, analyzed the composition and electron beam diffraction patterns, and measured electric characteristics. We confirmed our materials exhibited the CDW state and finally tested the STT made with NbS_3 single crystals.

2. Experiments and results

NbS_3 is obtained from Niobium and Sulfur powders in molar fraction 1:3 by chemical vapor transport method. Mixture of these powders was put in an evacuated quartz tube ($\sim 10^{-5}$ torr). The tube was placed in the furnace at 600 °C for two weeks with temperature gradient 1 °C/cm. Needle-shaped crystals were obtained near the low temperature end. Two polytypes of NbS_3 has been reported: type I has no Peierls transition near room temperature, whereas type II shows the CDW transition at 330 K [6]. Since polytypes of NbS_3 crystals could hardly be distinguished them from appearance, we measured the composition of synthesized materials, a transmission electron diffraction pattern, temperature dependence of resistance and $I-V$ characteristic of our sample.

Firstly, we examined compositions of synthesized materials (60 μm $\times$ 15 μm $\times$ 1 μm) that were put on a silicon substrate. Energy dispersion spectrometer (SEIKO SED-8000) was used to determine compositions. The peaks of Nb, S and Si were only seen and compared with those of NbS_2. Consequently, the molar fraction of Nb and S elements was determined to be 1:3 in our synthesized materials.

Secondly, we took diffraction patterns of NbS_3 at room temperature by a transmission electron microscope (JEOL JEM-2000FX). Fig. 1 shows the diffraction pattern of a NbS_3 crystal. We estimated plane distances that are $b = 3.50$ Å and $c = 18.83$ Å. We also discovered new satellites due to formation of the CDW that were defined as $q = (0, 1/4b^*, 1/4c^*)$.

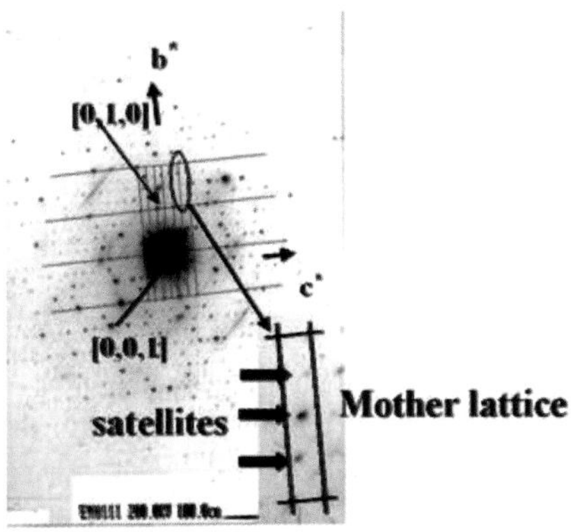

Figure 1. The diffraction pattern of the CDW in a NbS_3 crystal in the (b^*, c^*) plane is indicated at room temperature, Mother lattice shows square. In inset, satellites show black point in mother lattice(square).

Thirdly, we measured temperature dependence of resistance. We put a single crystal of NbS_3 on a silicon substrate and evaporated gold which a distance between electrodes was 60 μm. Silver wires of 50 μm in a diameter were attached by using a silver paste. The sample was put on a copper heat sink and the temperature was increased slowly up to 380 K and decreased slowly down to room temperature. The results of sample A and sample B are shown in Fig. 2 (a) and (b), respectively. Sample A exhibited the CDW transition at 330 K, whereas sample B is monotonic. The variation of R with the inverse of T is plotted in a logarithmic scale in inset of Fig. 2 (a). The activation energy is approximately $\Delta = 2320$ K.

Finally, the $I - V$ characteristic of sample A was shown in Fig. 3. We examined the contact resistance as 58 kΩ by comparing results of two and four-terminal measurements. The linear resistance at 295 K was 846 kΩ. ρ

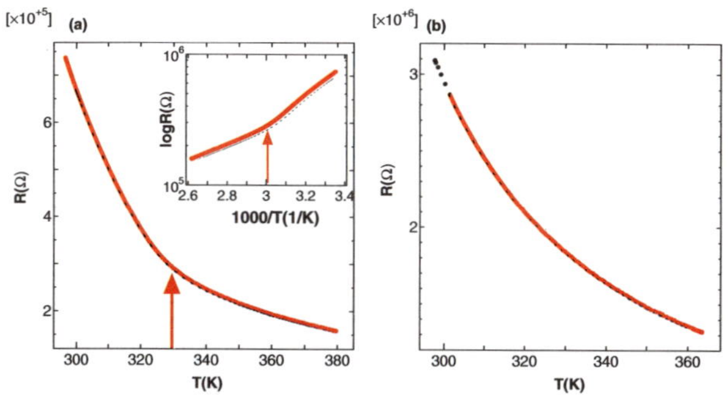

Figure 2. (a) exhibits the CDW transition at 330 K. In inset, (a) exhibits a resistance variation in a semilogarithmic plot as a function of $1/T$ for NbS_3. (b) exhibits monotonic.

$= 21.15\ \Omega \cdot \mathrm{cm}$ was obtained by sample size ($60\ \mu\mathrm{m} \times 15\ \mu\mathrm{m} \times 1\ \mu\mathrm{m}$). We also found sample A exhibited nonlinear transport. The threshold field at 13.3 V/cm was determined.

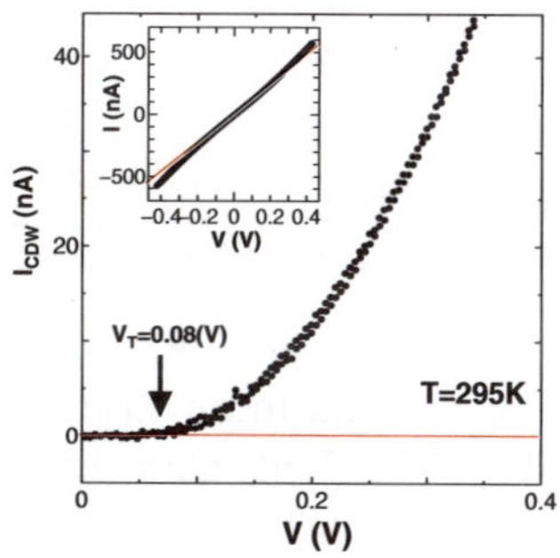

Figure 3. Variation of the CDW current as a function of the voltage at 295 K. Inset shows the current versus the voltage. A down arrow indicate a threshold value V_T.

NbS_3 (type I) was semiconductor with a resistivity at room temperature of $80\ \Omega \cdot \mathrm{cm}$. This resistivity is thermally activated with an energy gap of 2200 K. NbS_3 (type II) has a CDW transition at 340 K. Diffraction patterns of NbS_3 (type II) obtained by rotating the sample around b^*. The distortion wave vector was defined as $q_1 = (1/2a^*, 0.298b^*, 0)$, $q_2 = (1/2a^*, 0.352b^*, 0)$. The resistivity of it was at room temperature of $8 \times 10^{-2}\ \Omega \cdot \mathrm{cm}$ and exhibited nonlinear transport.

It should be important to compare our experimental results with previous reports by Z.Z. Wang [6]. Our plane distances that are $b = 3.50$ Å and $c = 18.83$ Å in the (b^*, c^*) plane is similar to theirs that are $b = 3.37$ Å and $c = 18.3$ Å of NbS_3. Although they reported that the distortion wave vector was defined as $q_1 = (1/2a^*, 0.298b^*, 0)$, $q_2 = (1/2a^*, 0.352b^*, 0)$, we discovered new satellites that were defined as $q = (0, 1/4b^*, 1/4c^*)$ in the (b^*, c^*) plane. Next, we discuss electric properties. We confirmed two kinds of temperature dependence of the resistivity as well as theirs. But in our sample A that showed the CDW transition at 330 K, the resistivity and the activation energy were similar to those of their NbS_3 (type I). This was a confusing result. Nevertheless, it is important for us to exhibit the CDW transition at 330 K and nonlinear transport. We plan to analyze their NbS_3 structures by a X-ray diffraction in the future.

We performed a preliminary study on the CDW device by our NbS_3 materials. It should be noted the cross relationship connected with the CDW and the JJ. In the JJ, the dynamical equation of the phase on fluxoid quantum is indicated by

$$\frac{\hbar c}{2e}\frac{d^2\theta}{dt^2} + \frac{\hbar}{2\pi R}\frac{d\theta}{dt} + I_c \sin\theta = I.$$

Whereas in the CDW, the dynamical equation of the phase on the CDW is indicated by

$$\frac{d^2\theta}{dt^2} + \Gamma\frac{d\theta}{dt} + \omega_0^2\theta = \frac{n_c \pi e}{m^*}E.$$

We used three terminals of the sample A that exhibited nonlinear transport. In addition to the source and drain electrodes, the center electrode was used for gate capacitor. a 0.15 μF gate capacitor was attached to it. The gate voltage source was coupled to the gate capacitor by way of a 1 MΩ resistor, which limited the current flowing through the crystal during changes in gate voltage when the gate capacitor was either partially charged or discharged. The sample was attached cryostat and put in a glass dewar vessel in the vacuum (10^{-3} torr) to prevent changing temperature. The measurement were carried out at room temperature (295 K). When the larger constant dc current around a threshold value E_T flowed, we observed that source-to-drain voltage was modulated by applied gate voltage in our system. Fig. 4 (a), (b), and (c) show source-to-drain voltage versus gate voltage for 0.12, 0.25, and 0.40 μA, respectively. The modulation was pronounced at 0.40 μA (Fig. 4 (c)). These results differ from the clear periodic modulation observed by J. H. Miller Jr. . We should continue on this study.

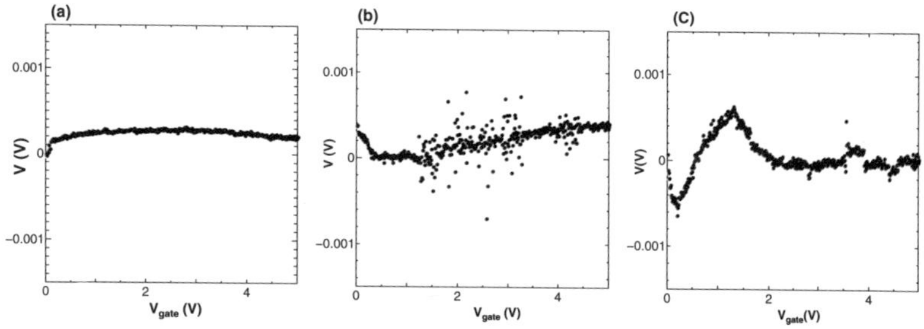

Figure 4. As three constant dc current around the threshold value E_T flowed, source-to-drain voltage versus gate voltage is indicated. (a), (b), (c) show source-to-drain voltage versus gate voltage for 0.12, 0.25, and 0.40 μA, respectively.

3. Summary

We synthesized NbS_3 crystals, determined that the molar fraction of Nb and S elements was 1:3, took diffraction patterns of NbS_3 at room temperature and discovered new satellites. The CDW transition and nonlinear transport in NbS_3 were shown by electric characteristics. We tried to make the prototype of room temperature CDW devices. Source-to-drain voltage was modulated by applied gate voltage, as the larger constant dc current around a threshold value E_T flowed. We need to continue on this study.

This work is supported by 21 century COE program "Topology of Science and Technology".

References

1. J. G. Bednorz and K. A. Muller, Z. *Phys. B* **64** 189 (1986).
2. J. Mannhart, J. G.Bednorz, K. A. Muller and D. Gschlom, Z. *Phys. B* **83** 307 (1991)
3. X. X. Xi, Q. Li, C. Doughty, C. Kwon, S. Bhattaharya, A. T. Findikoglu and T. Venkatesan, *Appl. Phys. Lett.* **59** 3470 (1991)
4. G. Grüner, *Rev. Mod. Phys.* **60** 1129 (1988)
5. J. H. Miller, Jr. G. Cardenas, A. Garcia-Perez, W. More and A. W. Beckwith, *J. Phys. A: Math. Gen.* **36** 9209 (2003)
6. Z. Z. Wang and P. Monceau, *Phys. Rev. B* **40** 11589 (1989)
7. T. L. Adelman, S. V. Zaitsev-Zotov, and R. E. Thorne, *Phys. Rev. Lett.* **74** 5264 (1995)

HOW DOES A SINGLE CRYSTAL BECOME A MÖBIUS STRIP?

T. MATSUURA, S. TANDA

Department of Applied Physics, Hokkaido University,
Kita 13, Nishi 8,
Sapporo 060-8628, Japan
E-mail: toru@eng.hokudai.ac.jp

T. TSUNETA

Low Temperature Laboratory, Helsinki University of Technology,
Helsinki FIN-02015 HUT, Finland

T. MATSUYAMA

Nara University of Education,
Takabataketyou,
Nara 630-8528, Japan

We propose a new general model explaining how the crystal becomes the Möbius strip geometry.

1. Introduction

The Möbius strip geometry of a niobium tri-selenide ($NbSe_3$) single crystal is discovered in 2002 by Tanda *et al.* [1] The Möbius strip is a topological object with only one side and only one boundary with a π twisting. Fig. 1b is a scanning electron microscope (SEM) image of the Möbius strip of the $NbSe_3$ crystal. Tanda *et al.* have found non-twisted ring crystals (Fig. 1a) and 2π twisted ring crystals (Fig. 1c). We can define a twist number T_w of each their geometry, such as, T_w of the non-twisted ring, the Möbius strip, and 2π twisted ring are 0, 1/2, and 1, respectively. It is easy to create such objects like the Möbius strip by giving a twist to a paper strip and connecting the ends. However, it is the amazing fact that the topology exists in nature.

Why can the crystals be topological? Tanda *et al.* have explained the mechanisms. [1] These crystals are synthesized using a chemical vapor trans-

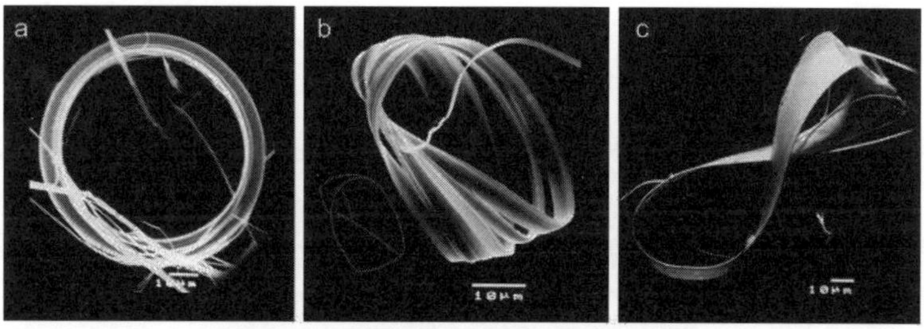

Figure 1. Topological crystals of $NbSe_3$.

port (CVT) method. Many small selenium (Se) droplets exist in admixture vapor. By the interfacial force, a seed crystal is absorbed on the droplet and restricted to grow only on the surface of it. Since the $NbSe_3$ crystal grows one-dimensionally, the two ends of the crystal approach each other on the droplet, and finally the ends are connected. The ring topology of the crystal remains after the droplet has disappeared due to re-evaporation or consumption by the crystal growing. If the crystal encircles 2 times around the droplet and connects the ends, it becomes a 2π twisted ring crystal. The number of encircling is called writhing number W_r. According to White's theorem, [3] summation of the twist number T_w and writhing number W_r of the closed loop is a topological invariant. Therefore, we can redefine more intrinsic parameter L_k, such as, $L_k = T_w + W_r = 1$ for the ring, and $L_k = 2$ for the 2π twisted ring, respectively. In general, while n is integer, the n winded ring ($T_w = 0$, $W_r = n$) is topologically equivalent to $n - 1$ twisted ring ($T_w = n - 1$, $W_r = 1$).

The π twisting of the Möbius strip crystal has not been explained by the mechanism because n is integer. Therefore, other mechanism has been needed. Tanda *et al.* have explained that the π twisting is caused by the low symmetric monoclinic crystal structure of $NbSe_3$. The monoclinic crystal or lower symmetric crystal is naturally twisted when bending, because the off-diagonal factors of the stress tensor are non-zero. It is a possible mechanism to explain the π twisting. However, some questions have remained. Actual NbSe3 crystal is twisted naturally by the bending? Why almost ring crystals are not twisted? Why the twisting is just π twisting? The origin of the π twisting has been unclear.

(a)

(b)

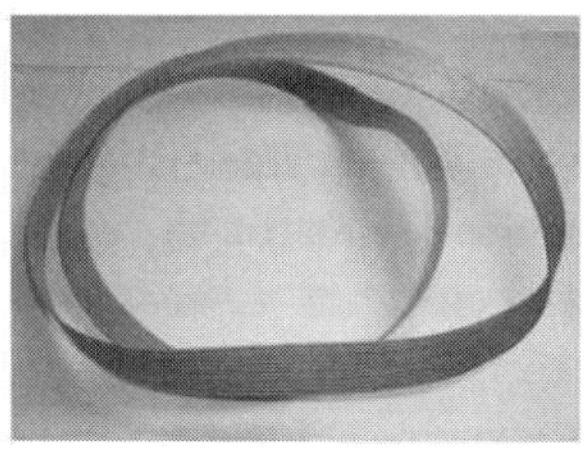

(c)

(d)

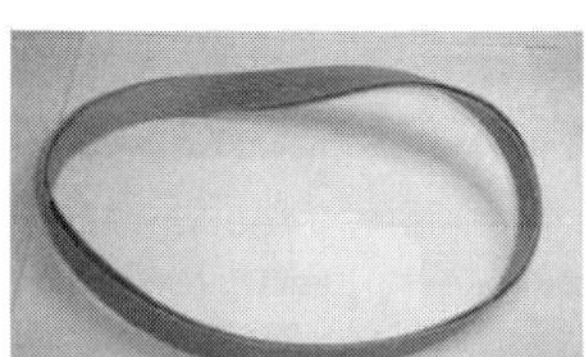

Figure 2. Pictures of the paper model of transformation from a 3 encircled ring to a Möbius strip. (a)The initial form of the 3 encircled ring ($W_r = 3$). (b)The ring transforms to find a metastable form under the conservation of $T_w + W_r$. (c) $T_w = 1, W_r = 2$. (d) After the connection between surface, it become Möbius strip crystal ($T_w = 1/2, W_r = 1$).

2. New mechanism

We will propose a new reasonable model to explain how the crystal has the Möbius strip topology. We assume that existence of a triple writhing ring ($W_r = 3$ and $T_w = 0$) like the paper model in Fig. 2(a). Actually, such crystal has not found yet but it is possible if the crystal encircled 3 times and connected its ends. Similar to the 2π twisted ring, the writhing number converts to the twist number under the conservation of the topological invariant. Therefore, it transforms to find a form with a minimum elastic energy after the droplet is disappeared (Fig. 2(b)). The form of $T_w = 1$ and $W_r = 2$ like the model in Fig. 2(c) must be the metastable form because the strain of twisting and bending is dispersed. Since the form is not entangled, the faces can be fitted. It looks very similar to the Möbius strip but topologically different. Here, we consider a topology change. The crystal is still in the growing process. One of the sides can adhere to other side because the crystal axes are parallel. Finally, it becomes really only one-face crystal. By the topology change, the topological invariant can be changed to $T_w = 1/2$ and $W_r = 1$. That is the Möbius strip crystal!

This new process to explain the π twisting is used only the encircling mechanism and crystal growing. Therefore, any exceptional mechanism is not necessary. Our idea is reasonable and general.

References

1. S. Tanda, T. Tsuneta, Y. Okajima, K. Inagaki, K. Yamaya, and N. Hatakenaka, *Nature* **417**, 397 (2002).
2. T. Tsuneta and S. Tanda, *J. Cryst. Growth* **264**, 223 (2004).
3. White, J. H. *Am. J. Math* **91**, 693-728 (1969).

DEVELOPMENT OF X-RAY ANALYSIS METHOD FOR TOPOLOGICAL CRYSTALS

K. YAMAMOTO,[1] T. ITO,[1] N. IKEDA,[2] S. HORITA,[1] N. IRIE,[1] Y. NOGAMI,[1,3] T. TSUNETA, [4,5] AND S. TANDA[4]

[1]*Department of Frontier and Fundamental Sciences, Graduate School of Natural Science and Technology, Okayama University, Okayama 700-8530, Japan.*
[2]*Japan Synchrotron Radiation Research Institute, SPring-8, Hyogo 679-5198, Japan.*
[3]*CREST, Japan Science and Technology Agency, Saitama 332-0012, Japan.*
[4]*Department of Applied Physics, Hokkaido University, Sapporo 060-8628, Japan.*
[5]*Low Temperature Laboratory, Helsinki University of Technology, Finland.*

To solve curved shape crystal, novel X-ray camera with two-axis sample rotator was developed. With use of this sample rotator, the preferred-orientation of micro-crystalline was nearly suppressed. The effect of two-axis rotation was effectively corrected by selecting integrated area of the imaging plate(IP). Its application to the topological crystals was presented briefly.

1. Introduction

In order to understand the physical and chemical properties of novel functional materials, the information of atomic arrangement in them is indispensable. To this end, we analyze crystal structure using X-ray diffraction. Usually, powder crystal method of X-ray structure analysis is used for inorganic compounds, because of their strong X-ray absorption and extinction. In this case of powder method, we need perfect disorder orientation, in other words, no preferred orientation to estimate X-ray diffraction intensity from the integration of a part of the Debye ring. On the other hand, single crystal method of X-ray structure analysis is often used for organic compounds. In this case we can calculate position of the Bragg reflection with the use of sample orientation matrix, thereby, we measure X-ray diffraction intensity with the integration of the observed intensity around the calculated position of the Bragg reflection.

Recently topological crystals of quasi one-dimensional compounds MX3 e. g. NbSe3 have been developed [1]. The new crystal forms are ring, Möbius strip and figure-of-eight strip. Figure 1 shows scanning electron microscope images of NbSe3 topological crystals [1].

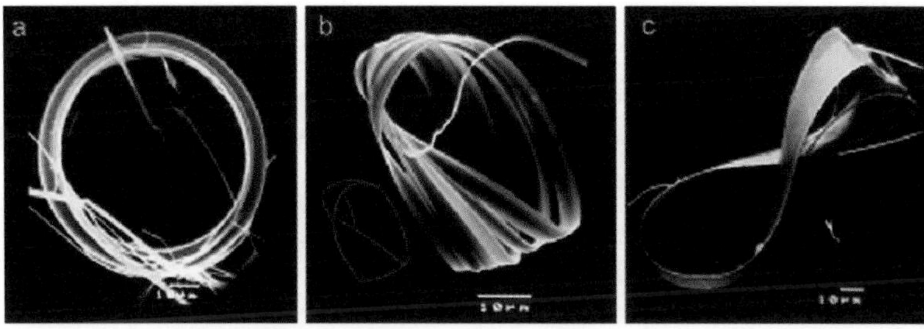

Figure 1. Topological Crystals of $NbSe_3$.

Since both ends of a crystal are bound and twisted, these crystals have entirely different topological class from ordinary crystals. These topological crystals collected much interest of scientists owing to their possible topology dependent effects on macroscopic quantum interference in the charge-density-wave (CDW) state which is specific quasi one-dimensional compounds.

Determination of atomic arrangement in the topological crystal seems very difficult owing to the curved shape. In other words, the orientation of each crystallite composing a topological crystal has positional dependence reflecting its shape. For this reason, we must abandon single crystal method applicable only to a crystal with constant crystal orientation. On the other hand, powder crystal method also faced the difficulty of strong positional variation of the crystal orientation depending the curved shape. If we divide or crash a topological crystal to average its crystal orientation and to obtain good powder crystal, we may lose important structural information originated from its bound and twisted shape.

Bearing this difficulty in our mind, we determined to develop new camera and analysis method to solve the structure of the topological crystal without any sample treatment. For the suppression of preferred orientation, we get some hint from traditional Gandolfi camera and developed new two-axis sample rotator. The original Gandolfi camera with the two-axis rotator was used in the mineralogy for qualitative analysis of constituting elements. However, their rotations were not independent nor incommensurate, because of the geared transmission between the two axis rotation. We developed a X-ray camera with independent two-axis sample rotator, principal(P) and secondary(S) rotators crossing at an angle of 45 deg. The sample is mounted on the top of S axis(see Fig. 2). With the rotation of the S, one reciprocal lattice point forms one ring trajectory. The assembly of ring trajectories reflecting the curved shape, covers the sphere partially in the reciprocal space. With the rotation of P, this part of the sphere intersects the Ewald sphere and diffraction pattern will be similar to Debye ring(see Fig. 3). Of course this is not ideal Debye ring which is found in powder crystals without preferred-orientation.

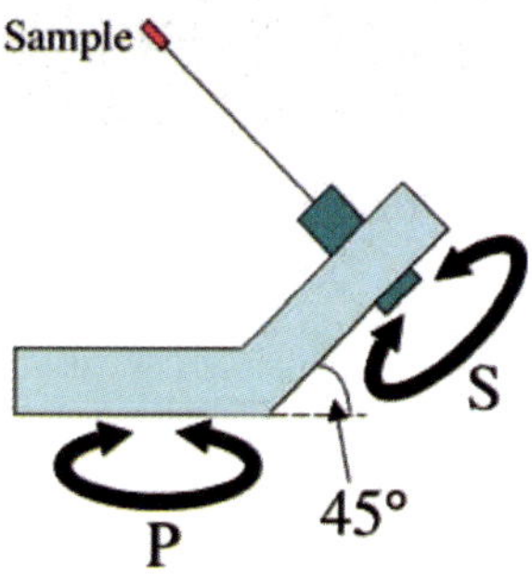

Figure 2. Schematic view of two-axis rotator.

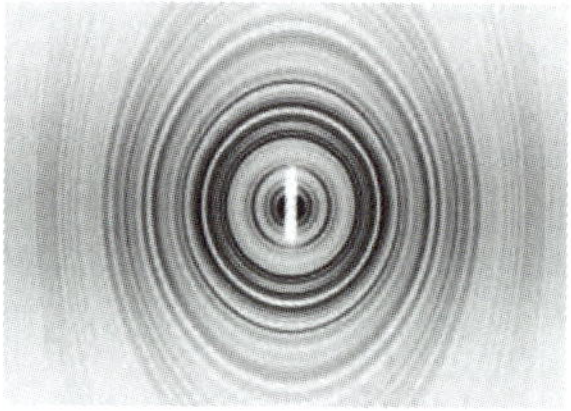

Figure 3. The diffraction pattern in the imaging plate(IP) from a $NbSe_3$ ring-shaped crystal using the new camera.

The observed X-ray intensity must be corrected with sample rotation effect to the Ewald sphere(= Lorentz factor). Since local crystal orientation is a function of the local bent and twist which is not the same among topological crystals, accurate estimation of Lorentz factor is not realistic. Hence, we determined to minimize correction of the Lorentz factor. The Lorentz factor depends chiefly on the distance between the rotation axis and the reciprocal point in the k space. With the rotation of the S, the reciprocal lattice point forms a ring trajectory in the k-space. The trajectorys of reciprocal lattice point by S and P rotations are shown in Fig. 4. When the speed ratio between P and S is incommensurate and X-ray exposure time is long, the reciprocal lattice point will be distributed net-area in Fig. 4. Figure 5 shows the calculated Lorentz factor as the function of the angle between the S axis and a specific recirocal point of 2θ=45 deg. With an increase in the angle between the point and the S axis, the nearest distance between the net trajectory and the Ewald sphere decreases until the ring touched the sphere surface, where the Lorents factor coming from another P axis rotation diverges. We removed the contribution from the above anomalous part by selecting the integration area of the IP (see inset figure of Fig. 5), and suppressed the singularity in the Lorentz factor. The corrected diffraction patterns by selecting integrations were analyzed by the Rietveld method(RIETAN2000). [2]

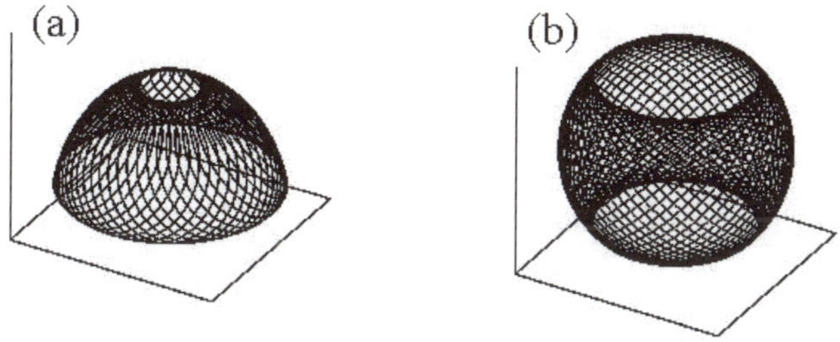

Figure 4. The trajectorys of reciprocal lattice point of single crystal in the k-space angle between point and S of (a) is 30 deg and that of (b) is 90 deg.

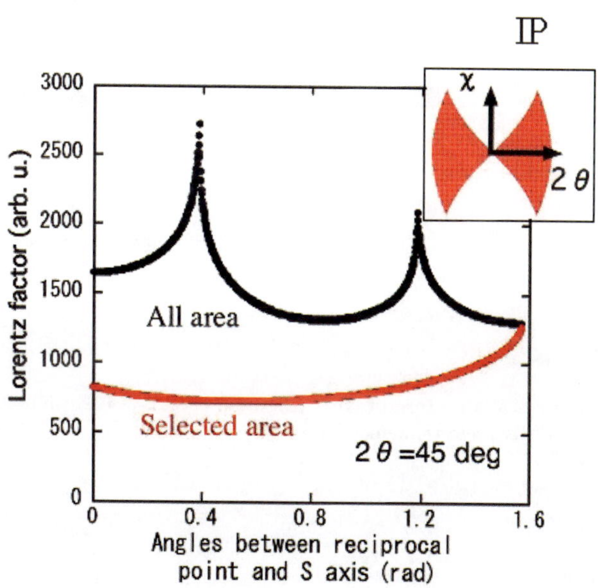

Figure 5. Lorentz factor for the new camera; inset, the selected integration area of the IP.

2. Experiment

We measured $NbSe_3$ figure-8 crystal with 1250 μm strip length, 35×60 μm^2 cross section. Considering the smallness(1000-100000μm^3) of the sample crystals, we used intensive synchrotron radiation at beam line BL02B1 of SPring-8 to enhance X-ray diffraction intensity. To suppress the X-ray noise of air scattering, the sample crystals measured in 10^{-4} Pa at room temperature using low temperature vacuum camera (LTVAC) surrounded by stainless-steel chamber.

3. Results and Discussion

The diffraction patterns of ring, and figure-8 can be analyzed using $P2_1/m$ space group, which agreed with that for whisker. The result of Rietveld refinement of crystal structure is indicated Fig. 6.

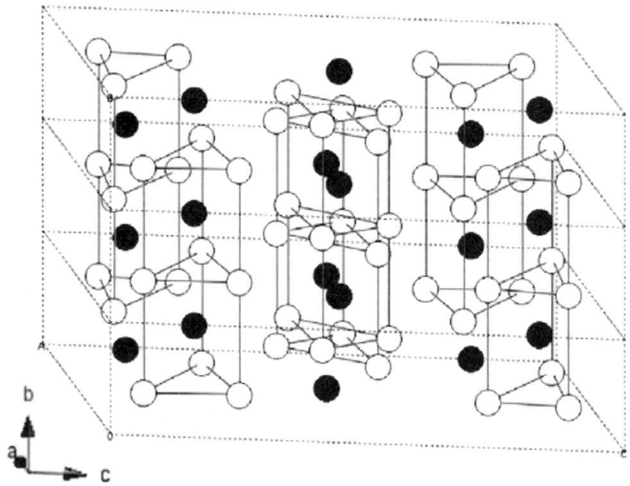

Figure 6. The crystal structure of $NbSe_3$ figure-8 crystal . • is a Nb atom, ∘ is a Se atom. R-factors of Rietveld refinement were $R_{\rm wp} = 4.10$ %, $R_{\rm wp} = 3.12$ %, $R_{\rm wp} = 0.91$ %, $R_{\rm wp} = 0.47$ %.

4. Summary

We developed the novel X-ray analysis method using two-axis rotator and by selecting integral area of IP. This method is applicable not only for topological crystals but also curved or twinned crystals.

Acknowledgments

This work has been partially supported by the 21COE program on "Topological Science and Technology" and by a grant-in-aid for the scientific research of priority areas "Novel Function of Molecular Conductors under Extreme Conditions" from the Ministry of Education, Culture, Sport, Science and Technology of Japan. We acknowledged a support by Okayama university priority research program on "Novel Quantum Effects and Phenomena in Materials with Structural Hierarchy. Integrated Approach to the Reorganization of the Material Structure Science".

References

1. S. Tanda *etal.*, *Nature* **417**, 397 (2002).
2. F. Izumi and T. Ikeda, *Mater. Sci. Forum* **321-324**, 198 (2000).

III

Topological Materials

FEMTOSECOND-TIMESCALE STRUCTURE DYNAMICS IN COMPLEX MATERIALS: THE CASE OF $(NbSe_4)_3I$.

D. DVORSEK AND D. MIHAILOVIC

Ultrafast relaxation dynamics in a quasi-one-dimensional semiconductor $(NbSe_4)_3I$ was investigated with time-resolved optical spectroscopy. New low frequency phonon modes and a central peak were exposed. We have identified a critical behavior connected with an eletronicaly driven structural phase transition at $T_c = 274K$.

1. Introduction

Time-resolved laser spectroscopy gives qualitatively new information on electronic structure as well as non-equilibrium electron and lattice dynamics in condensed matter. In superconductors, correlated electron systems and charge density wave systems the low energy electronic excitations are coupled to the lattice degrees of freedom, which can be effectively investigated on the femtosecond time scale using time resolved spectroscopy.

In previous studies of quasi-one-dimensional and quasi-two-dimensional charge-density wave (CDW) systems, as well as extensive time-resolved studies of cuprates, MgB_2, heavy fermion systems and manganites[1–6], a phenomenological understanding of the dynamics of both single particle and collective excitations has been developed. The systematic experimental studies have also enabled the development of quantitative theoretical models which have been successfully quantitatively tested.

The importance of these studies is twofold. Firstly, the information on the electronic structure on the short timescales is crucially important for understanding the functional properties of materials. Secondly the information on lifetimes of different excitations is very important from the point of view of applications.

In this paper, the dynamics of the electronically driven structural reordering in the quasi-1D Jahn-Teller system $(NbSe_4)_3I$ will be presented and discussed in detail. The structure of $(NbSe_4)_3$ consist of $(NbSe_4)$ chains parallel to the c axis separated from one another by iodine atoms. In the chains the Nb atoms alternate with Se_4 rectangles. This structure enables a directional overlap of Nb 4d$_{z^2}$ orbitals, which gives the material highly 1D character. Above T_c the chains are strongly distorted and have two long Nb-Nb distances of 3.25 Å and a short Nb-Nb distance of 3.06 Å inside a unit cell.[7] At the phase transition, which is ferrodistortive but non-ferroelctic, the space group changes from $P4/mnc$ (D^4_{4h}) above T_c to $P\bar{4}2_1c$ (D^4_{2d}) below T_c. Below T_c the chains are less distorted with following Nb-Nb distances 3.31 Å, 3.17 Å and 3.06 Å.[8] This material is unusual amongst the selenides in that instead of displaying a charge density wave Peirels transition at low temperature, it undergoes a cooperative pseudo-Jahn-Teller structural phase transformation at T_c=274 K.[10] Vibronic coupling between a valence A_{2g} band and a nearby conduction B_{2u} band by a B_{1u} transverse acoustic mode results in a structural phase transition which widens the gap for electronic excitations from 96.9 meV at room temperature to over 220 meV at low temperature (Fig. 1). The order parameter η in this case is related to the density difference between the highest valence band and the lowest conduction band.

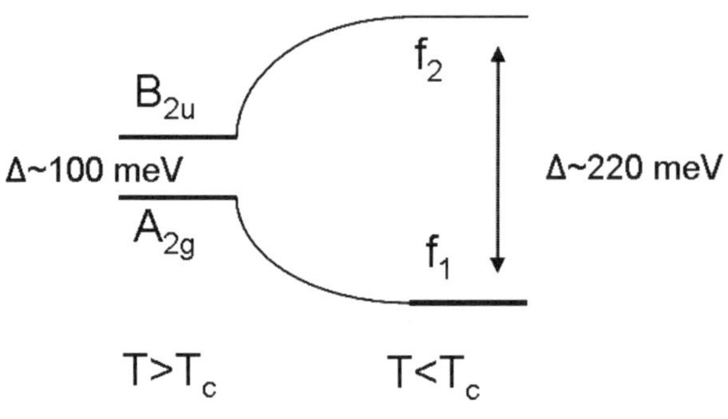

Figure 1. A shematic presentation of the change of the energy gap at the phase transition in $(NbSe_4)_3I$, due to the pseudo-Jahn-Teller effect. (f_1, f_2 are band occupation numbers.

Recent room temperature studies of $(NbSe_4)_3I$ have revealed unexpected spectral signatures in the optical conductivity and angle-resolved photoemission[11], showing that the ground state cannot be understood in terms of a simple 1D band insulator picture. The dynamic nature of the electronic excitations was found to be particularly unusual.

In the present experiments we investigate the electronic and structural dynamics by perturbing the system from its low temperature ordered state by a short pump laser pulse, creating non-equilibrium structural disorder. We then probe the relaxation back to the ordered state by monitoring the changes in the dielectric constant (reflectivity R) with suitably delayed probe laser pulses. The source for both pump and probe light pulses was an output of Ti:sapphire mode-locked laser, which was amplified with a regenerative amplifier at the repetition rate of 250 kHz. The duration of the obtained laser pulses was $\tau_p \approx 80$ fs. The wavelength of the pulses was centered at approximately $\lambda \approx 800$nm (1.58eV) and the intensity ratio of pump and probe pulses was approximately 100:5. The pump and probe beams were crossed on the sample's surface, where the angle of incidence of both beams was less than 10°. The diameter of both beams on the surface was $\sim$100 μm and the typical energy of pump pulses was 12 nJ

2. Experimental results

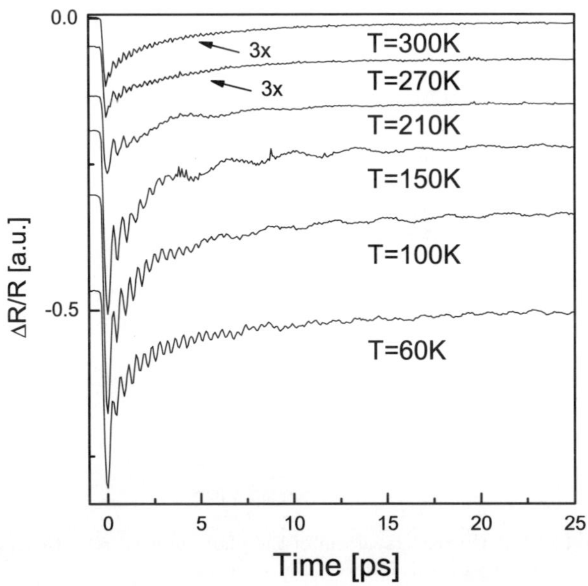

Figure 2. The photoinduced reflection $\Delta R/R$ from $(NbSe_4)_3I$ at different temperatures as a function of time, measured with the polarization of probe pulse in the direction parallel to the crystal axis c (the direction of $(NbSe_4)$ chains).

The photoinduced (PI) changes in reflectivity $\Delta R/R$ as a function of time are shown in Fig. 2. The measurements were done at different temperatures and at the direction of polarization of probe pulses along the $(NbSe_4)$ chains (c crystal axis). The presented data clearly shows oscillatory components, attributed to different coherently excited phonons, superimposed on the components that show exponential like decay. To better analyze the different components of the photoinduced reflection $\Delta R/R$ we performed a fast Fourier transform (FFT) on the data. In Fig 3 (a) we show the FFT of the data at 60 K from the Fig. 2. Actually we performed FFT only on the time dependent photoinduced reflection $\Delta R/R$ measured after the excitation. We can see a central peak and different phonons, whose temperature dependence can be better observed in Fig. 3 (b), where the central peak was fitted out from the original data. The central peak component was analyzed separately in the time domain.

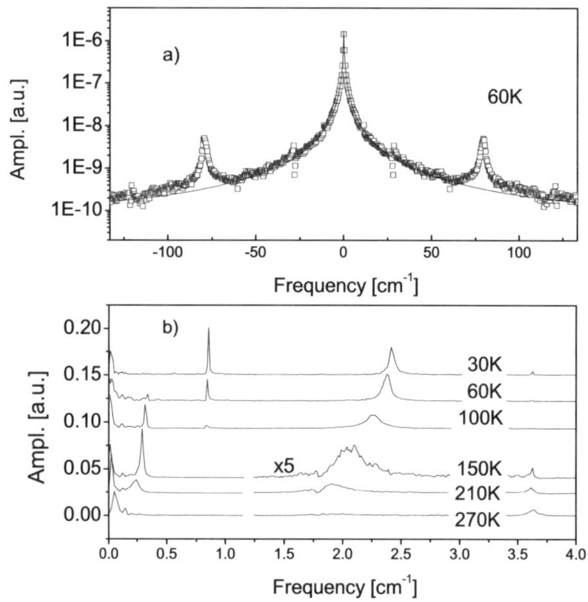

Figure 3. (a) The fast Fourier transform of the photoinduced reflection $\Delta R/R$ data at 60 K. (b) The phonon spectra at different temperatures.

Now we can compare the phonon spectra with Raman spectra obtained at the same material.[9,10] We can immediately see that our FTR spectroscopy enables us to observe low frequency phonons down to very low frequencies that are not accessible by ordinary Raman spectroscopy. In the range of frequencies where spectra of both techniques are overlapping we see the same phonon modes present. We can see a strong mode at around 80 cm^{-1} that softens and broadens with the increasing temperature. It also exhibits a dip when it crosses a stationary 60 cm^{-1} mode, which we couldn't see in Fig. 3 (b) before crossing. This behavior of the two modes was already observed with a Raman spectroscopy[9,10] and the prediction was made of the anticrossing behavior with the appearance of the lower energy phonon mode that would soften completely at the phase transition. We have observed two additional phonon modes in the low frequency range, but couldn't identify the behavior predicted in Ref. 9. The mode at 27 cm^{-1} is almost stationary, whereas the 12 cm^{-1} mode exhibits strong softening. The 12 cm^{-1} mode becomes also overdamped as T approaches T_c. The temperature dependences of the frequencies of the different phonon modes are presented in Figs. 4 (a) and (b). For comparison the Figs. 4 (a) and (b) include also the Raman data[9,10] and the neutron scattering data[12], where a low frequency mode with the same temperature dependence as our 12 cm^{-1} mode was observed.

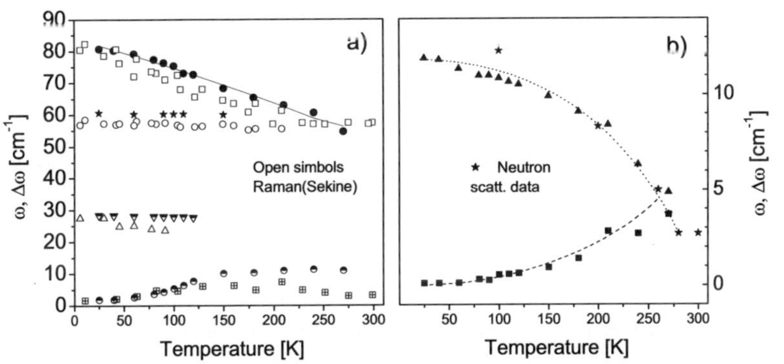

Figure 4. (a) The temperature dependence of the frequency of different phonon modes. The data obtained with the Raman spectroscopy[9,10] is presented with the open symbols for comparison. The temperature dependence of the line width of 80 cm^{-1} mode is shown by half open circles (FOS) and crossed rectangles (Raman). (b) The broadening and softening of the frequency of 12 cm^{-1} soft mode. The continues line through the date points is a guide for the eye. The data from neutron scattering[12] are presented by stars.

The non oscillatory decaying component of the photo induced signal was analyzed in the time domain. We had used the stretch exponential function of the form $\Delta R(\mathrm{t,T})/R = A(\mathrm{T})\exp(-(\mathrm{t}/\tau_A(\mathrm{T}))^{\mu})$ to fit measured signals at different temperatures. There is no direct physical reason for the use of the stretch exponential function, we have also tried to fit with the sum of two exponentially decaying functions but the fits were not as good. To have as few free parameters in our fit as possible, we have fixed the $\mu = 0.4$, which has turned out as good value for fitting the data at all temperatures. The quality of the fit is shown in Fig. 3 (a) by the straight line. The obtained temperature dependence of $1/\tau_A(\mathrm{T})$, which shows critical behavior near the phase transition is presented in Fig. 5 (a). The corresponding amplitude $A(\mathrm{T})$ sharply decreases as T increases toward T_c [Fig 5 (b)]. Similar behavior of relaxation dynamics near electronic phase transitions

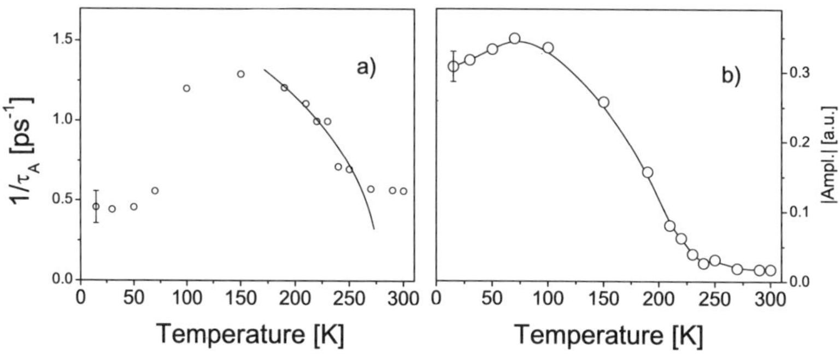

Figure 5. The temperature dependance of non-oscillatory component of photoinduced signal. (a) The relaxation rates $1/\tau_A$ at different temperatures and (b) the temperature dependance of the corresponding amplitude $A(\mathrm{T})$.

was previously observed in different high temperature superconductors and CDW systems.[1–3] The relaxation dynamics was in those cases attributed to the relaxation of the photoinduced quasiparticles across the energy gap in the excitation spectrum. The decrease of the amplitude and the relaxation rate was a result of the closing of the energy gap at the phase transition. In the case of the pseudo-Jahn-Teller phase transition the energy gap does not close completely at T_c as shown in Fig. 1, but nevertheless it does decrease substantially, which results in the change of relaxation dynamics. Bearing this in mind when considering the time-resolved experiments on $(NbSe_4)_3I$ system we can infer that the application of short pump laser pulse changes the occupation numbers of the lowest conduction band and the highest valence band from their equilibrium values which results in perturbation of the order parameter η. Due to the linear coupling between the order parameter η and the structural distortion, which follows from the theoretical analysis of the pseudo-Jahn-Teller type of phase transition[13,14], the different phonon modes can also be coherently excited.

3. Conclusions

We clearly observe real-time modulations due to phonons which are coupled to the order parameter η, and monitor their behavior over a wide range of temperatures. As T_c is approached from below the lowest-frequency recognizable phonon mode at 12 cm^{-1} shows full soft mode softening and broadening. In addition, we observe an anomalous intense zero-frequency (central peak) excitation. The temperature dependence of its width and intensity clearly show that it is directly coupled to the order parameter. Although by symmetry alone it is not possible to distinguish which mode is driving the transition, this low frequency excitation which is believed to be of electronic origin is more likely to be the driving force than the B_{1u} phonon, since it involves a much larger perturbation of the electronic system than the phonon mode.

The unusual behavior of the present quasi-1D system shows that electronically-driven structural dynamics may govern the behavior not only in metallic systems, but also for other semiconducting and semi-metallic one-dimensional materials, with implications that similar behavior might be observed in materials such as semiconducting carbon nanotubes, MoSI nanowires [15] and even DNA.[16]

References

1. V. V. Kabanov, J. Demsar, B. Podobnik, and D. Mihailovic, *Phys. Rev. B* **59,** 1497 (1999).
2. V. V. Kabanov, J. Demsar, and D. Mihailovic, *Phys. Rev. B* **61,** 1477 (2000); J. Demsar, B. Podobnik, V. V. Kabanov, Th. Wolf, and D. Mihailovic, *Phys. Rev. Lett.* **82,** 4918 (1999); C.J.Stevens, D.Smith, C.Chen, J.F.Ryan, B.Podobnik, D.Mihailovic, G.A.Wagner and J.E.Evetts, *Phys. Rev. Lett.* **78**, 2212 (1997).
3. D. Mihailovic, D. Dvorsek, V. V. Kabanov, J. Demsar, L. Forro and H. Berger, *Appl. Phys. Lett.* **80**, 871 (2002); D. Dvorsek, V. V. Kabanov, J. Demsar, S. M. Kazakov, J. Karpinski and D. Mihailovic, *Phys. Rev. B* **66**, 020510-1 (2002); J.Demsar, K.Biljakovic and D.Mihailovic, *Phys. Rev. Lett.* **83**, 800 (1999).
4. J. Demsar, R. D. Averitt, K. H. Ahn, M. J. Graf, S. A. Trugman, V. V. Kabanov, J. L. Sarrao and A. J. Taylor, *Phys. Rev. Lett.* **91**, 027401 (2003).
5. J. Demsar, R. D. Averitt, A. J. Taylor, V. V. Kabanov, W. N. Kang, H. J. Kim, E. M. Choi and S. I. Lee, *Phys. Rev. Lett.* **91**, 267002 (2003).
6. T. Mertelj, D. Mihailovic, Z. Jaglicic, A. A. Bosak, O. Yu. Gorbenko, and A. R. Kaul, *Phys. Rev. B* **68,** 125112 (2003).
7. A. Meerschaut, P. Palvadeau and J. Rouxel, *J. Solid State Chem.* **20**, 21 (1977).

8. P. Gressier, L. Guemas and A. Meerschaut, *Mater. Res. Bull.* **20**, 539 (1985).
9. T. Sekine, K. Uchinokura, M. Izumi and E. Matsuura, *Solid State Commun.* **52**, 379 (1984).
10. T. Sekine and M. Izumi, *Phys. Rev. B* **38**, 2012 (1988).
11. V. Vescoli ,F. Zwick, J. Voit, H. Berger, M. Zacchigna, L. Degiorgi, M. Grioni, and G. Grüner,. *Phys. Rev. Lett.* **84**, 1272 (2000).
12. P. Monceau, L. Bernard, R. Currat and F. Levy, *Physica B* **156**, 20 (1989).
13. N. Kristoffel, and P. Konsin, *Ferroelectrics* **6**, 3 (1973); N. Kristoffel and P. Konsin, *Phys. Status Solidi* **28**, 731 (1968).
14. N. Kristoffel and P. Konsin, *Phys. Status Solidi* **21**, K39 (1967).
15. M. Remskar, A. Mrzel, Z. Skraba, A. Jesih, M. Ceh, J. Demsar, P. Stadelmann, F. Lévy, and D. Mihailovic, *Science* **292**, 479 (2001).
16. A. Omerzu, M. Licer, T. Mertelj, V. V. Kabanov and D. Mihailovic, *Phys. Rev. Lett.* **93**, 218101 (2004).

ULTRAFAST DYNAMICS OF CHARGE-DENSITY-WAVE IN TOPOLOGICAL CRYSTALS

K. SHIMATAKE, Y. TODA,* T. MINAMI AND S. TANDA

Department of Applied Physics, Hokkaido University, Sapporo 060-8628, Japan.
E-mail: simatake@eng.hokudai.ac.jp

We have investigated the quasi-one-dimensional (quasi-1D) compound NbSe3 ring and whisker crystals by means of a time-resolved optical pump-probe measurement. Around a Peierls transition temperature at which the 3D CDW ordering is achieved by adjusting the coulomb correlation between individual 1D chains, the single particle relaxation show a remarkable difference between the crystals, suggesting a difference of the phase correlation depending on the crystal topology.

1. Introduction

Quasi-one-dimensional (quasi-1D) metals often undergo Peierls instability associated with the charge density wave (CDW) ordering. Below a Peierls transition temperature T_p, a three-dimensional (3D) ordering of the CDWs results in a well-defined single particle (SP) gap on Fermi surface. Therefore the phase correlation between the 1D chains gives rise to a striking influence on the Peierls transition. For example, if the system is really 1D, large fluctuations strongly decrease T_p below the mean field transition temperature. Indeed, a pronounced finite size effect has been observed in $NbSe_3$, in which the reduction of the number of parallel chains makes the Peierls transitions less visible and decreases T_p [1]. In this sense, fluctuation effects should also be expected in the changes of crystal topology since the topology may impose additional constraints on the phase correlation between the chains. However, despite the considerable interests on the size effects, effects of the crystal topology have received little attention to date. Recently, it has been found that several transition metal chalcogenides of the type MX_3 can have various types of topological structures. Especially

*Also at PRESTO Japan Science and Technology Agency, Saitama 332-0012, Japan

$NbSe_3$, which is regarded as a model material of CDW, forms a ring, a disk, or a Mobius strip [2,3].

In this work, we investigated, for the first time, the influences of the crystal topology on the Peierls transition by comparing the SP dynamics between crystals with different topology. When approaching to T_p from below, a remarkable increase of the SP relaxation time (τ_s) was observed in the $NbSe_3$ whisker. In contrast, the increase of τ_s in a ring is less pronounced than in the case of whisker. The result can be explained by the enhanced phase fluctuation in the closed-loop topology.

2. Experimental

For evaluating the CDW dynamics, we employed here the optical measurement, which is an ideal probe for the topological structures since the photon can act as a non-contact probe and preserve the crystal topology. The time-resolved measurement was achieved by a conventional pump-probe technique. The light source was a mode-locked Ti:sapphire laser (pulse width 130 fs, repetition rate 76 MHz). First, the pump pulse excites carriers via an interband transition. Then, these SPs relax down to an equilibrium state. In these processes, we can detect the transient response of the excited carriers as a reflectivity change in the probe pulse with a certain delay from the pump pulse. The spot size of pulses was estimated to be 10mm in diameter.

3. Results

Figure 1 shows various transient reflectivity changes below and above T_p (59K) in whisker and ring $NbSe_3$ crystals. Since the excitation photon energy of 1.56eV can excite the carriers into continuum states far above the CDW gap, the general trend of data can be attributed to a transient response of the photoexcited carriers associated with SP transitions. The observed SP response could be divided into three components according to the report of Demsar *et al.* [4]

(a) Abrupt increase within the time resolution corresponding to both the excitation of the SP over the gap and the relaxation down to the upper edge of the gap

(b) Subsequent decay corresponding to a inter-band relaxation across the gap

(c) Long-lived decay corresponding to the relaxation from a phason (PM) state pinned above the ground state.

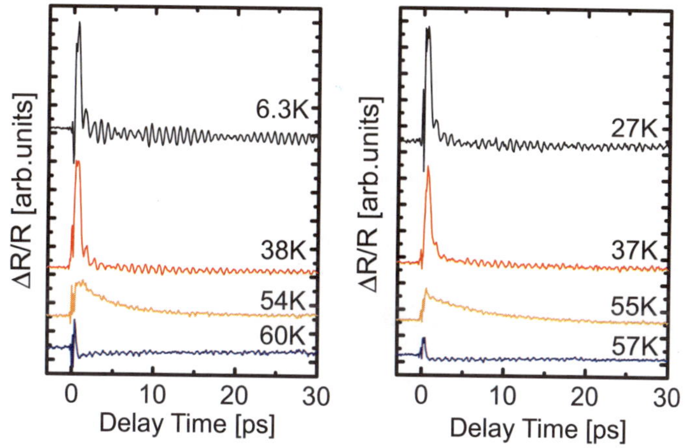

Figure 1. The typical transient reflectivity change measured across T_p (59K) in ring(left) and whisker(right) crystals.

Since the CDW gap depends on the sample temperature, the signal components from (ii) and (iii) will exhibit the remarkable temperature dependence. $T \ll T_p$, phonon emission is dominant. Therefore the signal shows fast decay. $T < T_p$, the process of phonon re-absorption becomes possible because of shrinkage of the gap. As a result, the increase of τ_s is observed. In addition, the PM can contribute the signal. $T > T_p$, the fast decay appears again.

Figure 2 shows the temperature dependence of τ_s in the ring and whisker crystals. In both crystals, a similar behavior is observed around T_p; τ_s shows increase when approaching to T_p from below. One of the remarkable differences between the data is the degree of the divergence of τ_s around T_p. τ_s in whisker diverges up to around 10 ps, which is more than 10 times longer than that in the lowest temperature. In contrast, τ_s in the ring reaches only 2 ps, which is three times faster than that in the whisker

For another qualitative analysis of ts, we analyzed the data by the mean field approximation (MFA). Since the SP relaxation reflects the gap energy, the temperature dependence of τ_s is described as

$$\tau_s(T) = \frac{1}{\Gamma}\sqrt{1 - \frac{T}{T_c}} \tag{1}$$

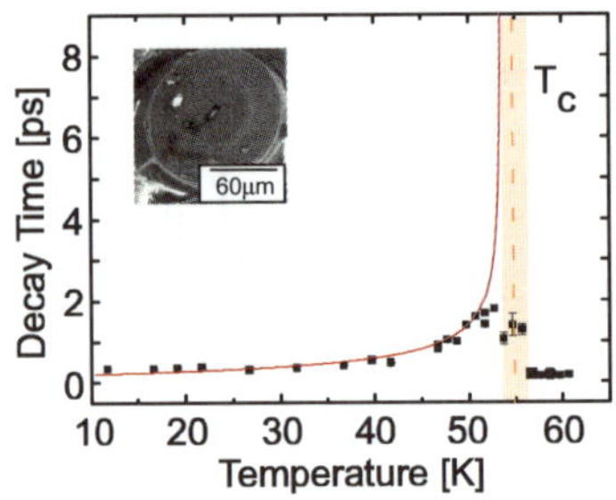

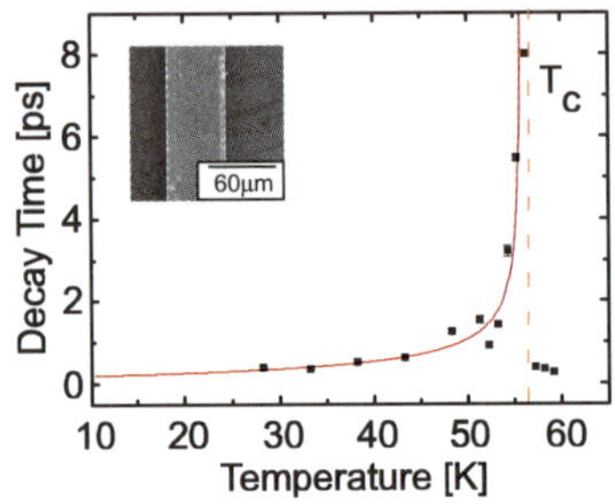

Figure 2. The temperature dependence of the SP relaxation time in a ring (up) and whisker (down).

where Γ is a fitting parameter reflecting the CDW fluctuation. From the fitting to the data, we obtained $\Gamma = 3.0 \pm 0.3$ in the whisker and 4.4 ± 0.2 in the ring.

4. Conclusion

We conclude that the difference of the temperature dependence of τ_s between the crystals is originated from the effects of topology on the CDW phase fluctuations. In the Peierls transition, the CDWs get 3D ordering by adjustment of the coulomb correlation between the 1D chains. Ring has a closed-loop crystal topology, which may impose additional constraints on the phase correlation between the chains. Therefore, the CDWs in the ring structure are difficult to obtain the coulomb correlation for the 3D ordering. Besides, the phase fluctuation strongly suppressethe divergence of the gap behavior represented by MFA. As a result, the fluctuation in the ring makes the critical relaxation clear, thus showing the suppression of the divergence around T_p.

In order to clarify the topological effect, we also investigated the homogeneities of the samples. For this purpose, we evaluated the dephasing time of collective excitations. The results indicate the similar dephasing time in each sample, thus we conclude that the contribution of the structural homogeneities to the phase transition is almost the same in each sample.

Acknowledgment

This work is partly supported by Grant-in-Aid for the 21st Century COE program "Topological Science and Technology".

References

1. E. Slot et al,. *Phys. Rev. Lett.* **93**, 176602 (2004).
2. S. Tanda et al,. *Journal. de. Physique.* **9**, 379 (1999).
3. T. Tsuneta et al,. *J. Crys. Grow.* **264**, 223 (2004).
4. J. Demsar et al,. *Phys. Rev. Lett.* **83**, 800 (1999).

TOPOLOGY IN MORPHOLOGIES OF A FOLDED SINGLE-CHAIN POLYMER

Y. TAKENAKA, D. BAIGL AND K. YOSHIKAWA

Department of Physics, Graduate School of Science, Kyoto University, Kyoto 606-8502, Japan

E-mail: takenaka.yoshiko@scphys.kyoto-u.ac.jp

E-mail: damien.baigl@ens.fr

E-mail: yoshikaw@scphys.kyoto-u.ac.jp

Folding transition of a single semiflexible polymer under poor solvent conditions can lead to a large variety of morphologies. We performed an analysis on these morphological changes by simple theoretical treatment and classified the different conformations by using the topological index called genus. In particular, we investigated the effects of stiffness and thickness of the chain on the toroidal conformation, a typical morphology of condensed semiflexible chains. Our results are in a semi-quantitative agreement with single-chain observation on long duplex DNA molecules.

1. Introduction

Polymers can be classified into flexible, semiflexible, or rigid polymers, depending on the relationship between the thickness δ, the contour length L, and the persistence length λ of the chain. Particularly interesting are semiflexible polymers, such as DNA, for which the inequality $\delta \ll \lambda \ll L$ is satisfied. Indeed, with a decrease in the solvent quality or in the presence of condensing agents, the single chain undergoes a first-order phase transition between an elongated coil state and a very dense, compact state[1]. As the result of folding transition, the compact chains can be of various morphologies and characteristic sizes[2−12]. In the case of DNA[2], the compact state has typically a toroidal morphology with an outer diameter of 60-100 nm. Under specific conditions, different morphologies and larger characteristic sizes can also be obtained[3]. Since the compact state is formed by a self-organization of the chain, we can expect that the various morphologies are generated spontaneously at the level of the single chain.

In this study, we have investigated the morphologies of a single polymer chain after the folding transition by using a simple theoretical approach.

Using a topological methodology, we classified the different single-chain morphologies according to an index called genus. Genus means the half of the linear connectivity, which is the number of the closed curves needed to cut off a curved surface. For instance, the genus of a rod is zero while that of a toroid is one. We will start from the situation of a neutral chain. In a second part, electrostatic effects will be taken into account to consider the situation of a charged chain (polyelectrolyte).

2. Neutral Chain

Figure 1 shows the various morphologies generated from a single neutral polymer chain as a function of solvent quality and stiffness of a chain[13], where stiffness means about the half of the persistence length. When the chain is flexible ($\delta \sim \lambda \ll L$), decreasing solvent quality induces the folding transition of the chain from the coil state to the spherical state (Index = 0). With an increase in the chain rigidity, structures such as a rod (Index = 0) or a toroid (Index = 1) appear. The difference between a rod and a toroid results from the strength of the pair-wise interaction between segments. Therefore, it can be concluded that stiffness of a chain can control the topology generated from a single neutral polymer chain; the index genus changes from 0 to 1 with a sufficient increase in the chain stiffness. Moreover, an index of 1 can correspond to various toroidal morphologies. Figure 2 shows the effects of chain stiffness and chain thickness on the toroidal morphology[14].

3. Charged Chain

In the case of a neutral chain, it was shown that only indices of 0 and 1 can be obtained. For example, in the case of a toroidal morphology, increasing chain length does not change the index (the single toroidal morphology is stable), but only induces an increase in the toroid thickness (difference between outer and inner diameter). The situation of a charged chain is much different. By using fluorescence microscopy observations on single-chain long DNA duplex molecules condensed by multivalent counterions[15], we observed that DNA condensates are made of one single chain that do not aggregate due to the remaining negative charge of the DNA condensate. Transmission electron microscopy (TEM) observations were used to resolve the detailed morphology of DNA condensates of various chain lengths. For DNA contour length up to ca. 10 μm, DNA condensates are always single toroids of 60-100 nm (Index = 1), in agreement with the classical

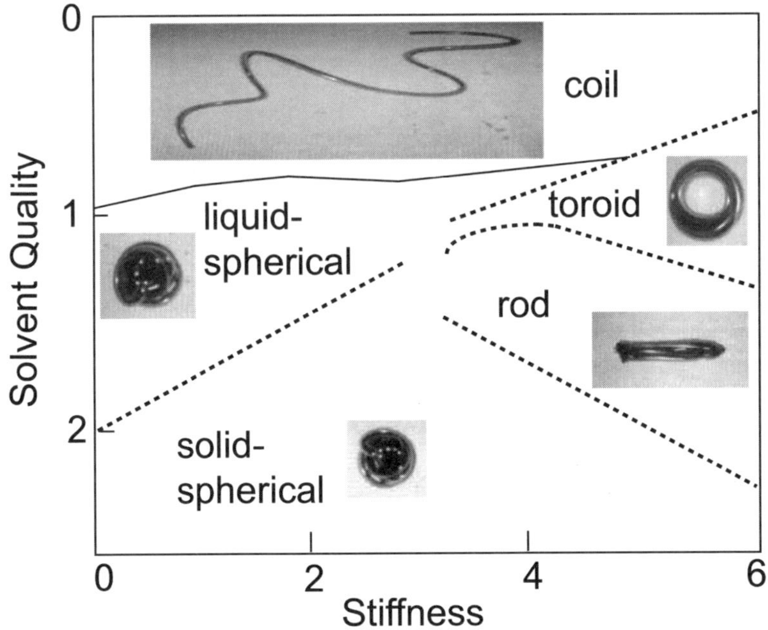

Figure 1. Schematic drawings of the phase diagram with solvent quality and stiffness of a chain. As increasing the stiffness, the index genus changes from 0 to 1.

observations[2]. However, for longer DNA chains (L > 10 μm), multi-tori structures (Index $\geq$ 2) appear. Figure 3 shows typical multi-tori structures in the case of T4 DNA (L = 57 μm).

Here, we consider a simple theoretical approach of the multi-tori structure made of one single charged chain with a fixed charge density. The free energy f of each toroid can be simply expressed as[16]

$$f = -ar^3 + br^2 + c\frac{q^2}{r} \tag{1}$$

where r is the average radius and q is the remaining charge on each condensate. The total free energy of multi-tori made of i toroids can be expressed as

$$F = if \sim -aV + bV^{\frac{2}{3}}i^{\frac{1}{3}} + cV^{\frac{5}{3}}i^{-\frac{2}{3}} \tag{2}$$

where V is the volume of multi-tori and a, b, and c are positive constants. By minimizing the free energy F, we obtained the relationship between total volume and the number of compact states i as plotted in Fig. 4. It

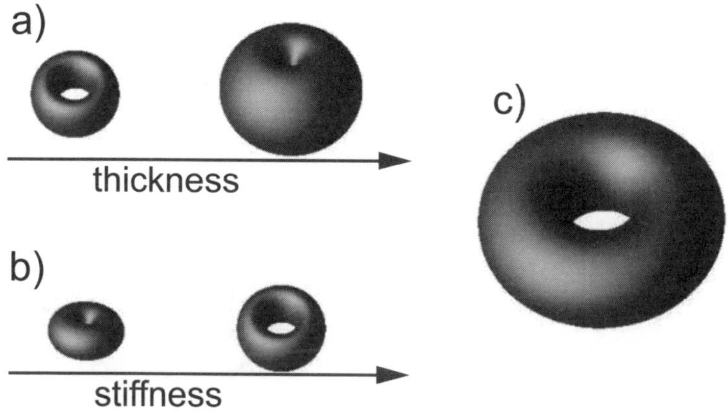

Figure 2. Morphological changes on toroid as a function of thickness and stiffness of a chain as obtained in ref.14. (a) Effect of thickness and (b) Effect of stiffness. Giant toroid appears with thick and stiff chain (c).

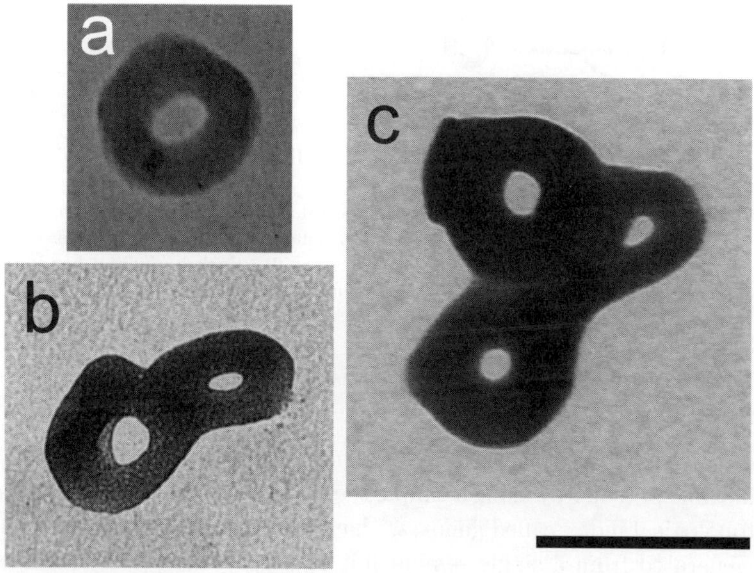

Figure 3. Transmission electron microscopic (TEM) images of condensates generated from a single T4 DNA (L = 57 μm, concentration 0.1 μM) in the presence of 10 μM of spermine, a tetracationic condensing agent. (a) Single toroid, (b) Two-tori and (c) Three-tori. The scale bar is 100 nm.

shows that the number of torii increases with an increase in chain length, in agreement with the calculations of the distribution function[17]. Therefore, in case of a charged chain, it can be concluded that multi-tori structures appear as increasing the chain length because of the instability due to the remaining charge on each condensate. It is to be noted that morphologies similar to the multi-tori such as rings-on-a-string chain structures have also been observed[18].

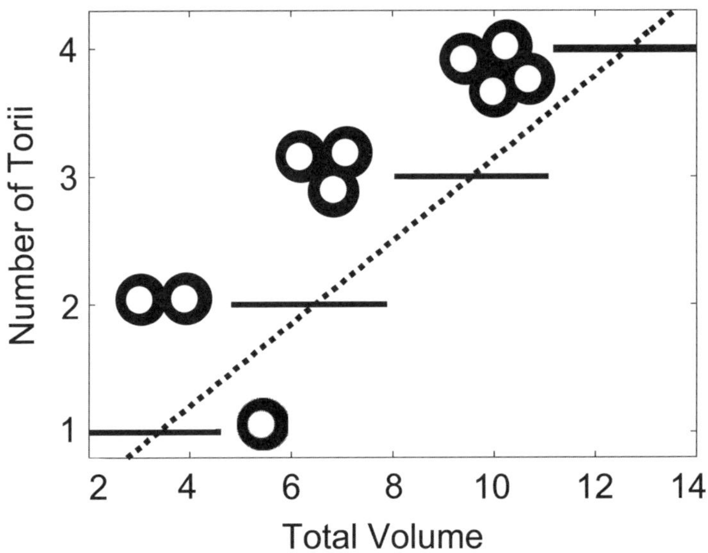

Figure 4. Number of torii as a function of the total volume, as the self-organized structure from a single polyelectrolyte. The dotted line shows i as a function of V as obtained from the minimization of the total free energy (Eq. 2). Each horizontal bar indicates the region where a given morphology is the most stable.

4. Summary

We have investigated the self-organization in the folding transition of a semiflexible polymer by using a simple theoretical approach. By introducing the topological index called genus, we have classified the various morphologies generated from a single semiflexible polymer chain. The relationship between the morphologies generated from a neutral/charged polymer and the indices can be summarized as follows:

(1) Neutral Chain. Indices 0 and 1 are generated by changing the stiff-

ness of a chain. In particular, for the index of 1, the morphological change on toroid is interpreted in terms of stiffness and thickness.

(2) Charged Chain (Polyelectrolyte). Indices 0 and 1 are generated similarly to the case of neutral chain. In addition, indices larger than 2 also appear in the case of sufficiently large chain, due to the surviving charge of the compact state.

Acknowledgments

The authors wish to thank Prof. Y. Yoshikawa (Nagoya Bunri College, Japan), Prof. Y. Koyama (Otsuma Women's University, Japan) and Prof. T. Kanbe (Nagoya Graduate School of Medicine) for experimental assistance. We also thank Dr. T. Sakaue (Kyoto University, Japan) for valuable discussions.

References

1. K. Yoshikawa, M. Takahashi, V. V. Vasilevskaya and A. R. Khokhlov, *Phys. Rev. Lett.* **76**, 3029 (1996).
2. C. C. Conwell, I. D. Vilfan and N. V. Hud, *Proc. Nat. Acad. Sci. USA* **100**, 9296 (2003).
3. Y. Yoshikawa, K. Yoshikawa and T. Kanbe, *Langmuir* **15**, 4085 (1999).
4. U. A. Laemmli, *Proc. Nat. Acad. Sci. USA* **72**, 4288 (1975).
5. L. C. Gosule and J. A. Schellman, *Nature* **259**, 333 (1976).
6. V. A. Bloomfield, *Biopolymers* **31**, 1471 (1991).
7. Y. Yoshikawa, N. Emi, T. Kanbe, K. Yoshikawa and H. Saito, *FEBS Lett.* **396**, 71 (1996).
8. K. Yoshikawa, Y. Yoshikawa, Y. Koyama and T. Kanbe, *J. Am. Chem. Soc.* **119**, 6473 (1997).
9. Y. Fang and J. H. Hoh, *FEBS Lett.* **459**, 173 (1999).
10. M. R. Shen,K. H. Downing, R. Balhorn and N. V. Hud, *J. Am. Chem. Soc.* **122**, 4833 (2000).
11. N. V. Hud and K. H. Downing, *Proc. Nat. Acad. Sci. USA* **98**, 14925 (2001).
12. V. Vijayanathan, T. Thomas, T. Antony, A. Shirahata and T. J. Thomas, *Nucleic. Acids, Res.* **32**, 127 (2004).
13. H. Noguchi and K. Yoshikawa, *J. Chem. Phys.* **109**, 5070 (1998).
14. Y. Takenaka, Y. Yoshikawa, Y. Koyama, T. Kanbe and K. Yoshikawa, *J. Chem. Phys.* **123**, 014902-1 (2005).
15. D. Baigl and K. Yoshikawa, *Biophys. J.* **88**, 3486 (2005).
16. Y. Yoshikawa, M. Suzuki, N. Chen, A. A. Zinchenko, S. Murata, T Kanbe, T. Nakai, H. Oana and K. Yoshikawa, *Eur. J. Biochem.* **270**, 3101 (2003).
17. T. Sakaue, *J. Chem. Phys.* **120**, 6299 (2004).
18. N. Miyazawa, T. Sakaue, K. Yoshikawa and R. Zana, *J. Chem. Phys.* **122**, 044902 (2005).

ONE TO TWO-DIMENSIONAL CONVERSION IN TOPOLOGICAL CRYSTALS

T. TOSHIMA, K. INAGAKI AND S. TANDA

Department of Applied Physics, Hokkaido University, Sapporo 060-8628, Japan

E-mail: toshima@eng.hokudai.ac.jp

We have studied topological crystals, metal trichalcogenides (MX_3) and metal dichalcogenides (MX_2). The topological crystals are defined by following property. Both ends of system are combined each together, such as ring shape or tube shape. The system loses translational symmetry and gain rotational symmetry. Such the properties are allowed by low dimensional structure of the systems. Therefore once material is chosen, its dimensionality restricts feasibility of shapes. We succeeded in synthesizing $TaSe_2$ ring crystal, quasi-topological crystal and nanoscale heterojunctions of $TaSe_3$-$TaSe_2$-$TaSe_3$ by the de-chalcogenide method.

1. Introduction

Topological crystals such as carbon nanotubes [1], ring, and Möbius strip crystals [2] have attracted much attention since they are keys for exploring unveiled character in quantum mechanics like Berry phase [3]. Synthesis of these topological crystals is ordinary attained in low dimensional materials due to their flexibility. For example, carbon nanotubes are roll of two-dimensional graphite sheets; the perpendicular to the sheet is flexible. Metal trichalcogenide MX_3 (M=Ta, Nb and X= Se, S), one-dimensional needle-like crystals sometimes form rings and Möbius crystals, which have curvature and twist; the perpendicular to the chain is also flexible. In Bravais ' s law [4], global shape of crystal is governed by its microscopic symmetry and dimensionality. If a material is chosen, its dimensionality restricts feasibility of shapes. It is hard for crystals with two and three dimension structure to form the topological shape.

Here we report new strategy to create topological crystals with higher dimensions. First step, we synthesize the needle shape crystals and some topological crystals such as rings, tubes and Möbius strip crystals of one-dimensional materials. ($TaSe_3$, TaS_3, $NbSe_3$, etc) The next step, we rearranged their chemical bonds to convert their dimensionality, as the result, topological crystals with higher dimension are obtained. Following the concept, we successfully synthesized niobium diselenide $NbSe_2$ nanotubes from $NbSe_3$ [5], $TaSe_2$ ring crystals from $TaSe_3$ ring crystal. In addition, the dimensional conversion method is also applicable for making up a new kind of heterojunctions, consisting of MX_3-MX_2-MX_3 structure, where n is less than three.

2. Experiment

We synthesized single crystals of $TaSe_3$ by the conventional chemical vapor transport method [6]. We choose a mixture of tantalum and selenium powders (purify 3N+) as starting materials. They are put in an evacuated quartz tube and are hearted to 700 °C with hearting rate of 10 °C/min and temperature gradient of 1 °C/cm. The chemical reaction time is 3 to 5h. After the reaction, one end of tube is quenched by liquid nitrogen. The other end is cut open in vacuum. Then $TaSe_3$ crystals are obtained. The size is approximately 10 nm–1.0 mm in width and thickness and 100 nm–10 mm in length. And some of topological crystals such as rings are also obtained.

We use these crystals as a template to convert from one-dimensional $TaSe_3$ to two-dimensional $TaSe_2$. They are put on quartz sheet and set in electric furnace. The sample chamber is evacuated to 10^{-5} Torr. And then $TaSe_3$ crystals are heated to approximately 300 °C or 1000 °C with a hydrogen gas. The chemical reaction is represented as

$$TaSe_3 + H_2 \text{ (gas)} \rightarrow TaSe_2 + H_2Se \text{ (gas)},$$
$$TaSe_3 + 3H_2 \text{ (gas)} \rightarrow Ta + 3H_2Se \text{ (gas)}.$$

These crystals were sonicated in isopropyl alcohol for 30 min and put it on the substrate to be observed by transmission electron microscope (TEM).

3. Result and discussion

Figure 1(a) is TEM image of $TaSe_2$ nanocrystal growing on the surface of $TaSe_3$ nanofiber. The $TaSe_2$ crystal is approximately 20 nm in width and less than 10 nm in thickness. Hexagonal crystal habit is due to two-dimensional hexagonal structure of $TaSe_2$. And needle shape is due to one-dimensionally $TaSe_3$ structure. This result shows that if one chose a template ($TaSe_3$) with nano scale, more small crystal of $TaSe_2$ can be

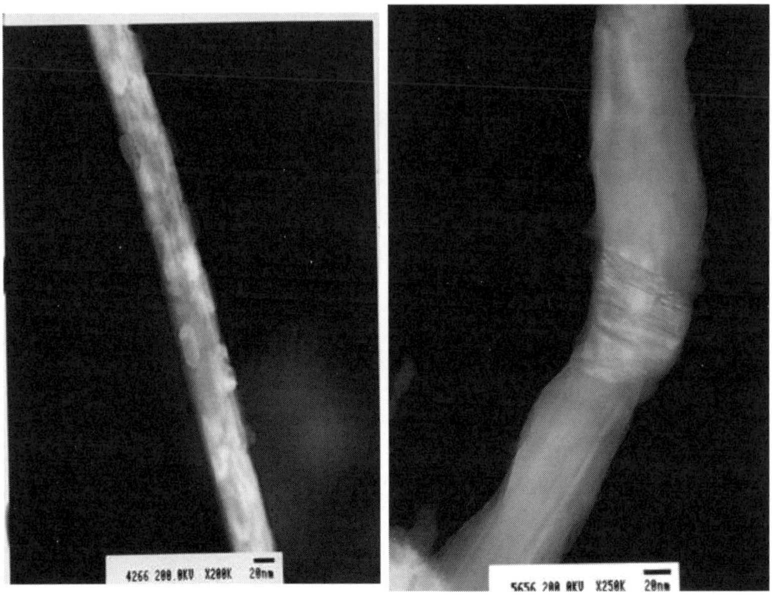

Figure 1. TEM image of $TaSe_2$ and $TaSe_3$ after the reaction. Fig 1(a) (left image) is $TaSe_2$ thin films are growing on $TaSe_3$. Fig 1(b) (right image) is $TaSe_2$ layers are sandwiched between $TaSe_3$ nanofibers.

synthesized, and global shape is kept over the chemical reaction. Figure 1(b) is $TaSe_2$ plate which is sandwiched between $TaSe_3$ nanofibers. This result shows that heterojunctions of MX_3-MX_2-MX_3, which is equal to superconductor-charge density wave-superconductor at the low temperature are realized in this system.

Figure 2 is TEM image of a ring crystal after the chemical reaction with moderate condition and the right image is transmission electron diffraction image of the ring crystal. The reaction temperature is 300 °C, heating rate is 300 °C/min and reaction time is 30 minutes with flowing hydrogen gas (1 sccm). During this reaction, the vacuum of the sample chamber is maintained at approximately 10^{-3} Torr. In this result, global shape of topological crystal is kept, but TED analysis shows that local chemical bonding is changed from $TaSe_3$ to $TaSe_2$. That is to say, dimension conversion from one-dimensionally system to two-dimensionally system in topological crystal. If one convert a part of MX_3 ring crystal, Josephson junction is realized. Moreover the convertion succeed only the surface of MX_3 ring crystal, MX_2 torus would be obtained.

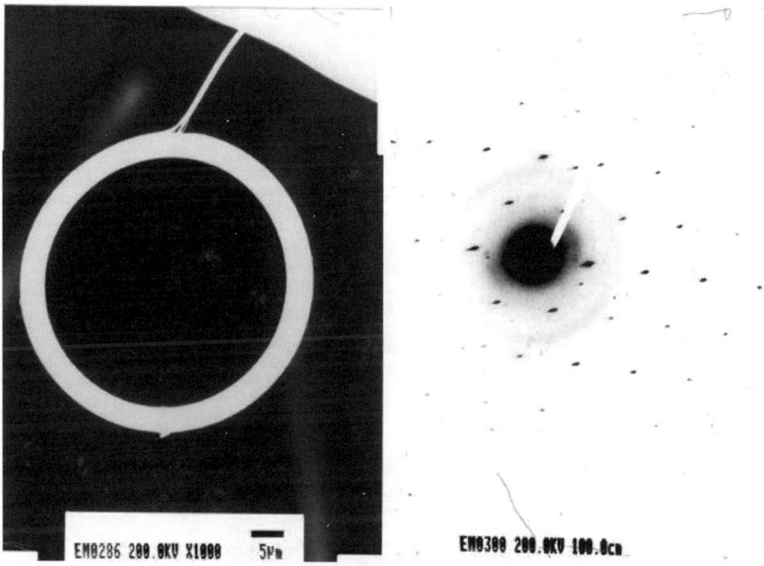

Figure 2. TEM (left) and TED (right) image of ring crystal after the reaction. Topological crystal keeps its shape during the chemical reaction, but local chemical bonding is changed.

Figure 3 is TEM image of quasi-topological crystals (center of figure, coil spring crystal), which are synthesized from $TaSe_3$ needle crystal with intense condition. The reaction temperature is 1000 °C, heating rate is 1000 °C/min and reaction time is 3 minutes. During this chemical reaction, the sample chamber is filled with hydrogen gas. These crystals have global rotational symmetry, but the ends of system still remain. That is to say "quasi-topological crystal" (Topological crystals have global rotational symmetry and both ends of system are combined each together). Such structure is introduced by difference of lattice constant between $TaSe_2$ and $TaSe_3$. When surface $TaSe_3$ crystal is converted to $TaSe_2$ layer crystal and inner $TaSe_3$ crystal is still not converted. Lattice defect is introduced between $TaSe_2$ and $TaSe_3$. It forces the system to form topological shape (with the curvature), in which the system gains rotational symmetry by losing translational symmetry. Moreover on some of TaSe2 ultrathin crystals, charge density wave reforms its wavelength from $\sqrt{13}$ to $\sqrt{4}$, $\sqrt{7}$, and $\sqrt{61}$. Such the transmutation is probably due to downsizing effect of system, which strengthens two-dimensionality.

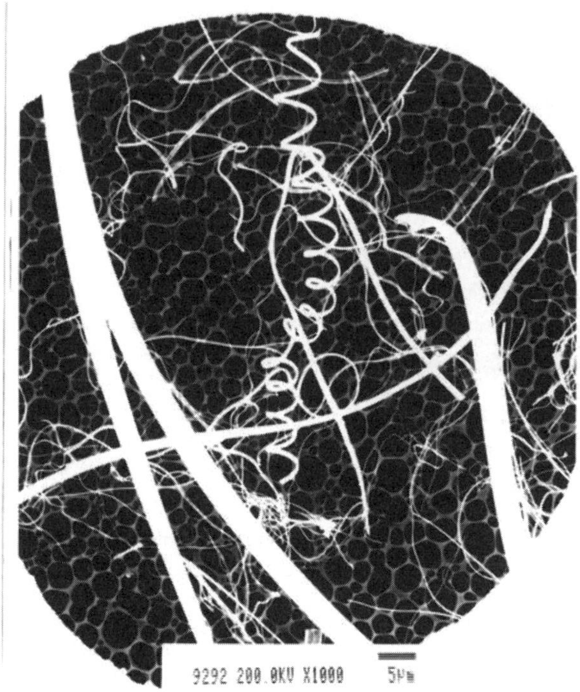

Figure 3. TEM image of quasi-topological crystal. The spring crystal has global rotational symmetry, but the system ends still remain.

Acknowledgment

This work has been partially supported by Grant-in-Aid for the 21st Century COE program on "Topological Science and Technology" from the Ministry of Education, Culture, Sport, Science and Technology of Japan.

References

1. S. Iijuma, *Nature* **354**, 56 (1991).
2. S. Tanda, T. Tsuneta, Y. Okajima, K. Inagaki, K. Yamaya, N. Hatakenaka, *Nature* **417**, 397 (2002).
3. M. V. Berry, *Proc. R. Soc. Lond.* **A392**, 45 (1984).
4. J. D. H. Donnay, D. Harker, *Amer. Mineral,* **22**, 446 (1937).
5. T. Tsuneta, T. Toshima, K. Inagaki, T. Shibayama, S. Tanda, S. Uji, M. Ashlskog, P. Hakonen, M. Paalanen, *Current Applied Physics* **3**, 473 (2003).
6. H. Schafer, *Chemical Transport Reactions. Academic Press.* New York (1964).

TOPOLOGICAL CHANGE OF FERMI SURFACE IN BISMUTH UNDER HIGH PRESSURE

M. KASAMI,* T. OGINO, T. MISHINA, S. YAMAMOTO AND J. NAKAHARA

Department of Physics, Graduate School of Science, Hokkaido University, Kita 6, Nisi 8, Sapporo, 060-0810, Japan.
E-mail: kasami@phys.sci.hokudai.ac.jp

The high pressure effect on the topological change of the Fermi surface in Bi is investigated using microscope pump and probe system. We have observed drastic changes of electronic response which is relaxation of photo-excited carriers at phase transition pressure. The electronic responses are strongly correlated with change in the electronic band structure, we propose the drastic changes of electronic responses can be topological change of the Fermi surface caused by the phase transition.

The recent advances of high pressure techniques make possible to study high pressure regimes in excess of the centre of Earth (360 GPa). Physical and chemical phenomena can be investigated under these extreme conditions. Particularly, pressure induced phase transitions are expected to cause drastic change in the lattice and electronic band structures. There have been several experiments at high pressure phase. Lotter *et al.* discover a sharp kink in the T_c dependence of superconducting Bi at phase transition pressure, they ascribe their results to topological change of the Fermi surface.[1] Bismuth is one of the most studied materials for unique electrical transport properties and variety of applications. Bismuth shows complicated phase diagram, which indicates several phases under high pressures at room temperature.[2] Bi is a semimetal at room temperature (RT) at ambient pressure, there is a band overlapping of 32-38 meV between the valence band at T point and the conduction band at L point.[3,4] As the pressure increases, the band overlapping gradually decreases,[5] structural phase transition occurs from rhombohedral into base centred monoclinic at

*Work partially supported by Hokkaido University, The 21st Century COE Program "Topological Science and Ttechnology"

2.5 GPa at RT.[2] Moreover, Bi transforms from semimetal to semiconductor near 2.5 GPa and the band overlapping vanishes.

Ultrafast laser pulses induce coherent lattice vibrations and relaxation of photo-excited carriers in time domain. Hase *et al.* have studied coherent phonons in Bi at low temperature and high excitation regimes.[6,7] However, there have not been studies extreme changes of lattice and electronic band structure in Bi at pressure induced phase transition in time-resolved spectroscopy.

In this work, the high pressure effect on the topological change of the Fermi surface in Bi is investigated using microscope pump and probe system. We have observed drastic change of coherent phonons and electronic response at 2.5 GPa where phase transition occurs. The electronic responses are strongly correlated with change in the electronic band structure, We propose the drastic changes of electronic responses can be topological change of the Fermi surface caused by the phase transition.

The single crystal of Bi is prepared by the zone melting method. Cleaved trigonal face whose thickness is 100 μm is used. The pressure generation and measurement technique were described in detail.[8] The screw type of diamond anvil cell (DAC) is used in this experiment as a pressure source. The sample is loaded in gasketed DAC together with a ruby chip. The diameter of the sample chamber is about 600 μm and its depth is about 150 μm. A mode-locked Ti:sapphire laser is operated with pulse duration of 130 fsec and a repetition rate of 80 MHz. Central wavelength is 790 nm. The beams are focused in the diameter of 10 μm. The excitation power density is 56 μJ/cm^2.

Figure 1 shows transient reflectivity change $\Delta R/R$ with the pump and probe measurements at pressures from ambient pressure to 3.0 GPa. The pump and probe beams pass diamond and pressure transmitting medium, however, large influences of the DAC for the signals are not observed. The damped oscillation and background components are coherent phonons and electronic responses, respectively. The frequency of the coherent phonons at ambient pressure at RT is 2.92 THz and it agrees well with phonon frequency of A_1g mode obtained by Raman scattering measurements.[9] However the pump and probe signals indicate drastic changes above 2.5 GPa, where the structural phase transition occurs. The amplitude of coherent phonons decreases and its frequency of coherent phonons shifts from 2.9 THz to 2.5 THz at phase transition pressure. The coherent phonon signals sensitive to change in the lattice structure, we attribute the behaviours of coherent phonons to structural phase transition.

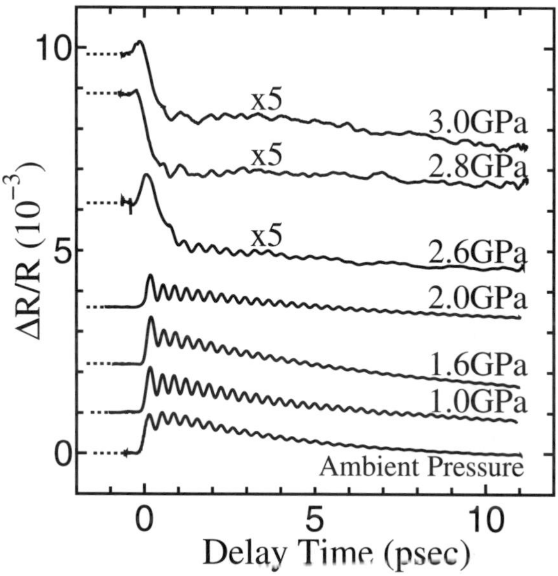

Figure 1. Pump and probe signals in Bi at pressure from ambient pressure to 3.0 GPa. Oscillation component of coherent phonons and background component of electronic responses are observed. The horizontal dotted lines indicate baseline of each signals.

Moreover, the electronic responses show drastic changes above 2.5 GPa. To investigate the change in the electronic response around phase transition, we subtract the electronic response from pump and probe signal. Figure 2 shows the pressure dependence of the electronic response after removal of coherent phonons from pump and probe signal. Sharp spike at 0 psec indicates a transient grating effect. In general, the electronic response represents the relaxation of photo-excited carriers. Therefore, it is considered that their relaxation process in semimetal and semiconductors is observed as a simple exponential decay. After sharp spike at 0 psec of delay time, the electronic responses rise positive direction and they are decaying with simple exponential below 2.5 GPa. The electronic responses, however, fall to the negative direction after the sharp spike at 0 psec and the decay rates become extremely small above 2.5 GPa. Bismuth transforms from semimetal to semiconductor above 2.5 GPa, and becomes an indirect semiconductor such as Si.

Sabbah *et al.* have observed the electronic response in Si with pump and probe measurement, and they have explained that the electronic response is caused by state filling effect.[10] Electronic responses in Si are similar to our

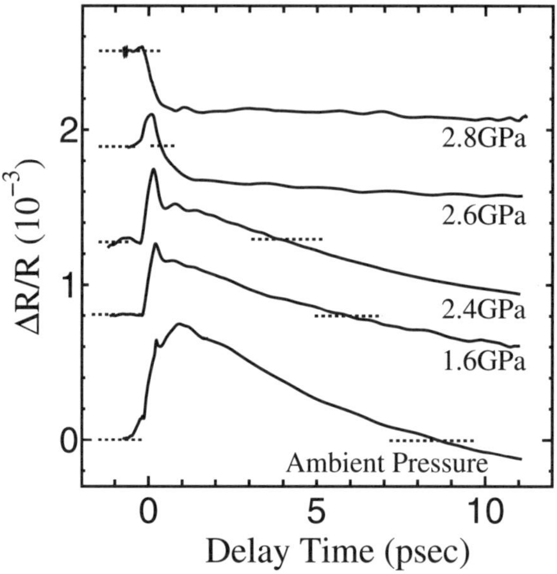

Figure 2. The pressure dependence of the electronic responses, after removal of the damped oscillation components of coherent phonons. The horizontal dotted lines indicate the baseline of each signals.

results obtained in semiconductor phase of Bi. We conclude that state filling effect is dominant to electronic responses in semiconductor phase. Since the decay rate of electronic responses in semimetal phase is too larger than that of the carrier recombination. We consider that the electronic responses in semimetal phase indicate a carrier scattering effect at the electronic band between the valence band at T point and the conduction band at L point.

The semimetal-semiconductor transition has an affect on the electronic band structure, we propose that the drastic changes of electronic response is indicative an electronic phase transition in connection with a topological change of the Fermi surface. Further experimental and theoretical studies are needed to understand our experimental results.

In conclusions, the high pressure effect on the topological change of the Fermi surface in Bi is investigated using microscope pump and probe system. We have observed drastic changes of coherent phonons and electronic response 2.5 GPa where phase transition occurs. The electronic responses are strongly correlated with change in the electronic band structure, We propose the drastic changes of electronic responses can be topological change of the Fermi surface caused by the phase transition.

Acknowledgements

The authors thank Mr. T. Kuwajima and Mr. H. Nomura of Technical Laboratory of lamina, Graduate School of Science, Hokkaido University. for sample preparation. The authors acknowledge OSAKA ASAHI METAL MFG. CO., LTD for supplying the 6N bismuth sample. This work was partially supported by Hokkaido University, The 21st Century COE Program "Topological Science and Technology".

References

1. N. Lotter and J. Wittig, *Europhys. Lett.* **6** (7), 659 (1988).
2. W. Klement Jr., A. Jayaraman, and G. C. Kennedy, *Phys. Rev.* **131**, 632 (1963).
3. J. A. vanHulst, H. M. Jaeger, and S. Radelaar, *Phys. Rev.* **B52**, 5953 (1995).
4. Y. A. Bogod and A. Libinson, *Sol. Stat. Commun.* **96**, 609 (1995).
5. K. G. Ivanov and V. M. Lopatkin, *Sov. Phys. Soild. Stat.* **20**, 1371 (1978).
6. M. Hase, K. Miozoguchi, H. Harima, S. Nakashima and K. Sakai, *Phys. Rev.* **B58**, 5448 (1998).
7. M. Hase, M. Kitajima, S. Nakashima and K. Mizoguchi, *Phys. Rev. Lett.* **88**, 067401 (2002).
8. M. Kasami, T. Mishina and J. Nakahara, *Phys. Stat. Sol.* **(b) 241**, No. 14, 3113 (2004).
9. J. S. Lannin, J. M. Calleja and M. Cardona, *Phys. Rev.* **B12**, 585 (1975).
10. A. J. Sabbah and D. M. Riffe, *Phys. Rev. Lett.* **66**, 165217 (2002).

TOPOLOGICAL CHANGE OF 4, 4'-BIS[9-DICARBAZOLYL]-2, 2'-BIPHENYL (CBP) BY INTERMOLECULAR REARRANGEMENT

K. S. SON, T. MISHINA, S. YAMAMOTO AND J. NAKAHARA

Department of Physics, Hokkaido University,
Sapporo, 060-0810, Japan.
E-mail: son@phys.sci.hokudai.ac.jp

C. ADACHI AND Y. KAWAMURA

Department of Photonics Materials Science,
Chitose Institute of Science and Technology (CIST)758-65 Bibi,
Chitose, Hokkaido 066-8655 and CREST Program, Japan Science and Technology Agency (JST), Japan
E-mail: c-adachi@photon.chitose.ac.jp

We discuss the behavior of the large change of photoluminescence (PL) spectrum at room temperature due to the change of the intermolecular arrangement of $4,4' - bis[9 - dicarbazolyl] - 2,2' - biphenyl$ (CBP). In this study, we report on PL characteristics in various aggregated morphologies such as deposited film and single crystal, and discuss its optical properties.

1. Introduction

Recently, CBP has been widely used as a host material in organic light emitting diodes (OLEDs) [1-10]. However, there are a few reports on the basic optical properties of CBP. We investigated PL characteristics of amorphous film, CBP in dichloromethane (solution), single crystal and aged film. Here, the aged film corresponds to the partially crystallization film due to storage in ambient air for two months. The structure of a CBP molecule is composed of biphenyl in both ends connected by two carbazoles, as shown in Fig. 1(a). According to the calculation (AM1 mothod) [11] of WinMOPAC, optimized structure of CBP molecule demonstrated the twisted structure between each carbazole and biphenyl, respectively [Fig.1(b)].

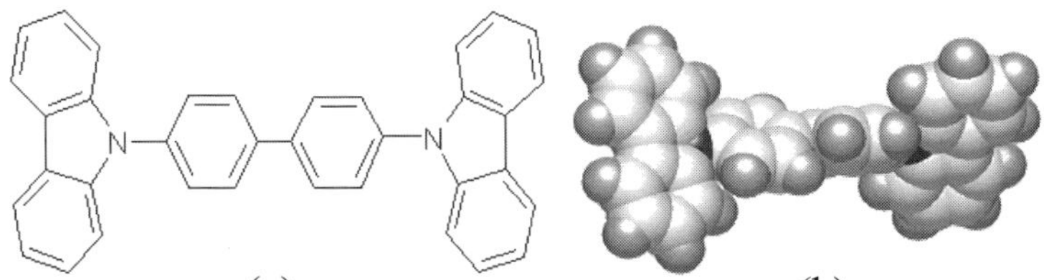

(a) (b)

Figure 1. (a) Molecule structure of $4,4'-bis[9-dicarbazolyl]-2,2'-biphenyl$ (CBP). (b) The structure of CBP optimized Hamiltonian of AM1 method by WinMOPAC.

2. Experimental Methods

We prepared a CBP thin film with a thickness of 100 nm by high-vacuum (1×10^{-3} Pa) thermal evaporation onto a pre-cleaned silicon substrate and the single crystal by using gas flow-train sublimation method. The PL spectrum and lifetime in the each condition were determined by using a streak camera (HAMAMATSU C1587, M1954). Figure 2 shows the scheme of optical measurement in our study. As a light source, we used mode-locked (80 MHz) Ti : Sapphire laser (790 nm). We used a BBO crystal (nonlinear crystal). It divides into fundamental beam (790 nm), second harmonic generation (SHG= 395 nm) and third harmonic generation (THG = 263 nm) light beams by a nonlinear effect when fundamental beam of Ti : Sapphire laser is irradiated to BBO crystal. We used only THG excitation light beam. The measurement of low temperature (23 K) was performed by using helium gas-flow system.

3. Results and discussion

As shown in Fig. 3, all of the PL spectra were integrated within the time range of 1.3 ns. The PL spectrum of CBP in dichloromethane (solution) showed only broad FWHM at room temperature. A change of the PL spectrum in the amorphous thin film was observed in the aged film. In addition, the film morphology gradually changed from a transparency film into a polycrystalline texture (white color) with periods of time due to the crystallization of the amorphous film. It is well known that the PL spec-

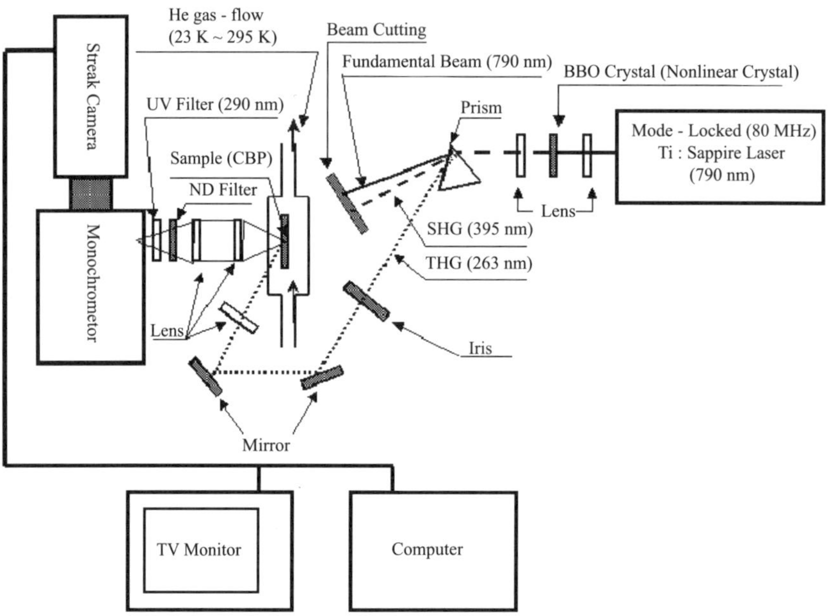

Figure 2. The setup of optical measurement

trum of an aged CBP film shows the fine structure and short lifetime of fluorescence which differs from a fresh amorphous CBP film. Similar change of the PL spectrum between the single crystal and aged CBP thin film indicates that the aged CBP is on the way of the crystallization. Presumably, this change is attributable to rearrange of the intermolecular structure of CBP. On the other hand, the PL spectra in the all of the conditions showed the characteristics fine structure at the low temperature (23 K). Relatively, each peak showed good agreement with energy range from 2.97 eV to 3.30 eV in the PL spectrum of the fresh amorphous film, aged film, and single crystal of CBP.

4. Conclusion

The PL spectrum of aged film of CBP was greatly changed by altering the intermolecular arrangement with the passage of the time while exposed in

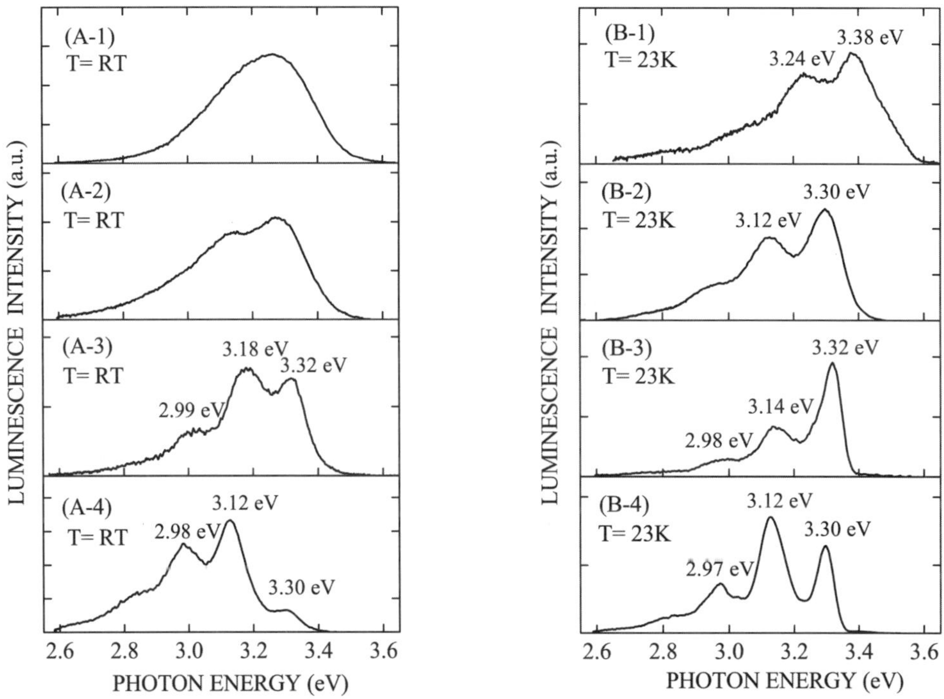

Figure 3. The PL spectra of CBP in dichloromethane (A- 1, B- 1), fresh amorphous film (A- 2, B- 2), aged film (A- 3, B- 3) and single crystal (A- 4, B- 4) room temperature and 23K, respectively.

air. This change be explained by a topological change in the intermolecular rearrangement.

Acknowledgments

This work was partially supported by the 21COE program on Topological Science and Technology at Hokkaido university. We thank Mr. Hiroyuki Yano and Mr. Takayuki Mimura for their help in this experiment.

References

1. C. Adachi, R. Kwong and S. R. Forrest, *Organ. Electro.* **2**, 37 (2001).
2. M. A. Baldo, S. Lamansky, P. E. Burrows, M. E. Thompson and S. R. Forrest, *Appl. Phys. Lett.* **75**, 4 (1999).
3. C. Adachi, M. A. Baldo and S. R. Forrest, *J. Appl. Phys.* **87**, 8049 (2000).
4. V. Adamovich, J. Brooks, A. Tamayo, A. M.Alexander, P. I. Djurovich, B.

W. Dandrade, C. Adachi, S. T. Forrest and M. E. Thompson, *New J. Chem.* **26**, 1171 (2002).

5. K. Goushi, R. Kwong, J. J. Brown, H. Sasabe and C. Adachi, *J. Appl. Phys.* **95**, 1 (2004).
6. S. Tokito, T. Iizima, Y. Suzuri, H. Kita, T. Tsuzuki and F. Sato, *Appl. Phys. Lett.* **83**, 569 (2003).
7. I. G. Hill, A. Rajagopal and A. Kahn, *J. Appl. Phys.* **84**, 3236 (1998).
8. R. J. Holmes and S. R. Forrest, *Appl. Phys. Lett.* **82**, 2422 (2003).
9. B. W. Dandrade, M. K. Thompson and S. R. Forrest, *Adv. Mater.* **14**, 147 (2002).
10. Y. Sakuratani, M Asai, M. Sone and S. Miyata, *J. Phys. D: Appl. Phys.* **34**, 3492 (2001).
11. M. J. S. Dewar, E. G. Zoebisch, E. F. Healy and J. J. P. Stewart, *J. Am. Chem. Soc.* **107**, 3902 (1985).

SPIN DYNAMICS IN HEISENBERG TRIANGULAR SYSTEM V15 CLUSTER STUDIED BY ^{1}H-NMR

Y. FURUKAWA, Y. NISHISAKA, Y. FUJIYOSHI AND K. KUMAGAI

Department of Physics, Hokkaido University,
Sapporo, 060-0810, Japan
E-mail: furu@phys.sci.hokudai.ac.jp

P. KÖGERLER

Department of Physics and Astronomy, Ames Laboratory,
Iowa State University, Iowa 50011, USA

We have investigated the spin dynamics in the V15 cluster at low temperature by measuring proton spin-lattice relaxation time (T_1) as a function of temperature (T) and external magnetic field (H). Both the H and T dependences of $1/T_1$ are well explained by a model in terms of the spin-phonon interaction above H_C where the ground state of the cluster is S=3/2. On the other hand, the T_1 data can not be explained by the model in the low magnetic field region where the ground state is two S=1/2 doubly degenerate states. The temperature independent behavior of $1/T_1$ is observed at low magnetic field. These results suggest that an existence of another contribution to $1/T_1$ in the low magnetic field region, which might be originated from peculiarities of S=1/2 triangle configuration in the V15 cluster.

1. Introduction

Recently much attention has been paid to nanoscale molecular magnets after discovery of a superparamagnetic behavior at low temperature combined with quantum phenomena such as quantum tunneling of the magnetization in so-called Mn12-ac [1,2] and Fe8 clusters [3,4]. Such an exciting discovery has triggered interests in exploring quantum effects in magnetic properties of other molecular magnets.

In particular, $K_6[V_{15}As_6O_{42}(H_2O)]_8H_2O$ (in short, V15) has attracted much interest, since the cluster is considered to be a typical s=1/2 Heisenberg triangular system [5]. The V15 cluster is made of fifteen V^{4+} ions with s=1/2 [5-7]. These vanadium ions are arranged in a quasi spherical layered structure formed of a triangle sandwiched between two hexagons as shown in Fig. 1. All exchange interactions between V^{4+} spins are antiferro-

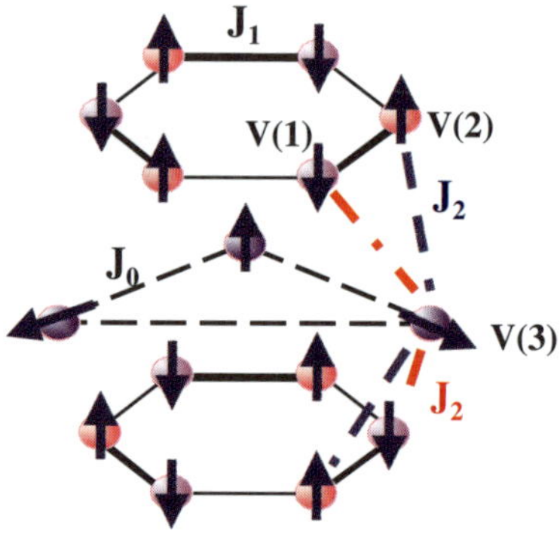

Figure 1. Configuration of V^{4+} (s=1/2) ions (solid circles) and exchange coupling scheme in V15 cluster.

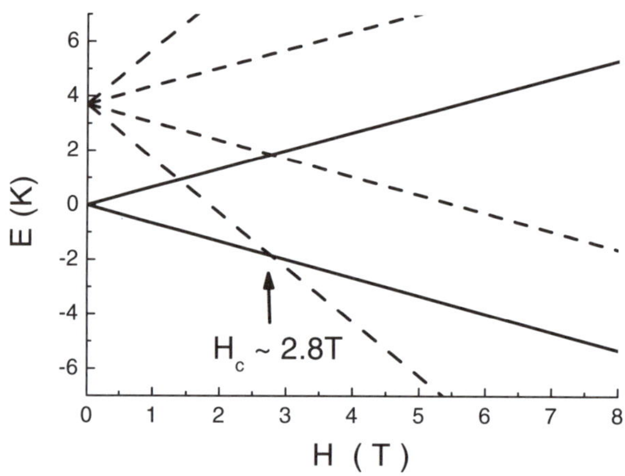

Figure 2. Level scheme of V15 cluster as a function of the external magnetic field. Solid and broken lines show nearly degenerated two S=1/2 branches and S=3/2 branches, respectively.

magnetic (AF) [5]. Each hexagon of the cluster has three pairs of strongly coupled spins with J_1~-800K [6,7]. Each spin of V^{4+} ions in the triangle is coupled with the spins in both hexagons with J_2=-150~-300K, resulting in a very weak exchange interaction between the spins within the triangle with J_0 ~-2.44K [8]. This pattern of couplings leads to a frustrated

s=1/2 triangle system. The ground state is formed by two S=1/2 doubly degenerate states separated by a small gap. The gap between the two degenerated states is reported to be ~0.1K by the magnetization measurements [9]. Since the three pairs in the hexagon have strong AF interaction of J_1~-800K, the magnetic properties of the V15 cluster at low temperature are determined only by the three V^{4+} spins on the triangle. The energy scheme at low temperature is given by the ground state of S=1/2 and the excited state of S=3/2 which lies ~3.8K above [10,11]. By the application of an external magnetic field, the ground state can be changed from two S=1/2 doubly degenerate states to a S=3/2 state, as shown in Fig. 2. Magnetization measurements at low temperature clearly exhibit below ~0.5K the magnetization step at a critical field $H_C \sim$ 2.8T where the two S=1/2 doubly degenerate ground state changes to S=3/2 ground state [11].

In order to shed right on the dynamical properties of the V^{4+} spins (s=1/2) on the triangle from a microscopic point of view, we have measured temperature (T) and external magnetic field (H) dependence of proton nuclear spin-lattice relaxa-tion time T_1 in the V15 cluster.

2. Experimental

Polycrystalline samples of $K_6[V_{15}As_6O_{42}(H_2O)]_8H_2O$ were prepared as described in Ref. 12. The temperature dependence of magnetic susceptibility measured in the samples agrees with the one reported previously [6,8]. The ^{1}H-NMR measurements were carried out utilizing a phase-coherent spin echo pulse spectrometer in the magnetic field range of 0.24-8.2 Tesla below 4.2K. The proton spin lattice relaxation rate ($1/T_1$) was measured by monitoring the recovery of the nuclear magnetization after saturating the NMR line with a short sequence of $\pi/2$ pulse, by irradiating at the Larmor frequency.

3. Results and discussion

Figure 3 shows the external magnetic field dependence of $1/T_1$ measured at T=1.5 and 4.2K. With increasing of magnetic field, $1/T_1$ decreases monotonically except for around $H \sim$ 2.7T. The enhancements of $1/T_1$ around H_C=2.8T are associated with the level crossings between S=1/2 and S=3/2. In order to analyze the $1/T_1$ results, we have calculated $1/T_1$ using a model in which thermal fluctuations of the magnetization among the different quantum number m sublevels of the total spin S state due to spin-phonon interaction induce the nuclear relaxation [13]. In this model, $(1/T_1)_{sp}$ is expressed as

$$(\frac{1}{T_1})_{sp} = \frac{A}{Z}\sum_{m=+10}^{-10}\frac{\tau_m \exp(-\frac{E_m}{k_B T})}{1+\omega_N^2\tau_m^2}, \tag{1}$$

where τ_m is the life-time of the magnetic m-th sublevels originating from the spin-phonon interactions. ω_N is resonance frequency, Z is the partition function and A is a parameter related to the hyperfine coupling constants. The life-time τ_m for each individual m state is determined by $1/\tau_m = P_{m\to m+1} + P_{m\to m-1}$. The transition probabilities due to the spin phonon interactions [14,15] can be expressed as $P_{m\to m+1} = C\Delta^3/[1-\exp(-\Delta/k_B T)]$ and $P_{m\to m-1} = C\Delta^3/[\exp(-\Delta/k_B T)-1]$ where C is the spin-phonon coupling constant and Δ is the energy difference between m and $m\pm1$ sublevels.

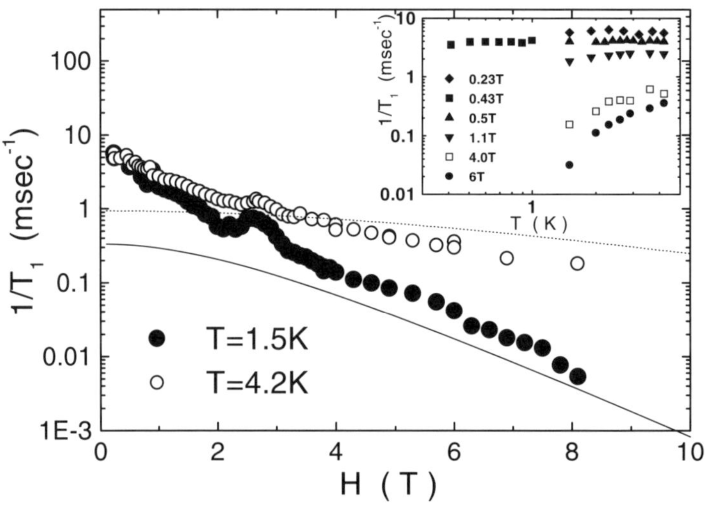

Figure 3. The external magnetic field dependence of 1/T1 measured at T=1.5K (solid circles) and 4.2K (open circles). Solid and broken lines show the calculated results for T=1.5K and 4.2K, respectively. The inset shows typical temperature dependence of $1/T_1$ measured at several external magnetic fields.

In the case of the V15 cluster, there are two total spin states, that is, S=1/2 and 3/2 so that $1/T_1$ can be expressed by sum of the two contributions due to the two S=1/2 and 3/2 states as,

$(1/T_1)_{sp}=(B_{1/2}(1/T_1)_{sp}^{S=1/2}+B_{3/2}(1/T_1)_{sp}^{S=3/2})/Z$ where $B_{1/2}$ and $B_{3/2}$ is the Boltzmann factor for the two states of S=1/2 and 3/2, respectively, including the two fold degeneracy for the S=1/2 states. The solid and broken lines in Fig. 3 show the calculated results for T=1.5 and 4.2K, respectively, with a set of parameters of $C=2.4\times10^4$ (Hz/K^3) and $A=4.0\times10^{14}$ (rad/s)^2. The calculated results seem to reproduce the experimental results above H_C where the ground state is S=3/2. On the other hand, below H_C where the ground state is two S=1/2 nearly degenerate states, $1/T_1$ values are much larger than the calculated values by the spin-phonon interaction model. It should be noted that $1/T_1$ shows almost constant with temperature at low magnetic field below ~1 Tesla as shown in the inset of Fig. 3. These results suggest another contribution to $1/T_1$ below H_C.

A possible origin for the contribution is due to quantum effects which originate from s=1/2 triangular lattice of three V^{4+} ions. According to Kawamura [16], the ground state of triangular spins has twofold degenerate according to the two chiralities (different spin configurations) due to the frustration effects. In this sense, two S=1/2 doubly degenerate states might be considered as the two chiralities in the V15 cluster. Since the fluctuations between two chiral states with different spin structures would give rise to a local field fluctuation at proton sites, this could be a relaxation mechanism. In contrast, such a fluctuation would not influence the magnetization measurements because both chiral states have the same S=1/2 states macroscopically. It would be very interesting if the T-independent $1/T_1$ were due indeed to the quantum fluctuation, although we can not establish this for sure from the present data. Further experiments under various magnetic fields at low temperature below 1K will give us information to elucidate the origin of peculiar behavior of $1/T_1$ at low magnetic field. Such measurements are now in progress.

Acknowledgments

The present work was in part supported by 21COE program on "Topological Science and Technology" from the Ministry of Education, Culture, Sports, Science and Technology of Japan.

References

1. J. R. Friedman, M. P. Sarachik, J. Tejada, and R. Ziolo, *Phys. Rev. Lett.* **76**, 3830 (1996).

2. L. Thomas, F. Lionti, R. Ballou, D. Gatteschi, R. Sessoli and B. Barbara, *Nature (London)* **383**, 145 (1996).
3. C. Sangregorio, T. Ohm, C. Paulsen, R. Sessoli and D. Gatteschi, *Phys. Rev. Lett.* **78**, 4645 (1997).
4. W. Wernsdorfer and R. Sessoli, *Science* **284**, 133 (1999).
5. A. Gatteschi, L. Pardi, A. L. Barra, A. Müller and J. Döring, *Nature* (London) **354**, 463 (1991).
6. A.L. Barra, D. Gatteschi, L. Pardi, A. Müller, and J. Döring, *J. Am. Chem. Soc.* **114**, 8509 (1992).
7. D. Gatteschi, L. Pardi, A. L. Barra and A. Müller, *Mol. Eng.* **3**, 157 (1993).
8. G. Chaboussant1, R. Basler, A. Sieber, S. T. Ochsenbein, A. Desmedt, R. E. Lechner, M. T. F. Tell-ing, P. Kögerler, A. Müller and H.-U. Güdel, *Europhys. Lett.* **59**, 291 (2002).
9. I. Chiorescu, W. Wernsdorfer, A. Müller, H. Bögge and B. Barbara, *Phys. Rev. Lett.* **84**, 3454 (2000).
10. V. V. Dobrovitski, M. I. Katsnelson and B. N. Harmon, *Phys. Rev. Lett.* **84**, 3458 (2000).
11. I. Chiorescu, W. Wernsdorfer, B. Barbara, A.Müller and H. Bögge, *J. Magn. Magn. Mater.* **221**, 103 (2000).
12. J. Choi, L. A. W. Sanderson, J. L. Musfeldt, A. Ellern and P. Kögerler, *Phys. Rev.* **B68** (2003) 064412.
13. A. Lascialfari, Z. H. Jang, F. Borsa, P. Carretta and D. Gatteschi, *Phys. Rev. Lett.* **81**, 3773 (1998).
14. J. Villain, F. Hartmann-Boutron, R. Sessoli and A. Rettori, *Europhys. Lett.* **27**, 537 (1994);
15. F. Hartmann-Boutron, P. Politi and J. Villain, *Int. J. Mod. Phys.* **10**, 2577 (1996).
16. H. Kawamura, *J. Phys.: Condens. Matter* **10** 4707 (1998).

STM/STS ON $NbSe_2$ NANOTUBES

K. ICHIMURA, K. TAMURA AND K. NOMURA
Division of Physics, Hokkaido University
Kita 10, Nishi 8,
Kita-ku, Sapporo 060-0810, Japan
E-mail: ichimura@sci.hokudai.ac.jp

T. TOSHIMA AND S. TANDA
Division of Applied Physics, Hokkaido University
Kita 13, Nishi 8,
Kita-ku, Sapporo 060-8628, Japan
E-mail: toshima@eng.hokudai.ac.jp

$NbSe_2$ nanotubes were studied by scanning tunneling microscopy (STM). Topographic images of $NbSe_2$ nanotubes with length of 300-200 nm were obtained at room temperature. The measured diameter of 2-20 nm suggests that the nanotubes are single-walled. The bundle structure and Y-junction were found similarly to carbon nanotubes.

1. Introduction

Carbon nanotube is the pioneer material which has the topological structure in nano scale. An important feature of nanotubes is that the electronic state can be controlled by helicity *i.e.* chirality. Carbon nanotube has given us an opportunity to study physics and technology for application in nano scale. Beside carbon, many nanotube materials such as BN and MoO_2 were synthesized. The fact tells us that nanotube might be formed from any layered material. Recently, $NbSe_2$ nanotube was synthesized independently by two groups[1,2]. Figure 1 shows a schematic view of $NbSe_2$ nanotube. Tsuneta *et al.*[1] reported that the nanotube is multi-walled and the stacking is determined as $2H$ from electron diffraction. On the other hand, the nanotube synthesized by Nath *et al.*[2] has $6H$a stacking. It should be emphasized that $NbSe_2$ nanotube is the first nanotube material, of which bulk phase undergoes the condensed state such as the density wave and superconductivity. We are interested in how the condensed state, which

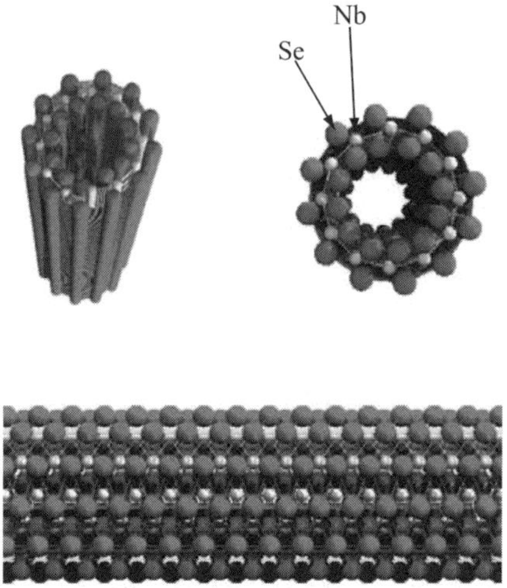

Figure 1. Schematic view of single-walled $NbSe_2$ nanotube.

are characterized with the coherence length of sub micrometer, survive in nano scale electronic system.

MX_2, where M is transition metal and X is chalcogen, has the layered structure. The band dispersion is two dimensional. Therefore, MX_2 compounds have the cylindrical Fermi surface. In $2H$-$NbSe_2$ (a=0.3444 nm, c=1.2552 nm) the incommensurate charge density wave (CDW) is formed below 33 K. The origin of the CDW with wave length of about $3a$ is understood by the nesting of a portion of the Fermi surface. Moreover, $2H$-$NbSe_2$ undergoes the superconductivity at 7.2 K.

In investigating the superconducting state, the electron tunneling is useful since the electronic density of states can be obtained directly with high energy resolution[3]. The tunneling spectroscopy using STM, *i.e.* scanning tunneling spectroscopy (STS), especially has an advantage because of non-contacting tip configuration. The local electronic density of states can be studied directly in superconductors. The microscopic structure of vortices in $2H$-$NbSe_2$[4] and the quasi-particle scattering resonance at impurity sites in $Bi_2Sr_2CaCu_2O_8$[5] were observed clearly with atomic resolution. Tunneling spectra along various crystal orientation are easily obtained by STS. The angle-resolved STS[6] was reported on the d-wave organic superconductor.

In this manuscript, we report STM images of $NbSe_2$ nanotubes and discuss about the diameter of tubes.

2. Experimental

$NbSe_2$ nanotubes were synthesized by the chemical vapor transport (CVT) method. After purifying, the nanotube sample for STM measurement was prepared by dropping $NbSe_2$ nanotubes agitated ultrasonically in 2-propanol on cleaved highly oriented pyrolytic graphite (HOPG). Mechanically sharpened Pt-Ir wire was used as an STM tip, which is attached to a tube type piezoelectric scanner.

3. Results and Discussion

Figure 2 shows a typical STM image at room temperature. We found linear structures with length of about 300-2000 nm. We conclude that the linear structure corresponds to $NbSe_2$ nanotube from dimension and linearity of the structure. At present, we cannot obtain the atomic resolution. The determination of the chirality is our future work.

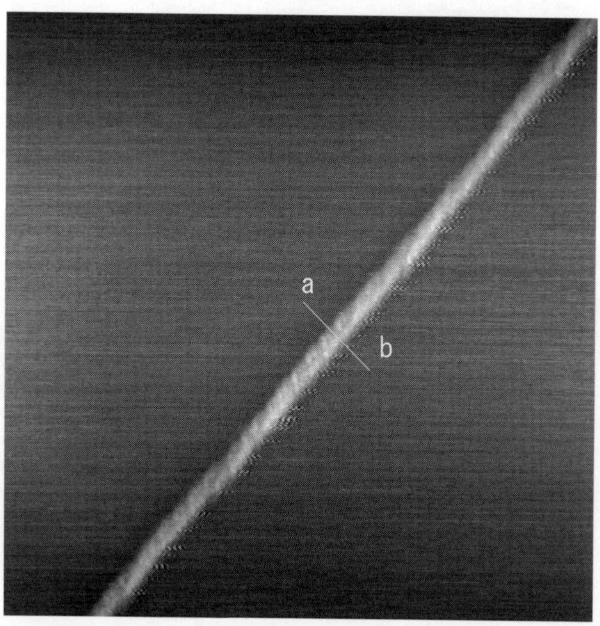

Figure 2. STM image of $NbSe_2$ nanotube on HOPG. Scan area is $200{\times}200$ nm^2.

Figure 3 shows the scan profile between (a) and (b) in Fig. 2. The peak structure in the profile suggests that the linear structure in Fig. 2 is

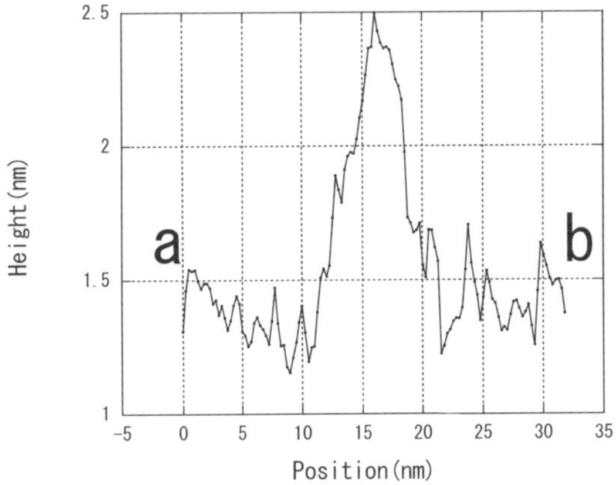

Figure 3. Scan profile between (a) and (b) in Fig. 1. The distance is measured from point (a).

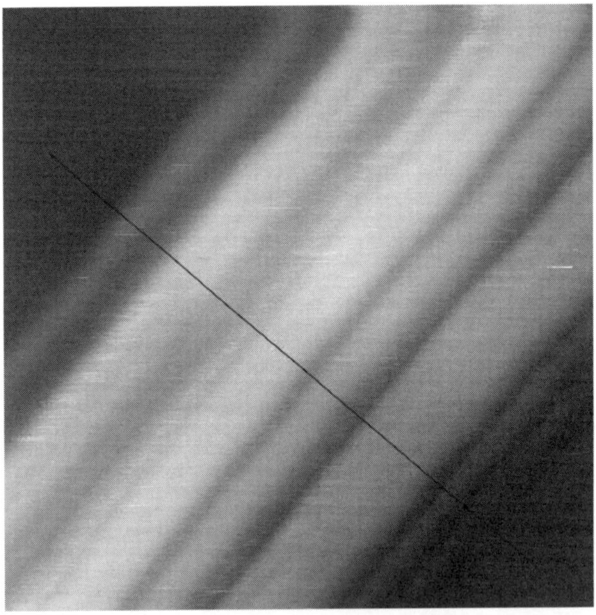

Figure 4. STM image of the bundle structure of $NbSe_2$ nanotubes. Scan area is 87.4×87.4 nm^2.

Figure 5. Y-junction of $NbSe_2$ nanotube. Scan area is 200×200 nm^2.

a nanotube. The width and height of the peak is 10 nm and 1 nm, respectively. Diameter of the tube shown in Fig. 2 is estimated as about 10 nm from the width. The height is much smaller than the width due to some reasons.[7] The diameter varies from sample to sample. We conclude that diameter of $NbSe_2$ nanotubes is ranging from 2 to 20 nm. The single-walled $NbSe_2$ nanotube consists of a set of three walls; one Nb and two Se walls. The smallest diameter is much smaller than that of other nanotubes such as carbon. Theoretical calculation[8] with respect to the structural energy gain predicted that single-walled $NbSe_2$ nanotubes are stabilized with diameter of several nm. We think that thiner $NbSe_2$ nanotubes observed in STM study are single-walled. Tsuneta *et al.*[1] reported that diameter of multi-walled $NbSe_2$ nanotube is estimated as 30-200 nm from transmission electron microscopy (TEM) study. We could find nanotubes with smaller diameter by STM. We think that thinner tubes we observed correspond to single-walled and thicker one correspond to multi-walled nanotubes.

The bundle structure of $NbSe_2$ nanotube was found similarly to carbon nanotube. It is well known that single-walled carbon nanotubes are assembled to a bundle. Figure 4 shows the STM image of the bundle with diameter of about 70 nm. Nanotubes are spaced closely and arranged in parallel. The scan profile across the bundle suggests that the bundle consists of several ten nanotubes with diameter of about 2 nm.

We also found Y-junction as shown in Fig. 5. The periodic structure on tubes is due to the system noise. Y-junction was often found in carbon nanotubes.[9] Y-junction of $NbSe_2$ nanotube suggests the possibility of an application to nano-scaled devices in the CDW and superconducting state.

Tunneling spectra are essential to elucidate the electronic structure. We are interested in the CDW and superconductivity in nano-sized sample. We are trying to carry out STS on $NbSe_2$ nanotubes at low temperature.

Acknowledgments

This work has been partially supported by the 21COE program on "Topological Science and Technology" from the Ministry of Education, Culture, Sport, Science and Technology of Japan.

References

1. T. Tsuneta, T. Toshima, K. Inagaki, T. Shibayama, S. Tanda, S. Uji, M. Ahlskog, P. Hakonen, M. Paalanen, *Curr. Appl. Phys.* **3**, 473 (2003).
2. M. Nath, S. Kar, A. K. Raychaudhuri, C. N. R. Rao, *Chem. Phys. Lett.* **368**, 690 (2003).
3. I. Gieaver, *Phys. Rev. Lett.* **5**, 147 (1960).
4. H. F. Hess, R. B. Robinson, R. C. Dynes, J. M. Valles, Jr., J. V. Waszczak, *Phys. Rev. Lett.* **62**, 214 (1989).
5. S. H. Pan, E. W. Hudson, K. M. Lang, H. Eisaki, S. Uchida, J. C. Davis, *Nature* **403**, 746 (2000).
6. T. Arai, K. Ichimura, K. Nomura, S. Takasaki, J. Yamada, S. Nakatsuji, H. Anzai, *Phys. Rev.* **B63**, 104518 (2001).
7. K. Ichimura, K. Tamura, K. Nomura, T. Toshima, S. Tanda, in preparation.
8. V. V. Ivanovskaya, A. N. Enyashin, N. I. Medvedeva, A. L. Ivanovskii, *Phys. Stat. Sol.* **B238**, R1 (2003).
9. L. P. Biro, Z. E. Horvath , G.I. Mark, Z. Osvath, A. A. Koos, A. M. Benito, W. Maser, Ph. Lambinc, *Diamond Relat. Mater.* **13**, 241 (2004).

NANOFIBERS OF HYDROGEN STORAGE ALLOY

I. SAITA[1]
T. TOSHIMA[2]
S. TANDA[2]
T. AKIYAMA[1]

[1]*Center for Advanced Research of Energy Conversion Materials, Hokkaido University*
Kita13 Nishi 8, Kita-ku, Sapporo, Hokkaido 060-8628, Japan
E-mail: itoko@eng.hokudai.ac.jp

[2]*Department of Applied Physics, Hokkaido University*
Kita13 Nishi 8, Kita-ku, Sapporo, Hokkaido 060-8628, Japan

This paper reports a novel methodology to produce metal hydrides and the product's remarkable structure. In this method, metal hydride was synthesized in gas phase and then sublimated into solid. High purity metal hydride of MgH_2 was synthesized by this method, in which raw material of magnesium was heated and evaporated in hydrogen atmosphere and then sublimate as the product of MgH_2. The product had fibrous shape, which is quite differ to conventional products. The fibers had uniform diameter of 500 nm and length ranging from 10 to 100 μm but no branches or curving. Transmission Electron Microscopy revealed that the fibers were formed with single crystalline.

1. Introduction

Crystal forms various structures by each growth process: For example, carbon constructs 2D sheets, nanotubes, and fullerene C_{60} by non-equilibrium growth conditions. In this study we synthesized nano-fiber crystals of a hydrogen storage alloy by new method of chemical reaction under pressurized H_2 gas. Hydrogen storage alloy, or so-called metal hydride involves chemical bonds between metals and hydrogen atoms, which is shorter than intermolecular distance of hydrogen. The fact has attracted a great deal of attention to its potential application as an energy carrier via hydrogen storage[1–2]. A lot of efforts has been done to develop new metal hydrides or to improve hydrogen storage properties. By way of example, mechanical milling presents the benefits of reducing the grain size, shortening the diffusion pass, and modifying kinetics of hydrogen charging at relatively

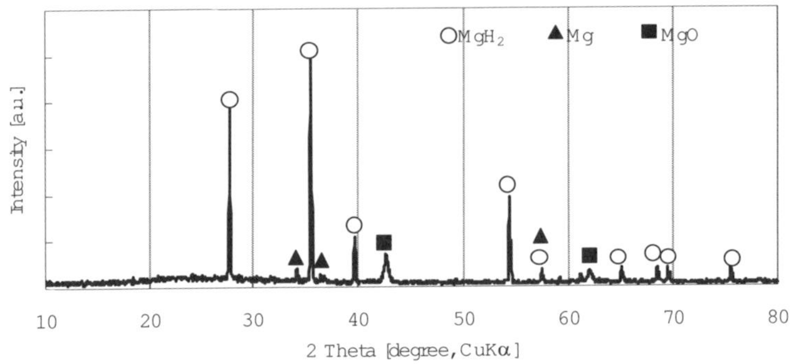

Figure 1. The XRD pattern of the product[3]. MgH_2 was the chief phase.

moderate temperature[4–7]. For the same purpose, we examined the synthesis of a nano-structured metal hydride by the method of vapor deposition.

2. Experimental

Magnesium hydride of MgH_2, which records high hydrogen density as much as 7.6 mass%, was selected as the final product. The raw material of magnesium (99.99 % in purity and less than 180 μm in size) evaporated by being heated at 900 K in 4 MPa hydrogen atmosphere. The precipitation from magnesium vapor on the substrate was investigated by X-ray diffraction

Figure 2. The SEM observations of synthesized MgH_2. The product showed fiber shapes without branches or curving. Length of the fibers were ranging from tens to hundreds μm, but the diameters were uniformly 500 nm.

(XRD) analysis, Scanning Electron Microscopy (SEM), and Transmitting Electron Microscopy (TEM).

3. Results and discussion

Figure 1 is the XRD pattern of the precipitation, whose visual appearance was white fine powder. The peaks were identified with MgH_2 as chief phase, less Mg and MgO phases. This result suggests fruitful methodology for metal hydride production with energy saving since it is difficult and needs enormous energy to hydrogenate bulk magnesium due to its less diffusivity of hydrogen atom. Figure 2 shows interesting SEM observation of the synthesized MgH_2. That is, the MgH_2 had whisker-like nano-structure longer than 10 μm and 500 nm across at the minimum. According to careful observation, each whisker had a straight longitudinal fissure. The TEM

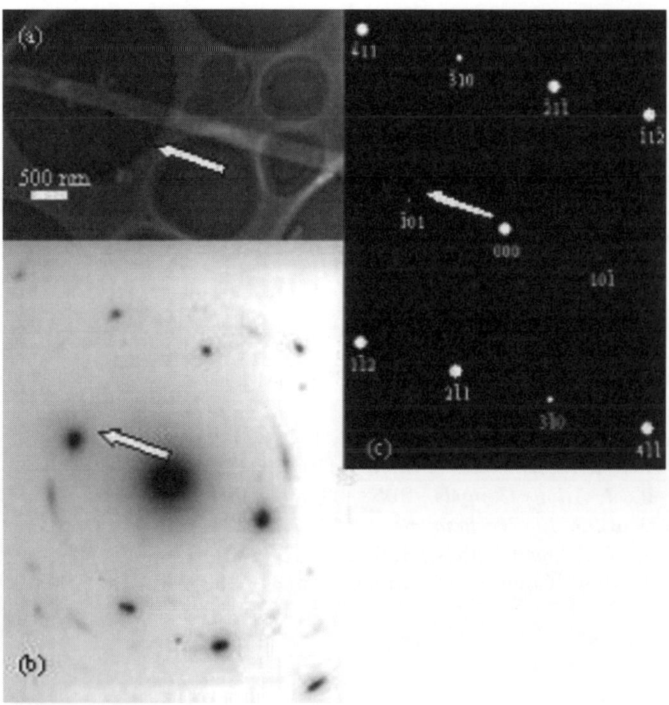

Figure 3. TEM observations: (a) is TEM image of the MgH_2 needle on a porous sheet (back net)[3]. (b) is the relative electron diffraction pattern[3] in which camera constant is 3.765 x 10^{-10} m^2. (c) is MgH_2 diffraction pattern calculated by using Crystal Studio© and the same camera constant and 1 3 1 zone axis.

analysis of the whisker (See Fig. 3) revealed that the whisker consisted of single-crystallized MgH_2. The hydrogen content of MgH_2 nano-structured by conventional mechanical milling is reduced because of the fact that the hydrogen at the grain boundary is unstable than in the crystallite[8]. In contrast, the obtained MgH_2 is expected to have stable hydrogen capacity since its nano-structure was highly crystallized.

4. Conclusion

This paper reports the synthesis of a nano-structured metal hydride based on vapor deposition at pressurized hydrogen. The product was fully hydrogen charged MgH_2 with straight fissure. The reactivity of the product is under investigation from the viewpoint of hydrogen charging/discharging and hydrogen generator using hydrolysis[9].

Acknowledgment

This work has been partially supported by the 21st century COE program on "Topological Science and Technology" from the Ministry of Education, Culture, Sport, Science and Technology of Japan.

References

1. "A national vision of America's transition to a hydrogen economy - to 2030 and beyond", United States Department of Energy, http://www.eere.energy.gov/hydrogenandfuelcells/pdfs/vision_doc.pdf, February (2002).
2. L. Schlapbach and A. Zuttel, *Nature,* **413**, 353-358 (2001).
3. I. Saita et al., *Mat. Trans.,* in printing.
4. A. Zaluska and L. Zaluski, *Appl. Phys.,* **A72**, 157 (2001).
5. E. Akiba and H. Iba, *Intermetallics,* **6**, 461-470 (1998).
6. T. Kuriiwa et al., *J. Alloys Compds.,* **295**, 433-436 (1999).
7. M. Tsukahara et al., *J. Electrochem. Soc.,* **147**, 2941-2944 (2000).
8. N. Hanada et al., *J. Alloys Compds.,* **366**, 269-273 (2004).
9. J. Huot et al., J. Alloys Compds., *J. Alloys Compds.,* **353**, L12-L15 (2003).

SYNTHESIS OF STABLE ICOSAHEDRAL QUASICRYSTALS IN Zn-Sc BASED ALLOYS AND THEIR MAGNETIC PROPERTIES *

S. KASHIMOTO AND T. ISHIMASA
Division of Applied Physics, Graduate School of Engineering,
Hokkaido University, Sapporo 060-8628, Japan
E-mail: kasimoto@eng.hokudai.ac.jp

We have measured magnetic properties of stable icosahedral quasicrystals Zn-M-Sc (M = Fe, Co and Ag). These quasicrystals were formed in Zn-Sc based alloys and possess high structural perfection which was confirmed by X-ray powder diffraction and selected-area electron diffraction. The magnetic susceptibility of the $Zn_{74.5}Ag_{9.5}Sc_{16}$ quasicrystal shows an increase with a rise in temperature over 80 K, which is accounted for by a temperature dependence of the Pauli paramagnetism. The temperature dependence of the magnetic susceptibility of $Zn_{77}Fe_7Sc_{16}$ and $Zn_{78}Co_6Sc_{16}$ quasicrystals follow Curie-Weiss law. The Fe in the $Zn_{77}Fe_7Sc_{16}$ quasicrystal has considerably large magnetic moments; the effective number of Bohr magneton per a Fe atom is estimated as 3.7 from Curie constant obtained by Curie-Weiss fitting, and shows spin glass behavior with the freezing temperature 7.0 K. This is the first case that most of Fe atoms have large magnetic moments in the stable icosahedral quasicrystals. On the other hands, the $Zn_{78}Co_6Sc_{16}$ dose not show large magnetic moments, and then, the singularity of the $Zn_{77}Fe_7Sc_{16}$ quasicrystal was clarified in this study.

1. Introduction

The remarkable property of "quasicrystals" is diffraction symmetry, $m\bar{3}\bar{5}$ (icosahedral), inconsistent with periodic translation symmetry and a quasiperiodic long-range order instead of a periodic one. Therefore, the quasicrystals are interesting from the view point of topological matter. The magnetic property of quasicrystals is worth investigating in regard to the magnetic order in the quasiperiodic structure. In quasicrystals, the local magnetic moments may arrange with a frustrated structure under an anti-ferromagnetic interaction because of aperiodic structure with the dodeca-

*This work is supported by the 21COE program on "Topological Science and Technology" from the Ministry of Education, Culture, Sports, Science and Technology of Japan.

hedral and icosahedral atomic arrangements. Recently, we confirmed that new thermodynamically stable icosahedral quasicrystals as a single phase are formed in $Zn_{77}Fe_7Sc_{16}$, $Zn_{78}Co_6Sc_{16}$, $Zn_{74}Ni_{10}Sc_{16}$, $Zn_{75}Pd_9Sc_{16}$ and $Zn_{74.5}Ag_{9.5}Sc_{16}$ alloys[1]. The remarkable feature of this series is the wide variety of additional M atoms ranging from $3d$ transition metals to noble metals in the periodic table. It is thought that these quasicrystals belong to the same structural type from the results of X-ray and electron diffraction experiments. In this study, we will present experimental results of the temperature dependence of the magnetization of Zn-M-Sc (M = Fe, Co and Ag) stable icosahedral quasicrystals.

2. Experimental procedure

Weighed high-purity materials of Zn (Nilaco, 99.99%), Sc (Shin-Etsu Chemical, 99.9%, according to the analysis of the material manufacturer, the Sc material includes some impurities; Al 0.01 wt%, Fe 0.21 wt%, O 0.65 wt%, F 0.01 wt%, Ca $<$ 0.01 wt%, Mg $<$ 0.01 wt%), Ag (Nilaco, 99.98%), Fe (Nilaco, 99.99%), Co (Nilaco, 99.9%) having the nominal composition ($Zn_{73.8}Ag_{9.7}Sc_{16.5}$, $Zn_{77.0}Fe_{7.0}Sc_{16.0}$ and $Zn_{77.8}Co_{6.2}Sc_{16}$) were put in an alumina crucible and sealed in a silica tube with an argon atmosphere at 3.4×10^{-2} Pa after evacuating to 1.3×10^{-4} Pa. The sealed samples were melted and annealed (1120 K $\times$ 1 h and 1026 K $\times$ 52 h for Zn-Ag-Sc, 1133 K $\times$ 3 h and 975 K $\times$ 12 h for Zn-Fe-Sc, 1133 K $\times$ 5 h and 943 K $\times$ 40 h for Zn-Co-Sc) using a computer controlled electric furnace. The structural characterization of the samples was carried out by powder X-ray diffraction using a X-ray diffractmeter (Rigaku, RINT2000-PC, Cu Kα radiation, 40 kV, 30 mA), and selected-area electron diffraction using an electron microscope (JEOL, JEM-200CX, an acceleration voltage of 160 kV). We confirmed that the composition of the quasicrystals are $Zn_{74.5}Ag_{9.5}Sc_{16}$, $Zn_{77}Fe_7Sc_{16}$ and $Zn_{78}Co_6Sc_{16}$ by a scanning electron microscopy using wavelength-dispersive X-ray spectroscopy (JEOL, JXA-8900M). The dc magnetic susceptibility was measured by a superconducting quantum interference device (SQUID) magnetometer (Quantum-Design, MPMS-2 and MPMS-7) in the temperature region between 2 and 300 K.

3. Results and discussion

Selected-area electron diffraction patterns of the $Zn_{74.5}Ag_{9.5}Sc_{16}$ are presented in Fig. 1. The symmetry of these patterns corresponds to the icosahedral symmetry $m\bar{3}\bar{5}$. All spots in the diffraction patterns are observed at the ideal icosahedral symmetric positions with very small deviations and could be indexed by six integers as an icosahedral quasicrystal using Elser's scheme[2]. For example, the reflections indicated by A and B in the pattern of the twofold axis have indices 0 0 0 0 0 1 and 1 1 $\bar{1}$ $\bar{1}$ 1 2 respectively. Their lattice spacing are estimated to be 1.014 and 0.2388 nm respectively from the analysis of the powder X-ray diffraction patterns.

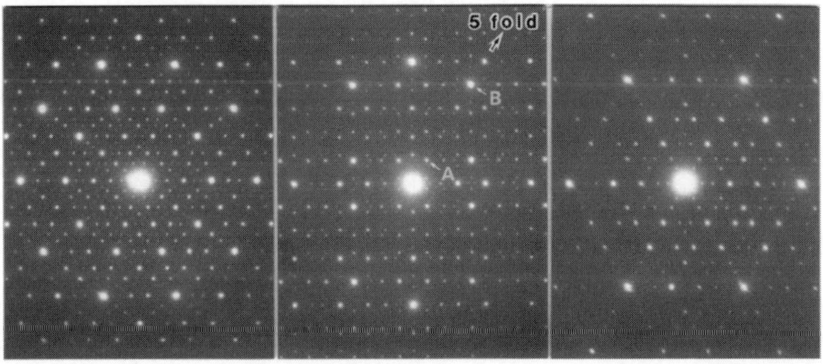

Figure 1. Selected-area electron diffraction patterns of the $Zn_{74.5}Ag_{9.5}Sc_{16}$ icosahedral quasicrystal taken with incident beams parallel to (a) fivefold, (b) twofold and (c) threefold axes.

Figure 2 shows the powder X-ray diffraction patterns of Zn-M-Sc (M = Fe, Co and Ag) stable icosahedral quasicrystals. All the reflections are very sharp and can be indexed as an icosahedral quasicrystal in the same scheme for the electron diffraction patterns. The peak widths, measured as full width at half maxima (FWHM), of the 0 2 $\bar{2}$ $\bar{3}$ 0 3 reflections are 0.0010 Å^{-1}, 0.0011 Å^{-1} and 0.0012 Å^{-1}, for $Zn_{77}Fe_7Sc_{16}$, $Zn_{78}Co_6Sc_{16}$ and $Zn_{74.5}Ag_{9.5}Sc_{16}$, respectively. These widths are nearly equal to that of the 4 0 0 reflection of the standard Si sample measured under the same condition, which is 0.00095 Å^{-1}. Judging from the results of electron diffraction and powder X-ray diffraction experiments, these icosahedral quasicrystals exhibit a high degree of structural perfection as a quasiperiodic order phase. Figure 3 shows the magnetic susceptibility of $Zn_{74.5}Ag_{9.5}Sc_{16}$ as a function of temperature at the magnetic field of 50 kOe. The sample shows diamagnetism at high temperatures and Curie-Weiss paramagnetism at low temperatures. The magnetic susceptibility χ ranges between -6.3×10^{-8} and -3.6×10^{-4} cgsemu g^{-1}. The origin of the Curie-Weiss paramagnetism may be due to magnetic impurities included in the material of Sc. The intrinsic property of the quasicrystal appears in high temperature region;

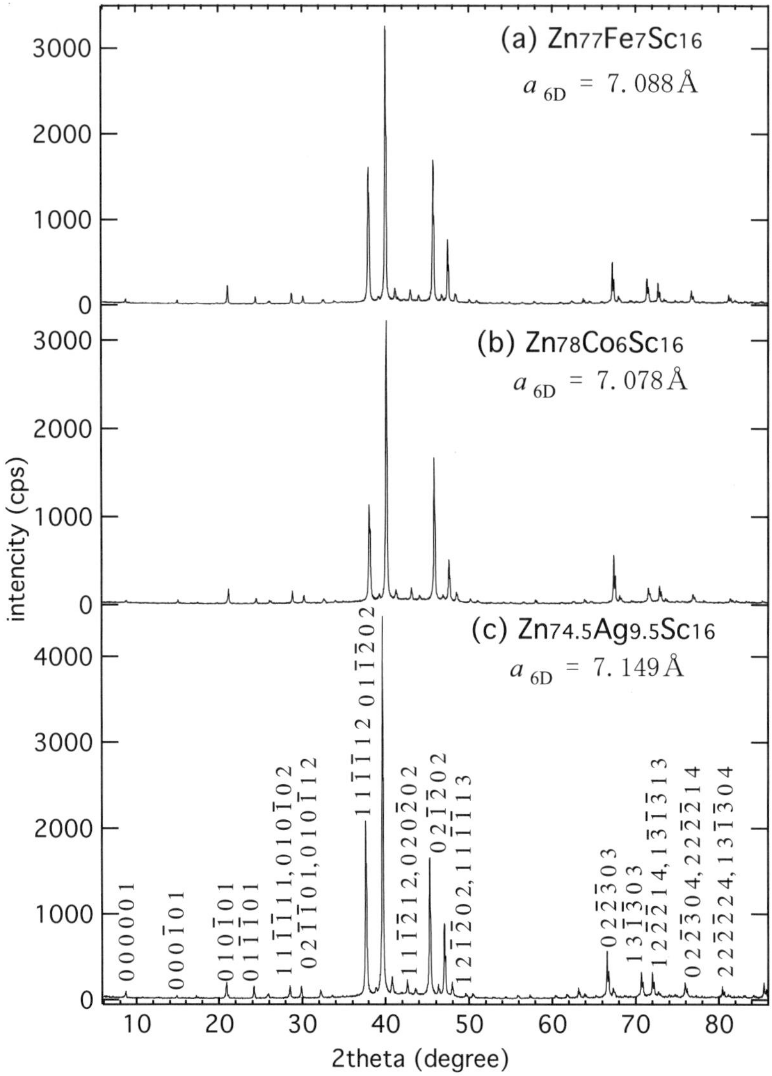

Figure 2. Powder X-ray diffraction patterns of Zn-M-Sc (M = Fe, Co and Ag) stable icosahedral quasicrystals. Six integers are Elser's indices as an icosahedral phase and a_{6D} is a six-dimensional lattice parameter[3].

the magnetic susceptibility distinctly increases with a rise in temperature. Such an increase in the magnetic susceptibility has been reported for some nonmagnetic quasicrystals such as Al-Cu-Fe and was interpreted by tem-

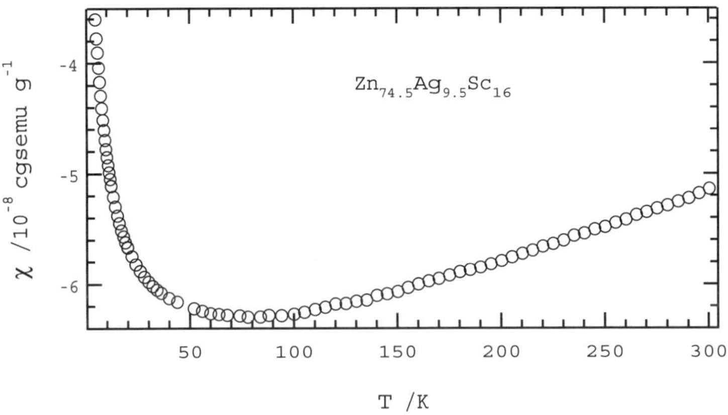

Figure 3. Magnetic susceptibility as a function of temperature of the $Zn_{74.5}Ag_{9.5}Sc_{16}$ icosahedral quasicrystal under 50 kOe.

perature dependence of Pauli paramagnetism of conduction electrons which varies as AT^2, where A is constant[3]. The temperature dependence is associated with the pseudogap in the electronic density of states at the Fermi surface. The T^2 term is given by the equation (2) in Ref. 3. If the coefficient A was obtained as positive, the second derivative of the density of states $d^2N(E_{\mathrm{F}})/dN(E_{\mathrm{F}})^2$ must be positive. Therefore, the results suggest a pseudogap in the density of states at Fermi energy due to the interaction between Fermi sphere and Brillouin zone.

The magnetic susceptibility of the $Zn_{77}Fe_7Sc_{16}$ and $Zn_{78}Co_6Sc_{16}$ at the magnetic field of 600 Oe is shown in Fig. 4. It seems that they obey Curie-Weiss law in the wide temperature region. The magnetic susceptibility of $Zn_{77}Fe_7Sc_{16}$ is considerably large (compare with the value of $Zn_{74.5}Ag_{9.5}Sc_{16}$ in Fig. 3); the effective magnetic moment is estimated to be $3.7\mu_{\mathrm{B}}$ from the Curie constant. Moreover, the temperature dependence of magnetic susceptibility exhibits a spin glass behavior with a freezing temperature 7.0 K. This is the first observation that most of the Fe atoms have magnetic moment in the icosahedral quasicrystal[4]. On the other hand, the $Zn_{78}Co_6Sc_{16}$ does not show large magnetic moments like the $Zn_{77}Fe_7Sc_{16}$ in spite of the same structure with $3d$ transition metal. It is thought that $3d$ states of Co do not satisfy the condition for local magnetic moment formation. An possible interpretation is the effect of the degree of occupancy of $3d$ states between Zn and Co on the overlapping of s- and d-states by the virtual bound state.

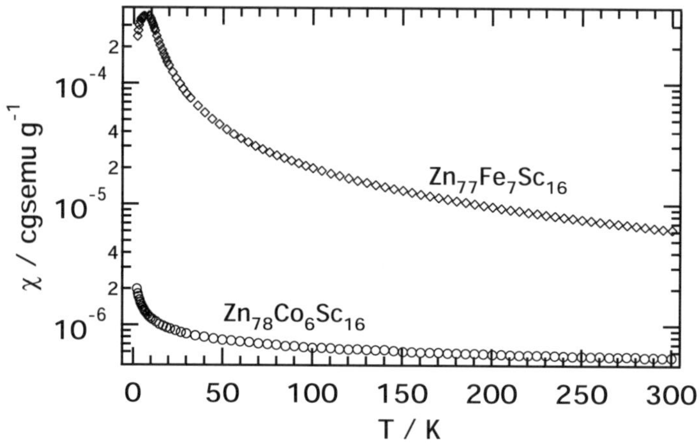

Figure 4. Magnetic susceptibility as a function of temperature of the $Zn_{77}Fe_7Sc_{16}$ and $Zn_{78}Co_6Sc_{16}$ icosahedral quasicrystals under 600 Oe.

In conclusion, we have shown that the variety of magnetism among the series of Zn-Sc based icosahedral quasicrystals. Especially, the $Zn_{77}Fe_7Sc_{16}$ is the singular case that *3d* transition metals have large magnetic moments in the stable icosahedral quasicrystal. The present results revealed that the $Zn_{77}Fe_7Sc_{16}$ is valuable to study of the magnetic behavior of the local magnetic moments of *3d* transition metals in quasiperiodic structure.

Acknowledgments

Thanks are due to the Research Center for Molecular-Scale Nanoscience, Institute for Molecular Science. We also thank C. Masuda and S. Francoual for their help in preparing one of the samples and acknowledge helpful discussions with Prof. S. Matsuo and S. Motomura.

References

1. T. Ishimasa, S. Kashimoto and R. Maezawa, *Mater. Res. Soc. Symp. Proc.* **805**, LL1.1.1 (2004).
2. V. Elser, *Acta. Cryst.* **A42**, 36 (1986).
3. S. Matsuo, H. Nakano, T. Ishimasa and Y. Fukano, *J. Phys.: Condens. Matter* **1**, 6893 (1989).
4. S. Kashimoto, S. Motomura, R. Maezawa, S. Matsuo and T. Ishimasa, *Jpn. J. Appl. Phys.* **42**, L526 (2004).

ONE-ARMED SPIRAL WAVE EXCITED BY RAM PRESSURE IN ACCRETION DISKS IN Be/X-RAY BINARIES

K. HAYASAKI

Department of Applied Physics, Graduate School of Engineering, Hokkaido University, Kitaku N13W8, Sapporo 060-8628, Japan
E-mail: hayasaki@topology.coe.hokudai.ac.jp

ATSUO T. OKAZAKI

Faculty of Engineering, Hokkai-Gakuen University, Toyohira-ku, Sapporo 062-8605, Japan
E-mail: okazaki@elsa.hokkai-s-u.ac.jp

A non-axisymmetric structure of accretion disks around the neutron star in Be/X-ray binaries is studied, analyzing the results from 3D SPH simulations performed by Hayasaki & Okazaki (2004)[1]. It is found that ram pressure due to the phase-dependent mass transfer from the Be-star disk excites a one-armed, trailing spiral structure in the accretion disk around the neutron star. The spiral wave has a transient nature; it is excited around the periastron, when the material is transferred from the Be disk, and gradually damped afterwards. The disk changes its topology from circular to eccentric with the development of the spiral wave, and then from eccentric to circular with the decay of the siral wave during one orbital period. We also find that the orbital phase-dependence of the mass-accretion rate is mainly caused by the inward propagation of the spiral wave excited on the accretion disks.

1. INTRODUCTION

The Be/X-ray binaries represent the largest sub-class of high-mass X-ray binaries. These systems generally consist of a neutron star and a Be star with a cool ($\sim 10^4 K$) equatorial disk, which is geometrically thin and nearly Keplerian. Be/X-ray binaries are distributed over a wide range of orbital periods ($10\,\mathrm{d} \leq P_{\mathrm{orb}} \leq 300\,\mathrm{d}$) and eccentricities ($e \leq 0.9$).

The majority of the Be/X-ray binaries show only transient activity in the X-ray emission and are termed Be/X-ray transients. Be/X-ray transients show periodical (Type I) outbursts, which are separated by the orbital period and have the lumiocity of $L_X = 10^{36-37}\mathrm{erg\,s^{-1}}$, and giant (Type II) outbursts of $L_X \geq 10^{37}\mathrm{erg\,s^{-1}}$ with no orbital modulation. These outbursts have features that strongly suggest the presence of an accretion disk around the neutron star.

Recently, Hayasaki & Okazaki (2004)[1] studied the accretion flow around the neutron star in a Be/X-ray binary with a short period ($P_{\mathrm{orb}} = 24.3\,\mathrm{d}$) and a moderate eccentricity ($e = 0.34$), using a 3D SPH code and the imported data by Okazaki *et al.* (2002)[2]. They found that a time-dependent accretion disk is formed around the neutron star. They also discussed the evolution of the azimuthally-averaged structure of the disk, in which a one-armed spiral structure is seen. It is important to note that the Be/X-ray binaries are systems with double circumstellar disks (the Be decretion disk and the neutron star accretion disk), which interacts mainly via the mass transfer, and that this gives a new point of view to understand the interactions in Be/X-ray binaries.

In this paper, we show that the ram pressure due to the material transferred from the Be disk around periastron temporarily excites the one-armed spiral wave in the accretion disk around the neutron star in Be/X-ray binaries.

2. ACCRETION DISKS DEFORMED BY RAM PRESSURE

Our simulations were performed by using the same 3D SPH code as in Hayasaki & Okazaki (2004)[1], which was based on a version developed by Bate *et al.* (1995)[4].

In order to inspect the effect of the ram pressure on the accretion disk, we compare results from model 1 in Hayasaki & Okazaki (2004)[1] (hereafter, model A) with those from a simulation (hereafter, model B) in which the mass transfer from the Be disk is artificially stopped for one orbital period. Except for this difference, two simulations have the same model parameters: The orbital period P_{orb} is 24.3 d, the eccentricity e is 0.34, and the Be disk is coplanar with the orbital plane. The inner radius of the simulation region r_{in} is $3.0 \times 10^{-3}a$, where a is the semi-major axis of the binary. The polytropic equation of state with the exponent $\Gamma = 1.2$ is adopted. The Shakura-Sunyaev viscosity parameter $\alpha_{\mathrm{SS}} = 0.1$ throughout the disk.

Fig. 1 gives a sequence of snapshots of the accretion disk around the neutron star for $7 \leq t \leq 8$ in model A, where the unit of time is the orbital period P_{orb}. The left panels show the contour maps of the surface density, whereas the non-axisymmetric components of the surface density and the velocity field are shown in the right panels. Annotated in each left panel are the time in units of $P_{\rm orb}$ and the mode strength S_1, a measure of the amplitude of the one-armed spiral wave, which is defined by using the azimuth Fourier decomposition of the surface density distribution, details of which are described by Sec 2.2 of Hayasaki & Okazaki (2005)[3]. It is noted from the figure that the one-armed, trailing spiral is excited at periastron and is gradually damped towards the next periastron. The disk is topologically changing from circular to eccentric with the development of the spiral wave, and then the process reverses to move from eccentric to circular with the decay of the wave during one orbital period.

For comparison purpose, we present the results for model B, in which the mass transfer is artificially turned off for $7 \leq t \leq 8$. Figure 2 shows the surface density (the left panel) and the non-axisymmetric components of the surface density and the velocity field (the right panel) at the time corresponding to the middle panel of Fig. 1. The format of the figure is the same as that of Fig. 1. It should be noted that the disk deformation due to the one-armed mode is not seen in model B. The disk is more circular and has a larger radius in model B than in model A. This strongly suggests that the excitation of the one-armed spiral structure in the accretion disk is induced by the ram pressure from the material transferred from the Be disk at periastron.

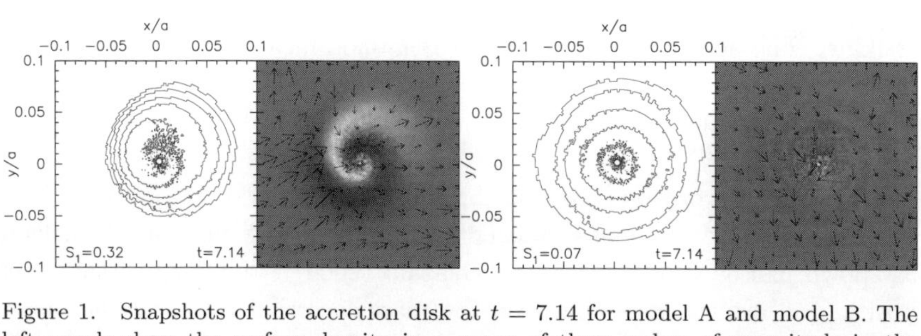

Figure 1. Snapshots of the accretion disk at $t = 7.14$ for model A and model B. The left panels show the surface density in a range of three orders of magnitude in the logarithmic scale, while the right panels show the non-axisymmetric components of the surface density (gray-scale plot) and the velocity field (arrows) in the linear scale. In the right panels, the region in gray (white) denotes the region with positive (negative) density enhancement. The periastron is in the x-direction and the disk rotates counterclockwise. Annotated in each left panel are the time in units of $P_{\rm orb}$ and the mode strength S_1.

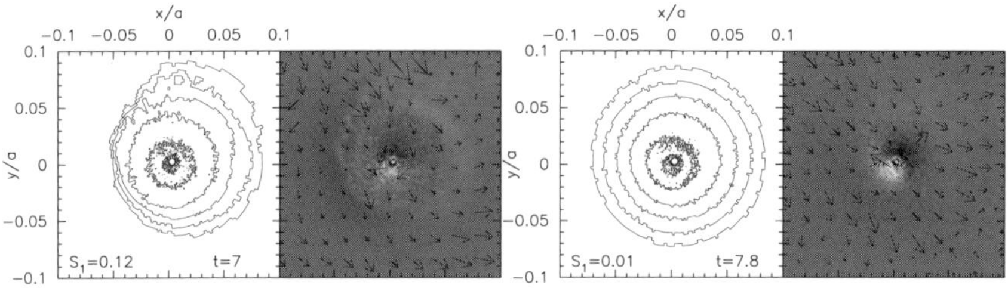

Figure 2. Same as Fig 1, but at $t = 7.0$ and $t = 7.8$ for model A.

2.1. *Phase dependence of the mass-accretion rate*

After the accretion disk is developed ($t \geq 5$), the mass-accretion rate has double peaks per orbit, a relatively-narrow, low peak at periastron and a broad, high peak afterwards [see Fig. 15(a) of Hayasaki & Okazaki (2004)[1]]. While the first low peak at periastron could be artificial, being related to the presence of the inner simulation boundary, the origin of the second high peak was not clear. Below we show that the one-armed spiral wave is responsible for the second peak in the mass-accretion rate.

Figure 3 shows the time dependence of the mass-accretion rate for $7 \leq t \leq 8$. The thick line denotes the mass-accretion rate in model A, in which the mass transfer from the Be disk is taken into account. For comparison, the mass-accretion rate in model B, in which the mass transfer from the Be disk is artificially turned off at $t = 7$, is also shown by the thin line. The difference between the accretion rate profiles for these two models is striking. The accretion rate in model B monotonically decreases over one orbital period, whereas that of model A shows a broad peak centred at $t \sim 7.32 - 7.35$, which corresponds to the second peak found in Hayasaki & Okazaki (2004)[1].

Although it is obvious that the above peak is caused by the mass transfer from the Be disk, the mass-transfer rate has a narrow peak at periastron as shown in Fig. 2 of Hayasaki & Okazaki (2004)[1]. This lag of the peak position on orbital phase between the mass-accretion rate and the mass-transfer rate results from the inward propagation of the wave from the disk outer radius to the inner simulation boundary.

3. SUMMARY AND DISCUSSION

We investigate a non-axisymmetric structure of accretion disks in Be/X-ray binaries, analyzing the results from 3D SPH simulations performed by

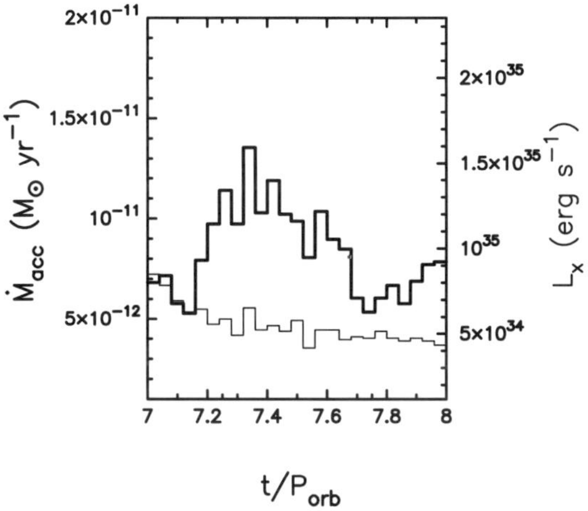

Figure 3. Time dependence of the mass-accretion rate for $7 \leq t \leq 8$. The thick and thin lines are for model A and model B, respectively. The right axis shows the X-ray luminocity corresponding to the mass-accretion rate.

Hayasaki & Okazaki (2004)[1].

We find that the phase-dependent mass transfer induces an eccentric deformation of the disk due to an one-armed spiral wave propagation. The ram pressure from the mass transfer beats on an outer part of the disk around periastron, by which the wave is excited. The inward propagation of the wave leads to an enhancement of the X-ray luminosity around apastron. This suggests that the observed X-ray luminosity might have some information about the topological change due to the disk deformation.

Acknowledgments

This work has been supported by Grant-in-Aid for the 21st Century COE Scientific Research Programme on "Topological Science and Technology" from the Ministry of Education, Culture, Sport, Science and Technology of Japan (MECSST).

References

1. K. Hayasaki, and A T. Okazaki, *MNRAS*,**350**, 971, (2004).
2. A T. Okazaki, M R. Bate, G.I. Ogilvie & J.E. Pringle , *MNRAS*, **337**, 967, (2002).
3. K. Hayasaki, and A T. Okazaki, *MNRAS*, **360L**, 15, (2005).
4. M R. Bate, I A. Bonnell & N M. Price *MNRAS*, **285**, 33, (1995).

IV

Topological Defects and Excitations

TOPOLOGICAL EXCITATIONS IN THE GROUND STATE OF CHARGE DENSITY WAVE SYSTEMS

P. MONCEAU

Centre de Recherches sur les Très Basses Températures,
laboratoire associé à l'Université Joseph Fourier,
CNRS, BP 166, 38042 Grenoble cedex 9, France

Most of strongly correlated electronic systems show various types of symmetry breaking at the origin of degenerate ground states. The degeneracy allows for special topologically non trivial perturbations, the most known forms being domain walls, lines of vortices or dislocations. Of special interest are classes of correlated systems, the so-called Electronic Crystals including Wigner crystals, charge and spin density waves, ... Many of these systems have a continuous degeneracy like in incommensurate charge density waves (CDW). Contrary to usual crystals, the number of sites is not fixed and can be readjusted to absorb transferred electrons. Locally, the addition of electrons to the "condensate" of the crystalline order goes *via* topologically excitations such as solitons. Experiments performed on the charge density wave compound $NbSe_3$ will be described in which it is shown that solitons play a central role in the current conversion between CDW current and normal current *via* the promotion of phase slip processes.

The dynamics of periodically modulated elastic media in the presence of quenched disorder has been the subject of intensive studies during the past years. Systems which exhibit such periodic superstructures include charge density waves (CDW) and spin density waves (SDW),[1−5] vortex lattices in type-II superconductors,[6] and Wigner crystals.[7] All these systems experience pinning forces originating from structural defects or impurities, but sliding of the superstructure is made possible by the application of sufficiently large applied forces. It is generally accepted that random pinning induces continuous elastic distortions of the ordered phase while plastic deformations occur at strong pinning defects. Phase slippage is the general phenomenon which releases phase gradients in the condensate through periodic 2π phase jumps.[8,9].

Quasi-one-dimensional materials with a CDW ground state, the prototype of which being $NbSe_3$, are model systems for the study of the interaction between structural disorder and the internal degrees of freedom of a condensed phase. Below the Peierls transition temperature T_P, a CDW is characterized by a modulated electronic density $\rho(x) = \rho_0\{1 + \alpha\cos(Q_0 x + \varphi)\}$ together with a periodic lattice distortion of the same wave number $Q_0 = 2k_F$ where k_F is the generally incommensurate Fermi momentum. The new periodicity opens a gap in the electron density of states and leads to new satellite Bragg peaks. CDW pinning by host defects fixes the local phase φ and destroys the translational invariance of the CDW ground state. Application of an electric field E above a threshold value E_T sets the CDW in motion and gives rise to a collective current.

Phase slippage is required at the electrodes for the conversion from free to condensed carriers. When the CDW is depinned between current contacts, CDW wave fronts are created near one electrode and destroyed near the other, leading to CDW compression at one end and stretching at the other end. In a purely 1D channel, the order parameter is driven to zero at a certain distance from the pinning ends.[8] For samples of finite cross section, phase slippage develops as dislocation loops (DLs) (see Refs. 10 and 11, which climb to the crystal surface, each DL allowing the CDW to progress by one wavelength. Associated with the dislocation loop nucleation and growth process, the sliding CDW phase is slightly distorted. The phase gradient is observable by means of X-ray diffraction as a longitudinal shift, $q(x) \propto \delta_x\phi$, of the CDW satellite peak position in the reciprocal space, $Q = Q_0 + q$.[12]

$NbSe_3$ has a chain-like structure, with chains parallel to the monoclinic b-direction (a = 10.006 Å ; b = 3.478 Å ; c = 15.626 Å ; β =109.30°). It undergoes two independent Peierls transitions at 145 K and 59 K with modulation wave vectors $(0,Q_1,0)$; $Q_1 = 0.241\ b^*$ and $(0.5,Q_2,0.5)$; $Q_2 = 0.260\ b^*$, respectively. All measurements were carried out on the $(0.1{+}Q_1,0)$ CDW satellite on the diffractometer TROIKA I (ID 10A) at ESRF (Grenoble/France) using an incident wavelength of 1.127 Å (E = 11 keV). The sample orientation was such that the $(a^*{+}c^*,b^*)$-plane coincided with the horizontal scattering plane, with a^*, b^*, c^* the $NbSe_3$ reciprocal vectors ($b^* \parallel b$). In order to obtain a high spatial resolution, the beam width (from 10 to 30 μm) was controlled by a slit, placed before the sample (see Fig. 1). The samples, of typical cross-sections $10 \times 2\ \mu m^2$, were mounted on sapphire substrates of 100 μm thickness to provide homogeneous sample cooling together with suitable beam transmission (50 %). Two 2 μm thick gold electrodes, of 1 mm width, have been evaporated, leaving an uncovered section of a few mm between electrodes.

In Figure 2 we have drawn the "double shift" $q_\pm = Q(+I) - Q(-I)$

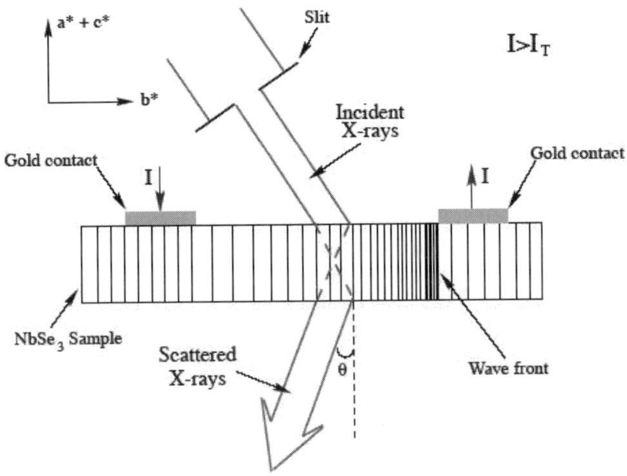

Figure 1. Sketch of the X-ray scattering geometry.

as a function of x for the same half of the sample for applied (pc and dc) currents of $|I| = 4.6$ mA $= 2.13 I_T$. I_T is the threshold current above which the CDW, depinned from impurities, starts to move. For dc currents, the spatial variation of $q_{\pm}(x)$ can be fitted with an exponential decay near the electrodes ($0 < x < 0.7$ mm) with a characteristic length of 375 ± 50 μm and a linear variation for 0.7 mm<x<2 mm with a slope of $\partial q_{\pm}/\partial x = -(2.0 \pm 0.1) 10^{-4} b^*$ mm^{-1}.[13] It should be noted in Fig. 2 that in the middle part of the sample there is no observed difference between direct and pulsed currents. On the contrary the pc shift is nearly zero at the electrode position and reaches a maximum at a distance of about 100 μm away from the contact boundary. These differences suggest a spatially dependent relaxational behavior for the CDW deformations, the fastest relaxation occurring at the contact position. In the pulsed current experiments the observed decay of a q near the contacts during the pulse delay, measured from 10^{-3} s to 10 s, resembles the Kim-Anderson law ($-\ln t$) for the supercurrent decay in superconductors *via* the creep of pinned vortices. Here we refer to the decay of the DW deformations near the contacts due to the backward creep of DLs during the long silent period between active pulses.

The semi-microscopic model by Brazovskii *et al.*[12], describing the normal $\leftrightarrow$ condensed carrier conversion by nucleation and growth of phase dislocation loops in a highly rigid CDW electronic crystal is based on the assumption of the local equilibrium between the electrochemical potentials

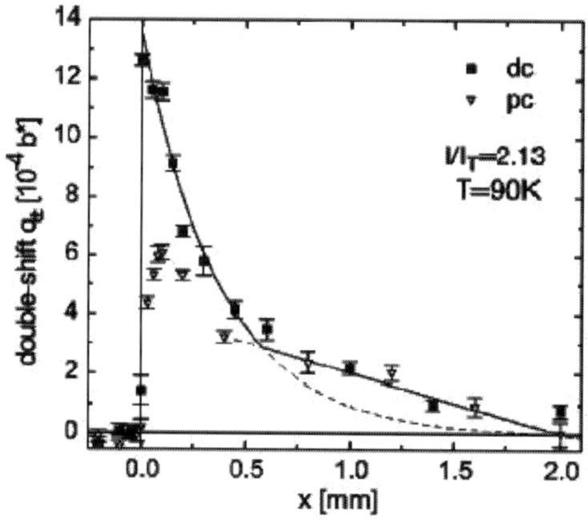

Figure 2. Double-shift $q_\pm(x) = Q(I) - Q(-I)$ (in units of b^*) for direct (full squares) and pulsed (open triangles) current ($I/I_T = 2.13$). The full line shows the exponential decay near the contact ($0 < x < 0.5$) and a linear dependence for $0.7 < x < 2$, $NbSe_3$, $T = 90$ K.

of the phase dislocations, U, and of the free carriers, μ_n. The quantity $\eta \equiv U - \mu_n \propto q$, which is directly measured by X-ray diffraction, measures the electrochemical potential unbalance between U and μ_n and hence the excess or lack of normal carriers. The stationary distribution of the CDW deformation obeys the equation (see Eq. (1) in Ref. 12

$$\frac{\partial \eta}{\partial x} = F_r(J_c) - \frac{e(J_n - J_T)}{\sigma_n}. \tag{1}$$

Here J_n, J_c, σ_n and J_T are, respectively, the normal carrier and CDW current densities, the normal carrier conductivity, and the current density at threshold. $F_r(J_c)$, the friction force, is approximated for high current values ($J_{\rm tot} \le 2J_T$) as $F_r(J_c) \approx eJ_c/\sigma_c$, where σ_c is the high-field CDW conductivity. The gradient of the CDW current density is determined *via* the conversion rate, R, between normal carriers and CDW condensate (see Eq. (2) in Ref. 12):

$$\partial J_c/\partial x \propto R(\eta). \tag{2}$$

In the case of heterogeneous passive nucleation of DLs, for which the conversion rate can be approximated as $R \propto \tau_{\rm cnv}^{-1}\eta$, the solution of Eqs. (1) and (2) yields $q(x) \propto \sinh(x/\lambda_0)$. q decreases exponentially near the contacts and vanishes in the central part of the sample. $\lambda_0 \propto \sqrt{\tau_{\rm cnv}}$, typically a few hundred μm, characterizes the length scale of the phase slip distribution.

$T_{\rm cnv}$, here treated as an adjustable parameter, is the lifetime of an excess carrier with respect to its conversion to the condensate, *i.e.* the mean free carrier lifetime before absorption by a DL.

CDW phase discontinuities may also occur along the sample due to defects which obstruct the CDW current.[14]

Figure 3 shows the shift, $q(x) = Q_I(x) - Q_0$ at $T = 90$ K for negative ($\circ$) and positive ($*$) polarities as a function of beam position, x, along a $NbSe_3$ sample. The vertical lines show the boundaries of the gold-covered contacts.

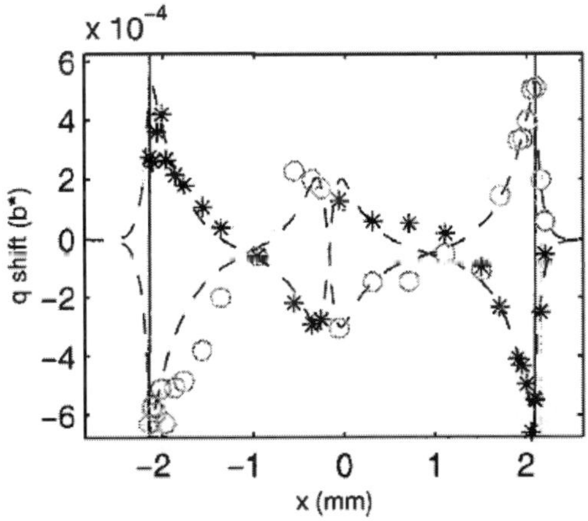

Figure 3. Shift $q(x) = Q_I(x) - Q_0$ of the CDW satellite peak position as a function of beam position for positive ($*$) and negative ($\circ$) polarities; $I = \pm 4.6$ mA ($I/I_T = 2.13$). The vertical lines show the boundaries of the gold-covered contacts, beam width: 30 μm; $NbSe_3$; $T = 90$ K.

This sample exhibits, between electrodes, one localised defect at the position $x_d \propto -0.15$ mm. The sliding-CDW satellite shift changes abruptly at this position, with maxima on either sides of the defect position (as well as at the electric contacts). In contrast, for defect-free samples, the phase gradient is large near the electrodes, where the conversion between normal carriers and CDW condensate occurs, and vanishes half-way between the electrodes.[12,13]

In conclusion, we have observed the influence of localised defects on the longitudinal variation, $q(x)$, of the CDW satellite position at fixed dc current as a function of beam position, x, between electrical contacts, in $NbSe_3$. We have shown that the sliding-CDW phase distortion, due to conversion between normal carriers and CDW condensate, is not uniquely a near-contact effect but also occurs in the vicinity of structural defects or regions intentionally damaged by X-ray radiation. Using the semi-microscopic model by Brazovskii *et al.*,[14] with an enhanced pinning force at the defect position, we have given a coherent description of the spatial dependence of the CDW phase gradient near both types of defects. The phase gradient reflects the fact that a fraction of the CDW current is transformed into normal current at the defect position.

Acknowledgments

This contribution is based on the papers Refs. 12, 13, and 14. I acknowledge the co-authors of these works: S. Brazovskii, R. Currat, C. Detlefs, G. Grübel, N. Kirova, J.E. Lorenzo, F. Nad, H. Requardt, D. Rideau.

References

1. Proceedings of the International Workshop on Electronic Crystals, edited by S. Brazovskii and P. Monceau, *J. Phys. France IV*, Colloque **C2**, 3 (1993).
2. Proceedings of the International Workshop on Electronic Crystals, edited by S. Brazovskii and P. Monceau, *J. Phys. France IV*, **9** (1999).
3. Proceedings of the International Workshop on Electronic Crystals, edited by S. Brazovskii, N. Kirova, and P. Monceau, *J. Phys. France IV*, **12** (2002).
4. G. Grüner, in "Density Waves in Solids" (Addison-Wesley, Reading, MA) (1994).
5. P. Monceau, in "Electronic Properties of Quasi-One-Dimensional Compounds", edited by Reidel Publ. Co (1985).
6. G. Blatter *et al.*, *Rev. Mod. Phys.* **66**, 1125 (1994).
7. E.Y. Andrei *et al.*, *Phys. Rev. Lett.* **60**, 2765 (1988).
8. L.P. Gorkov, *JETP Lett.* **38**, 87 (1983).
9. N.P. Ong and K. Maki, *Phys. Rev.* B **32**, 6582 (1985).
10. J. Dumas and D. Feinberg, *Europhys. Lett.* **2**, 555 (1986).
11. D. Feinberg and J. Friedel, *J. Phys. France* **49**, 485 (1988).
12. S. Brazovskii *et al.*, *Phys. Rev.* B **61**, 10640 (2000).
13. H. Requardt *et al.*, *Phys. Rev. Lett.* **80**, 5631 (1998).
14. D. Rideau *et al.*, *Europhys. Lett.* **56**, 289 (2001).

SOLITON TRANSPORT IN NANOSCALE CHARGE-DENSITY-WAVE SYSTEMS

KATSUHIKO INAGAKI, TAKESHI TOSHIMA, AND SATOSHI TANDA

Department of Applied Physics, Graduate School of Engineering
Hokkaido University, Kita 13 Nishi 8 Kita-ku, Sapporo 060-8628 Japan
E-mail: kina@eng.hokudai.ac.jp

We report studies of o-TaS_3 nanocrystals, including sample preparation of nanocrystals, fabrication of electrodes attached to the nanocrystals, and transport properties. Single crystals of o-TaS_3 were synthesized by chemical transport method. The observed temperature dependence of the nanoscale o-TaS_3 crystal did not show clear Peierls transition. The resistance was well described in terms of one-dimensional (1D) variable-range-hopping conduction over the wide range of temperature (100 K to 4.2 K). These features suggest transport properties of the nanoscale o-TaS_3 crystals were governed by soliton nucleation even at the highest observed temperature. We found finite-size effect of sloition transport in nanoscale o-TaS_3 crystals. One kind of the samples exhibit the standard activation formula, $\exp[-E_0/E]$, whereas other kind of the samples show a modified form, $\exp[-(E_0/E)^2]$, where E_0 is a constant, and E is an applied field. We interpreted it as the dimensional crossover of soliton transport, in the framework provided by Hatakenaka et al. The system dimension relevant to soliton transport was either 1D or 2D, which depends on the system size. The sample of the smallest effective cross section, determined by nominal bulk resistively, exhibits the 1D behavior, whereas the larger samples showed the 2D one. This result is consistent with the idea that electric transport phenomenon of the nanoscale o-TaS_3 crystals should be attributed to soliton nucleation.

1. Introduction

The soliton in a charge-density-wave (CDW) system is a typical kind of topological defects[1,2,3], accompanied with a charge of $\pm 2e$. A CDW system is, in principle, an insulator because of Peierls gaps opened at Fermi surfaces, but there some possibilities of electric transport, such as quasiparticle excitation, collective motion of CDW and soliton transport. A quasi-one-dimensional conductor TaS_3 is known for existence of soliton transport under Peierls transition[4,5].

Nucleation of soliton is of importance to understand low temperature behaviour of the soliton transport phenomenon. Early studies gave a formula

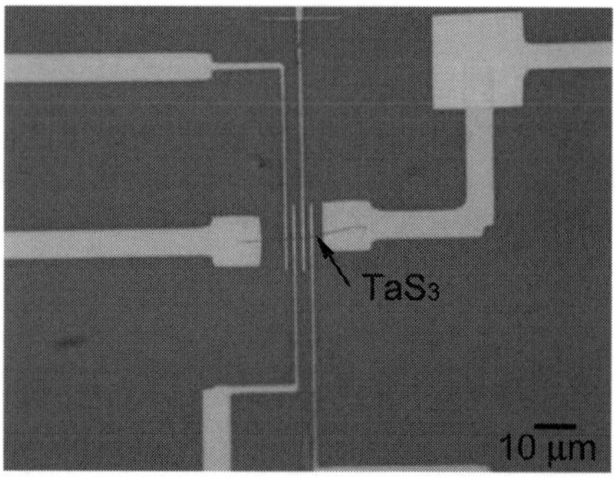

Figure 1. An o-TaS_3 nanocrystal attached with six gold electrodes. The scale bar shows 10 μm.

similar to that of Arrhenius-type activation[1], however, other possibility was pointed out both by an experiment[5], soon followed by a theory[2]. Discrepancy between these two theories has been unified in terms of dimensional crossover of soliton nucleation[3]. We have been developing a versatile technique for fabrication of nanoscale CDW systems[6]. Here we report studies of o-TaS_3 nanocrystals, including sample preparation of nanocrystals, fabrication of electrodes attached to the nanocrystals, and transport properties.

2. Experimental

Single crystals of o-TaS_3 were synthesized by chemical transport method. A pure tantalum sheet and sulfur powder were put in a quartz tube. The quartz tube was evacuated to 1×10^{-6} Torr and heated in a furnace at 530 °C for five hours. The crystals were sonicated in toluene for 15 minutes, kept from perturbation for several hours, and then the dispersion was deposited on a silicon substrate with thermal oxide layer of 1 μm. After blow-drying, the crystals were inspected with an optical microscope. The obtained crystals were typically 0.2 μm in width and 10-100 μm in length.

Electrodes were fabricated by standard electron beam lithography with a scanning electron microscope (SEM) equipped with a homemade writing system. The electrode was heated by irradiation of electron beam with another SEM, which can produce more intense electron beam, to obtain ohmic electric contact between the nanocrystal and the electrode[6]. Figure 1 is a micrograph of a typical result of the o-TaS_3 nanocrystal fabrication. It is shown that the gold electrodes were well defined and located precisely on the nanocrystal.

The samples were cooled down to 0.4 K by a ^{3}He cryostat. All wires are carefully shielded and filtered so that buring out of the samples should be avoided by spurious noises and static. Resistance was measured by a two-probe configuration because the contact resistance is thought to be negligiblly smaller than that of the sample. Noise floor of the measurement system is around 10^{-13} A.

3. Results and Discussions

It is known that the temperature dependence of the resistance of conventional bulk samples is complex and consists of several regimes. At room temperature a bulk o-TaS_3 is metallic. Below Peierls temperature ($T_P \sim 220$ K), the conductivity decreases exponentially, $\exp(-\Delta/k_BT)$, where $\Delta \sim 800$ K. At lower temperatures than ~ 100 K, the potential Δ changes to a small energy, ~ 200 K. This change in the potential is described in terms of soliton transport[4]. Below 20 K, the formula of the temperature dependence changes from the Arrhenius form to the variable-range-hopping (VRH) conduction, represented as

$$\sigma(T) = \sigma_0 \exp[-(T_0/T)^{1/(d+1)}], \tag{1}$$

where T_0 is a characteristic temperature and $d = 1$ is dimension of the system[7,8].

On the other hand, the observed temperature dependence of the nanoscale o-TaS_3 crystal was significantly different from that of bulk crystal. Figure 2 shows that temperature dependence of resistance for Sample A is different from that of the bulk samples. Most significant feature is the lack of Peierls transition in the nanocrystal. Moreover, the temperature range where the resistance was described in terms of 1D-VRH conduction was remarkably wide (100 K to 4.2 K). We believe in this regime the carrier is topological dislocation of CDW, or charged soliton. Assume that a soliton propagates in a random medium. Hopping energy of a soliton to move to a next possible site should differ to each position due to randomly distributed impurities. Nevertheless, possibility of VRH of topological dislocations in a superconductor, namely fluxons, has already proposed theoretically[9].

We also found that the nonlinear conduction due to soliton transport in the charge-density-wave system became independent of temperature below 2 K. Figure 3 shows I-V characteristics of Samples A and B measured at 389 mK.The observed I-V curves were classified to two kinds. One kind of the samples exhibit the standard activation formula, $I \propto \exp[-E_0/E]$, whereas

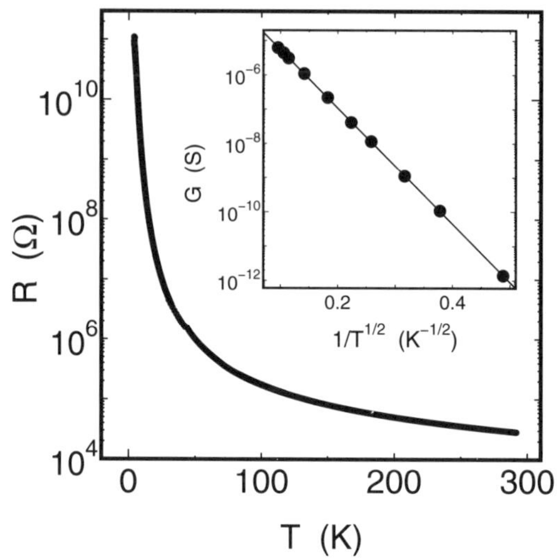

Figure 2. Temperature dependence of resistance for the TaS_3 nanocrystal. Inset shows the linear conductance at 4.2 K to 100 K as the function of $T^{-1/2}$. The data well fit to a linear line, suggesting that the variable-range-hopping conduction: $\sigma \propto \exp[-(T_0/T)^{1/2}]$.

other kind of the samples show a modified form, $I \propto \exp[-(E_0/E)^2]$, where E_0 is a constant, and E is an applied field. Since both kinds of the samples were synthesized in a same batch, it is plausible to attribute the difference of the I-V curves to that of the sample sizes.

We interpreted it as the dimensional crossover of soliton transport, in the framework provided by Hatakenaka *et al.*[3]. They predicted the quantum nucleation possibility of the solutions Γ, which is proportional to the current, obeys a dimensionality-dependent formula,

$$\Gamma \propto \exp(const./E^D), \tag{2}$$

where D is the system dimension. According to this, our observation suggests that the system dimension relevant to soliton transport was either $D = 1$ or $D = 2$, which depends on the system size. This interpretation is justified when the system dimension D relates to the effective cross section of each sample. The sample of the smallest effective cross section (1.1×10^{-4} μm^3) exhibits the 1D behavior, whereas the larger samples showed the 2D one. This result strongly suggests that electric transport phenomena of the nanoscale o-TaS_3 crystals should be attributed to soliton nucleation, nevertheless, at low temperatures.

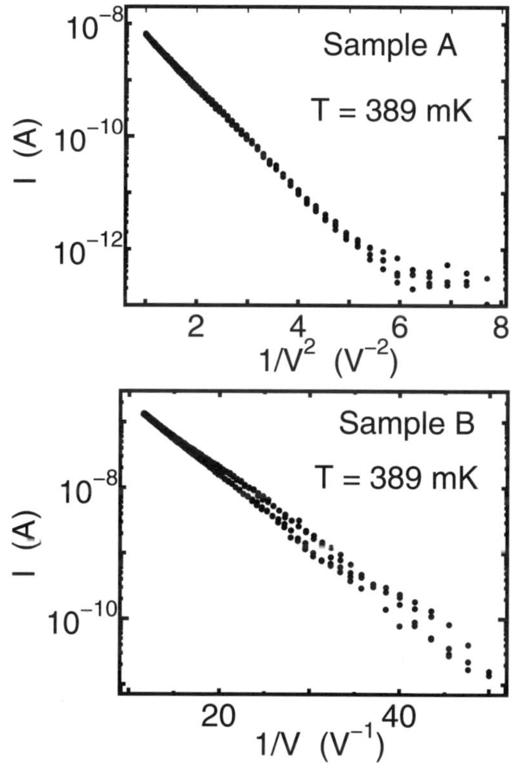

Figure 3. *I-V* characteristics of Samples A and B measured at 389 mK. Sample A exhibits $I \propto \exp[-(V_0/V)^2]$, while B shows $I \propto \exp[-(V_0/V)]$, though both samples come from a same batch.

Promising applications of the soliton transport in the charge density wave system are three-terminal devices such as field effect transitors[10]. Nucleation of soliton is either promoted or supressed according to the internal electric field as a periodic function of $Q_0 = eN_{ch}/\epsilon A$, where N_{ch} is the number of chains, ϵ is the dielectric constant, and A is the cross section. We believe our technique of electrode fabrication is suitable for this application. The gold finger on the sample does not conduct electric current until it is irradiated by a certain amount of electron beam. Hence an untreated finger works as the gate of a field effect transistor. We now plan to start preliminary study of the soliton field effect transistor made of nanoscale o-TaS_3 crystals.

Acknowledgment

This work is partly supported by Grant-in-Aid for Scientific Research (K. I. and S. T.), and the 21st Century COE Program.

References

1. K. Maki, *Phys. Rev. Lett.* **39**, 46 (1977).
2. J.-M. Duan, *Phys. Rev.* **B 48**, 4860 (1993).
3. N. Hatakenaka, M. Shiobara, K. Matsuda, and S. Tanda, *Phys. Rev.* **B 57**, 2003 (1998).
4. T. Takoshima, M. Ido, K. Tsutsumi, T. Sambongi, S. Honma, K. Yamaya, and Y.Abe, *Solid State Commun.* **35**, 911 (1980).
5. S. V. Zaitsev-Zotov, *Phys. Rev. Lett.* **71**, 605 (1993).
6. K. Inagaki, T. Toshima, S. Tanda, K. Yamaya, and S. Uji, *Appl. Phys. Lett.* **86**, 073101 (2005).
7. S. K. Zhilinskiĭ, M. E. Itkis, I. Yu. Kal'nova, F. Ya. Nad', and V. B. Preobrazhenskiĭ, *Sov. Phys. JETP* **58**, 211 (1983).
8. M. E. Itkis, F. Ya. Nad', and P. Monceau, *J. Phys.: Condens. Matter* **2**, 8327 (1990).
9. M. P. A. Fisher, *Phys. Rev. Lett.* **65**, 923 (1990).
10. J. H. Miller, Jr., G. Cárdenas, A. García-Perez, W. More, and A. W. Beckwith, *J. Phys. A: Math. Gen.* **36**, 9209 (2003).

TOPOLOGICAL DEFECTS IN TRIPLET SUPERCONDUCTORS UPT$_3$, SR$_2$RUO$_4$, ETC.

K. MAKI

S. HAAS

D. PARKER

Department of Physics and Astronomy, University of Southern California, Los Angeles, CA 90089-0484 USA

H. WON

Department of Physics, Hallym University, Chuncheon 200-702, South Korea

After a brief introduction on nodal superconductors, we review the topological defects in triplet superconductors such as UPt$_3$, Sr$_2$RuO$_4$, etc. This is in part motivated by the surprising discovery of Ana Celia Mota and her colleagues that in some triplet superconductors the flux motion is completely impeded (the ideal pinning). Among topological defects the most prominent is Abrikosov's vortex with quantum flux $\phi_0 = \frac{hc}{2e}$. Abrikosov's vortex is universal and ubiquitous and seen in both conventional and unconventional superconductors by the Bitter decoration technique, small angle neutron scattering (SANS), scanning tunneling microscopy (STM), micromagnetometer and more recently by Lorentz electron micrograph. In order to interpret the experiment by Mota et al a variety of textures are proposed. In particular, in analogy to superfluid ^{3}He-A the $\hat{\ell}$-soliton and $\hat{d}$-soliton play the prominent role. We review these notions and point out possible detection of these domain walls and half-quantum vortices in some triplet superconductors.

1. Introduction

We shall first survey the new world of nodal superconductors, which appeared on the scene in 1979. Indeed nodal superconductors are a child of the 21st century [1,2,3]. Although the presence of nodal superconductors in heavy-fermion superconductors like CeCu$_2$Si$_2$, UPt$_3$, UBe$_{13}$ and others was found in the late eighties [1], the systematic study of the gap function $\Delta(\mathbf{k})$ began only after the discovery of the high-T$_c$ cuprates La$_{2-x}$Ba$_x$CuO$_4$ by Bednorz and Müller [4] in 1986. The d-wave symmetry of high-T$_c$ cuprates YBCO, Bi-2212, etc. was established circa 1994 through the elegant Josephson interferometry [5,6] and the powerful angle resolved photoemission spectrum (ARPES) [7] among others. In 1993 Volovik [8] derived the quasiparticle density of states of the vortex state in nodal superconductors within the semiclassical approximation. The surprising $\sqrt{H}$ dependence of the specific heat has been seen in YBCO [9], LSCO [10], and Sr_2RuO_4[11,12]. Later Volovik's approach was extended in a variety of directions: a) the study of the thermal conductivity [13,14]; b) for an arbitrary field direction [15]; and c) for different classes of $\Delta(\mathbf{k})$ [16]. These are summarized in Ref. 17. Until now the powerful ARPES and Josephson interferometry have not been applied outside of high-T_c cuprate superconductors.

Since 2001 Izawa *et al.* have determined the gap function $\Delta(\mathbf{k})$ in Sr_2RuO_4[18],$CeCoIn_5$ [19], $\kappa-(ET)_2Cu(NCS)_2$ [20], YNi_2B_2C [21], and $PrOs_4Sb_{12}$ [22,23] via the angle-dependent magnetothermal conductivity. These $|\Delta(\mathbf{k})|$'s are shown in Fig. 1. In addition, the gap function of UPt_3

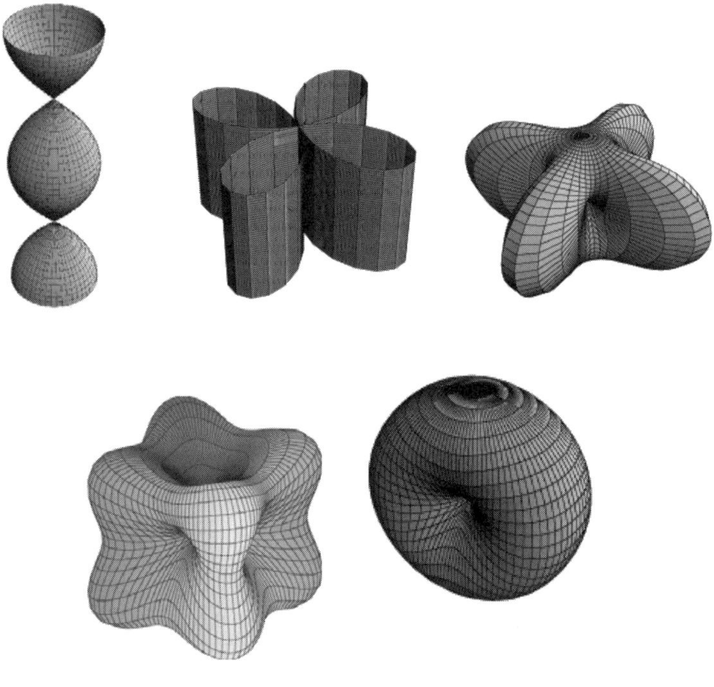

Figure 1. From top left, order parameters for Sr_2RuO_4, $CeCoIn_5$ and $\kappa-(ET)_2Cu(NCS)_2$, YNi_2B_2C, $PrOs_4Sb_{12}$ (A and B phases)

was established around 1994-6 as E_{2u} through the anisotropy in the thermal conductivity [24] and the constancy of the Knight shift in NMR [25]. Somewhat surprisingly all these superconductors are nodal and their quasiparticle density of states increases linearly in $|E|$ for $|E|/\Delta \ll 1$:

$$N(E) \sim |E|/\Delta \quad (1)$$

For example, this implies that the p-wave superconductivity in Sr_2RuO_4 as proposed in Ref. 26 is not consistent with the specific heat data [11]. Also as discussed elsewhere [27,28], the two gap model is of little help in this matter.

More recently the quasiparticle density of states in the vortex state in Sr_2RuO_4 has been reported [29]. Indeed the observed quasiparticle density of states is very consistent with that predicted for an f-wave order parameter [30]. Also many of these superconductors are triplet: UPt_3, Sr_2RuO_4, $(TMTSF)_2PF_6$, $U_{1-x}Th_xBe_{13}$, URu_2Si_2, $PrOs_4Sb_{12}$, UNi_2Al_3 and $CePt_3Si$, for example.

2. Textures in triplet superconductors

Here we consider possible textures in triplet superconductors. For simplicity we concentrate on two f-wave superconductors: UPt_3 (3-dimensional) and Sr_2RuO_4 (quasi 2-dimensional), which we understand well. Mota *et al.* [33,34] have discovered the ideal pinning in the B phase of UPt_3 and below T=70 mK in Sr_2RuO_4. Also these systems are characterized by $\hat{\ell}$ and $\hat{d}$ similarly to superfluid ^{3}He-A [35,36]. Here $\hat{\ell}$ is the direction of the pair angular momentum. But unlike superfluid ^{3}He-A $\hat{\ell}$ is fixed to be parallel to one of the crystal axes: $\hat{\ell} \parallel \pm\hat{c}$. $\hat{\ell}$ can be called the chiral vector. On the other hand $\hat{d}$ describes the spin configuration of the pair and is perpendicular to the pair spin $\mathbf{S_{pair}}$. In the equilibrium configuration $\hat{d} \parallel \pm\hat{\ell}$ in superfluid ^{3}He-A, UPt_3 and Sr_2RuO_4 [26]. Also we believe that some triplet superconductors may not have $\hat{\ell}$ and $\hat{d}$ vectors, as observed in superfluid ^{3}He-B. So it is very important to know if the superconductivity breaks the chiral symmetry, as observed in the experiments by Mota *et al.* [33,34] on the B phase in $U_{1-x}Th_xBe_{13}$ (see Fig. 2). But it is possible that the superconductor in UBe_{13} does not have $\hat{\ell}$ and $\hat{d}$. Also we conjecture that the superconductivity in the B phase of $PrOs_4Sb_{12}$ has $\hat{\ell}$ and $\hat{d}$ vectors [23] and that probably all other quasi 2-dimensional systems such as URu_2Si_2, UNi_2Al_3 and $CePt_3Si$ do as well. Therefore we can think of a variety of domain walls as in superfluid ^{3} He-A [38,39]. Indeed the $\hat{\ell}$-soliton in Sr_2RuO_4 has been considered by Sigrist and Agterberg [40]. The $\hat{\ell}$-soliton is created when in one side of the wall $\hat{\ell} \parallel \hat{c}$, while in the other side $\hat{\ell} \parallel -\hat{c}$. The

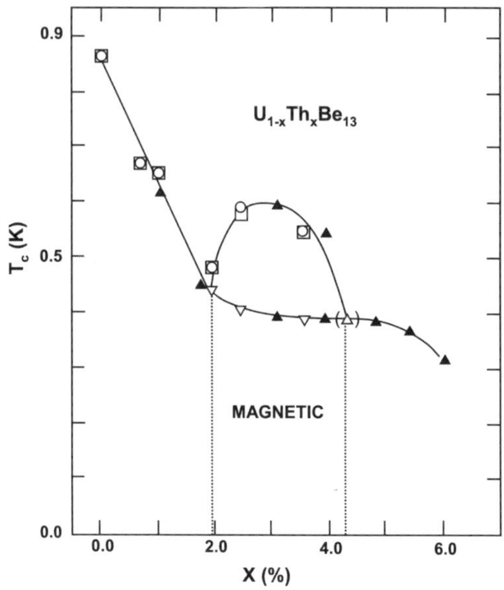

Figure 2. Phase diagram of $U_{1-x}Th_xBe_{13}$ from Heffner *et al.*[41]

chirality changes across the $\hat{\ell}$-soliton. However, unlike in superfluid ^{3}He-A $\hat{\ell}$ is practically fixed to be parallel to $\pm\hat{c}$. For example, when $\hat{\ell}$ is in the ab-plane there will be little superconducting order parameter left. Therefore we can estimate the $\hat{\ell}$-soliton energy per unit area as

$$f_{\hat{\ell}} \simeq \frac{1}{4} N_0 \Delta^2(T)\xi(T) \tag{2}$$

$$\simeq \frac{1}{4} N_0 v_F \Delta(T) \tag{3}$$

where $\Delta(T)$ is the maximum value of the energy gap and v_F is the Fermi velocity.

On the other hand the $\hat{d}$-soliton may be much more easily created [42]. The crucial element here is the spin-orbit energy which binds $\hat{d}$ parallel to $\hat{\ell}$. The relevant energy can deduced from the NMR data of UPt_3 [43] and Sr_2RuO_4 [44]. We estimate $\Omega(T)/\Delta(T)$ for UPt_3 and Sr_2RuO_4: $\Omega(T)/\Delta(T) \sim 0.5 \times 10^{-3}$ (B phase) and 0.2×10^{-4} respectively. Here $\Omega(T)$ is the characteristic frequency associated with the $\hat{\ell}$ and $\hat{d}$ coupling [42]. Then the areal energy for the $\hat{d}$-soliton is given by

$$f_{\hat{d}} \simeq \frac{1}{4} N_0 v_F \Omega(T). \tag{4}$$

This is smaller than $f_{\hat{\ell}}$ for UPt_3 and Sr_2RuO_4 by factors of 0.5×10^{-3} and 0.2×10^{-4}, respectively. Also $\xi_d = v_F/\Omega(T)$ gives the spatial extension of the domain well. This ranges from 10μm $\sim$ 1 mm.

In the presence of $\hat{\ell}$-solitons vortices may enter into the superconducting state as observed in the vortex sheet of superfluid ^{3}He-A [45,46]. Otherwise the motion of the vortex across the $\hat{\ell}$- soliton is completely impeded as discussed in [40]. When an Abrikosov vortex encounters a $\hat{d}$-soliton, the vortex appears to split into two half-quantum vortices (HQV) in the vicinity of $T \simeq T_c$. At lower temperatures it appears that Abrikosov's vortex should tunnel through the $\hat{d}$-soliton. This type of HQV was first predicted in the context of superfluid ^{3}He-A [47,48,49]. However, these HQV's have not yet been observed in superfluid ^{3}He-A [50]. Therefore half-quantum vortices may be first observed in triplet superconductors.

3. Half-quantum vortices

Here we shall consider a pair of half-quantum vortices (HQV) attached to a $\hat{d}$-soliton [42]. The texture free energy is given by

$$f_{pair} = \frac{1}{2}\chi_N C^2 \int dy\, dz (K(\nabla\phi)^2 + \sum_{i,j} |\partial_i \hat{d}_j|^2 + \xi_d^{-2} \sin^2(\Psi)) \tag{5}$$

where C is the spin wave velocity, $\xi_d = C(T)/\Omega(T)$ and

$$K = \frac{\rho_s}{\rho_{s,spin}} = \frac{1+\frac{1}{3}F_1}{1+\frac{1}{3}F_1^a} \frac{1+\frac{1}{3}F_1^a(1-\rho_s^0)}{1+\frac{1}{3}F_1(1-\rho_s^0)} \tag{6}$$

where ρ_s and $\rho_{s,spin}$ are the superfluid density and the spin superfluid density respectively. Here F_1 and F_1^a are the Landau parameters and $\rho_s^0(T)$ is the superfluid density when $F_1 = F_1^a = 0$. The superfluid density for UPt_3 and Sr_2RuO_4 are shown in Fig. 3. In this analysis we assumed that $|\Delta(\mathbf{k})| \sim |\cos\theta \sin^2\theta|$ and $\sim |\cos\chi|$ in UPt_3 [51] and Sr_2RuO_4 respectively. We note that ρ_s^0 in Sr_2RuO_4 is the same as in d-wave superconductivity [52]. In particular, for T in the vicinity of T_c we obtain

$$\rho_s^0 \simeq -\frac{4}{3}\ln(\frac{T}{T_c}) \text{ for } Sr_2RuO_4 \tag{7}$$

$$\simeq -\frac{143}{105}\ln(\frac{T}{T_c}) \text{ for } UPt_3 \tag{8}$$

respectively. Here we assumed that $\mathbf{H} \parallel \hat{a}$ and that the domain wall extends in the y-z plane. Here we consider 2 typical cases as shown in Fig. 4a) and b). In Fig. 4a) the $\hat{d}$-soliton runs parallel to the c axis while in Fig. 4b) it

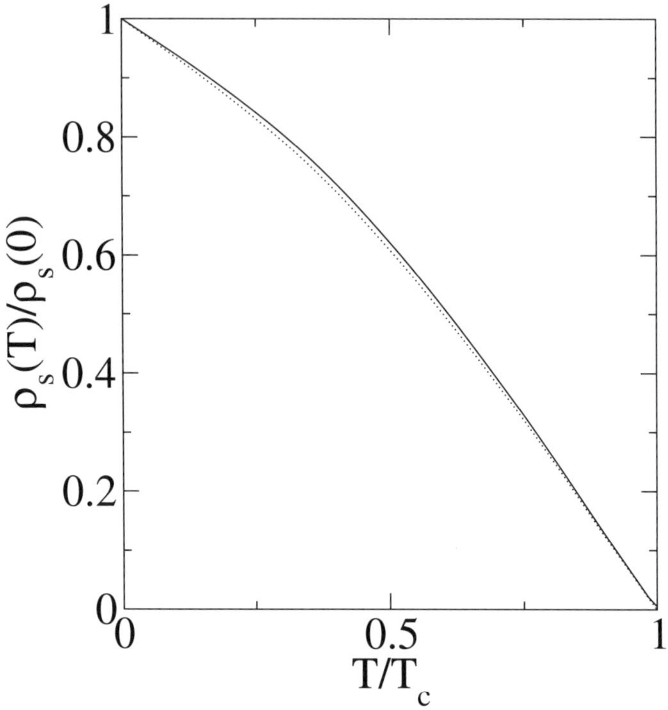

Figure 3. The superfluid density for the B phase of UPt_3 (solid line) and Sr_2RuO_4 are shown

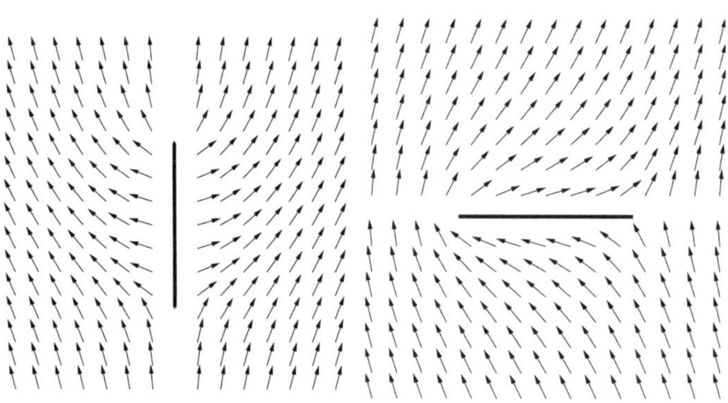

Figure 4. The spatial orientation of the $\hat{d}$-vector

runs parallel to the b-axis. For the first configuration (i.e. a) we parametrize $\hat{d} = \cos\psi\hat{z} + \sin\psi\hat{y}$ with

$$\psi(y,z) = \frac{1}{2}(\arctan(\frac{z+(R/2)}{y}) - \arctan(\frac{z-(R/2)}{y})) \qquad (9)$$

with 2 HQV located at $(y,z) = (0, R/2)$ and $(0, -R/2)$ while in the second configuration (b)

$$\psi(y,z) = \frac{1}{2}(\arctan(\frac{y+(R/2)}{z}) + \arctan(\frac{y-(R/2)}{z})) \qquad (10)$$

Also Φ in Eq.(7) is the phase of the order parameter $\Delta(\mathbf{k})$. Then the total free energy reduces to

$$f_{pair} = \frac{1}{2}\chi_N C^2(\pi K \ln(\lambda/R) + \pi\ln(R/\xi) + \pi(\frac{R}{2\xi_d})^2 \times \ln(4\xi d/R) \qquad (11)$$

where λ and ξ are the magnetic penetration depth and the coherence length respectively.

By minimizing f_{pair} with respect to R, we obtain

$$R_0^2 = 2\xi_d^2(K-1)/[\ln(\frac{4\xi f}{\sqrt{e}R_0})] > 0 \qquad (12)$$

where R_0 is the optimal distance of a pair of HQVs. First in order to have a pair of HQVs we need $K > 1$, which is guaranteed when $F_1 > F_1^a$ and $T \simeq T_c$. Also it is necessary to have

$$R_0 < \frac{4}{\sqrt{e}}\xi_d \qquad (13)$$

In particular when $K - 1 \ll 1$, we obtain

$$R_0/\xi_d \simeq \frac{4}{\sqrt{e}} - \frac{\sqrt{e}}{2}(K-1) \qquad (14)$$

where $e = 2.71828\ldots$. Also the separation betweeen a pair of HQV should be of the order of $\xi_d \sim 10\mu$m ~ 1 mm.

4. Concluding remarks

We have described an abundance of triplet superconductors. Many of their order parameters possess the $\hat{\ell}$ and $\hat{d}$ vectors: $\Delta(\mathbf{k})$ in the B phase of UPt_3, Sr_2RuO_4, both the A and B phase of $PrOs_4Sb_{12}$ and perhaps many other systems. In these systems the $\hat{d}$-soliton is the most common domain wall. The presence of the $\hat{d}$-soliton can impede the flux motion in a variety of ways. The most intriguing is the splitting of an Abrikosov vortex into a pair of half-quantum vortices as discussed in [42]. We expect that some of the techniques used to observe Abrikosov's vortex can be used in the present circumstances. These techniques include the Bitter decoration technique [53], small angle neutron scattering (SANS) [54], scanning tunneling microscopy (STM)[55], micromagnetometer [56] and more recently Lorentz electron micrograph [57]. We expect the exploration of these topological defects in triplet superconductors will enhance our understanding of these exotic superconductors. Also they will provide ideal laboratories to check rich field-theoretical concepts at moderately low temperatures from 10 $\sim$ 100 mK. Therefore the future of topological defects in nodal superconductors is still wide open.

Acknowledgments

We have benefitted from collaborations with Balazs Dora, Hae-Young Kee, Yong Baek Kim, András Ványolos and Attila Virosztek on related subjects.

References

1. M. Sigrist and K. Ueda, Rev. Mod. Phys, **63**, 239 (1991).
2. H. Won, S. Haas, D. Parker, K. Maki, Phys. Stat. Sol. (6) **242**, 363 (2005).
3. K. Maki, S. Haas, D. Parker, H. Won, Chinese Journal of Physics **43**, 532 (2005).
4. J.G. Bednorz and K.A. Müller, Z. Phys. B **64**, 189 (1986).
5. D. J. van Harlingen, Rev. Mod. Phys. **67**, 515 (1995).
6. C. C. Tsuei and J. R. Kirtley, Rev. Mod. Phys. **72**, 969 (1995).
7. A. Damascelli, Z. Houssain and Z.X. Shen, Rev. Mod. Phys. **45**, 473 (2003)
8. G. E. Volovik, JETP Lett. **58**, 496 (1993).
9. K.A. Moler *et al.*, Phys. Rev. Lett. **72**, 2744 (1994).
10. S.J. Chen et al, Phys. Rev. B **58** 14753(R), (1998).
11. S. Nishizaki, Y. Maeno and Z. Mao, J. Phys. Soc. Jpn. **69**, 573 (2000).
12. H. Won and K, Maki, Europhys. Lett. **52**, 427 (2000).
13. C. Kübert and J. P. Hirschfeld, Phys. Rev. Lett. **80**, 4963 (1998)
14. H. Won and K. Maki, cond-mat/0004105.
15. I. Vekhter, J.P. Carbotte and E.J. Nicol, Phys. Rev B **69**, 7123 (1999).
16. T. Dahm, K. Maki and H. Won, cond-mat/0006307.
17. H. Won, S. Haas, D. Parker, S. Telang, A. Ványolos and K. Maki, in "*Lectures on the physics of highly correlated electron systems IX*" edited by A. Avella and F. Mancini, API conference proceeding 789 (Melville,NY 2005) pp3-43.
18. K. Izawa et al, Phys. Rev. Lett. **86**, 2653 (2001).
19. K. Izawa et al, Phys. Rev. Lett. **87**, 57002 (2001).
20. K. Izawa et al, Phys. Rev. Lett. **88**, 027002 (2002).
21. K. Izawa et al, Phys. Rev. Lett. **89**, 137006 (2002).
22. K. Izawa et al, Phys. Rev. Lett. **90**, 117001 (2003).

23. K. Maki, S. Haas, D. Parker, H. Won, K. Izawa and Y. Matsuda, Europhys. Lett. **68**, 720 (2004).
24. B. Lussier, B. Ellman and L. Taillefer, Phys. Rev. Lett. **23**, 3294 (1994).
25. H. Tou, Y. Kitaoka, K. Asayama, N. Kimura, Y. Onuki, Y. Yamamoto and K. Maegawa, Phys. Rev. Lett. **77**, 1374 (1996).
26. A.P. Mackenzie and Y. Maeno, Rev. Mod. Phys. **75**, 657 (2003).
27. B. Dora, K. Maki, A. Virosztek, Europhys. Lett. **62**, 426 (2003).
28. H-Y. Kee, K. Maki, C.H. Chung, Phys. Rev. B **67** 86534(R) (2003).
29. C. Lupien, S.K. Dutta, B.I. Barker, Y. Maeno and J.C. Davis, cond-mat/0503317.
30. M. Kato, H. Suematsu and K. Maki, Physica C **408-410**, 53 (2004).
31. E. Bauer et al, Phys. Rev. Lett. **92**, 027003 (2004).
32. P.A. Frigeri et al, Phys. Rev. Lett. **92**, 097001 (2004).
33. A. Amann, A.C. Mota, M.B. Maple and H.v. Löhneysen, Phys. Rev. B **57**, 3640 (1998).
34. E. Dumont and A.C. Mota, Phys. Rev. B **65**, 144519 (2000).
35. D. Vollhardt and P. Wölfle, "The Superfluid Phases of Helium-3", (Taylor and Francis, New York, 1990).
36. G.E. Volovik, "Exotic Properties of Superfluid Helium", (World Scientific Pub. Co., Singapore, 1991).
37. R.H. Haffner et al, Phys. Rev. Lett. **65**, 2816 (1990).
38. K. Maki and P. Kumar, Phys. Rev. B **17**, 1088 (1978).
39. K. Maki in "Solitons", edited by S.E. Trullinger, V.E. Zakharov and V.L. Pokrovskii (North-Holland, Amsterdam, 1986).
40. M. Sigrist and D.F. Agterberg, Prog. Theor. Phys. **102**, 965 (1999).
41. R.H. Heffner et al, Phys. Rev. Lett. **65**, 2816 (1990).
42. H.-Y. Kee, Y.B. Kim and K. Maki, Phys. Rev. B **62** R 9275 (2000).
43. H. Tou et al, Phys. Rev. Lett. **80**, 3129 (1998).
44. H. Murakami et al, Phys. Rev. Lett. **93**, 167004 (2004).
45. M.T. Heinilä and G.E. Volovik, Physica B **210**, 300 (1995).
46. Ü. Parts et al, Physica B **210**, 311 (1995).
47. M.M. Salomaa and G.E. Volovik, Phys. Rev. Lett. **55**, 1184 (1985).
48. K. Maki, Phys. Rev. Lett. **56**, 1312 (1986).
49. C.-R. Hu and K. Maki, Phys. Rev. B **36**, 6871 (1987).
50. G.E. Volovik in "Vortices in Unconventional Superfluids and Superconductors", edited by G.E. Volovik, N. Schopohl and P.R. Huebner (Springer, Berlin 2002).
51. G. Yang and K. Maki, Euro. Phys. J. B **21**, 61 (2001).
52. H. Won and K. Maki, Phys. Rev. B **49**, 1397 (1994).
53. P.L. Gammel, D.J. Bishop, J.P. Rice and D.M. Ginsburg, Phys. Rev. Lett. **68**, 3343 (1992).
54. B. Keimer et al, Phys. Rev. Lett. **73**, 3459 (1994).
55. I. Maggio-Aprile et al, Phys. Rev. Lett. **75**, 2764 (1995).
56. J.R. Kirtley et al, Appl. Phys. Lett. **66**, 1138 (1995).
57. A. Tonomura et al, Nature **397**, 308 (1999).

MICROSCOPIC STRUCTURE OF VORTICES IN TYPE II SUPERCONDUCTORS

K. MACHIDA, M. ICHIOKA, H. ADACHI, AND T. MIZUSHIMA
Department of Physics, Okayama University, Okayama 700-8530, Japan

N. NAKAI
Yukawa Institute for Theoretical Physics, Kyoto University, Kyoto 606-8502, Japan

P. MIRANOVIĆ
Department of Physics, University of Montenegro, Podgorica 81000, Serbia and Montenegro

We analyze the low temperature specific heat and magnetization in the mixed state to extract signatures which are able to discriminate various pairing symmetries. Microscopic calculations based on quasi-classical and Bogoliubov-de Gennes formalisms are performed. We emphasize the importance of low-energy excitations induced around a vortex core to understand the vortex physics.

1. Introduction

There has been much attention focused on various unconventional superconductors, ranging from high temperature cuprates, organic conductors to heavy Fermion systems. Some of these systems are characterized by higher orbital angular momentum of the relative coordinates for a Cooper pair, such as p-, d-, or f-wave pairings. Unconventional Cooper pairing can also occur when the center-of-mass coordinates acquires a finite momentum, leading to the so-called Fulde-Ferrell-Larkin-Ovchinnikov (FFLO) state. This state has not been firmly identified yet in real materials, but certain materials are now examined in light of this possibility[1].

Here we undertake a study to help identify the unconventional superconductivity through the vortex core structure. The electronic excitation spectrum around a vortex core sensitively reflects the underlying pairing symmetry, in particular the nodal topology of the energy gap. The low lying excitations around a core are bounded if the energy gap is non-vanishing everywhere on the Fermi surface. This is quite contrasted with the nodal gap case where the low lying excitations are extended along the nodal directions. Thus the local density of states (LDOS) behaves differently, depending on the nodal gap structure. These behaviors can be detected experimentally by a variety of methods. In particular LDOS is directly measured by STM-STS experiment. But other thermodynamic measurements can probe these excitation spectrum directly and indirectly.

The purpose of this paper is to provide unified view of series of our works related to these tasks[2–7]. Namely, the field (H) or angle (θ) dependence of the low temperature specific heat under applied field, or the linear specific heat coefficient $\gamma(H,\theta)$, is calculated for various pairing symmetries and various field orientations between the nodal position and the field direction[2–4]. The orientational dependence of the magnetization $M(H,\theta)$ is also evaluated to give valuable information on the pairing symmetry[5].

We also consider the Pauli paramagnetic effect on the vortex structure[6,7], which turns out to give important modification for the above quantities and also provide important information on the underlying pairing form. Since the Pauli effect ultimately leads to FFLO state, we touch upon the vortex structure in FFLO state also[7].

2. Microscopic calculations

The basic strategy to attack the above tasks is to perform the microscopic calculation. The quasi-classical Eilenberger formalism is best suited for these purposes because it is valid for $k_F\xi \gg 1$ with k_F Fermi wave number and ξ the coherent length, where usual superconductors satisfy this requirement. This formalism is easy to take into account various kinds of anisotropic gaps, including point or line nodes.

In the clean limit quasiclassical equations read as

$$\left[2\hbar\omega_n + \hbar\boldsymbol{v}\left(\nabla + \frac{2\pi i}{\Phi_0}\boldsymbol{A}\right)\right] f = 2\Psi(\boldsymbol{r})\Omega(\phi,\theta)g, \tag{1}$$

$$\left[2\hbar\omega_n - \hbar\boldsymbol{v}\left(\nabla - \frac{2\pi i}{\Phi_0}\boldsymbol{A}\right)\right] f^\dagger = 2\Psi^*(\boldsymbol{r})\Omega(\phi,\theta)g. \tag{2}$$

Here $\hbar\omega_n = \pi T(2n+1)$ with integer n are Matsubara frequencies, $\boldsymbol{v}$ is Fermi velocity, Φ_0 is flux quantum, and f, $f^\dagger$, g are Green's functions integrated over energy normalized so that $ff^\dagger + g^2 = 1$. The anisotropic pairing is expressed within the separable model of pairing potential $V(\boldsymbol{k},\boldsymbol{k}') = V_0\Omega(\boldsymbol{k})\Omega(\boldsymbol{k}')$. The order parameter takes the following form: $\Delta(\boldsymbol{r},\boldsymbol{k}) = \Psi(\boldsymbol{r})\Omega(\boldsymbol{k})$. The Fermi surface is assumed to be sphere. Order parameter $\Psi(\boldsymbol{r})$ and vector-potential $\boldsymbol{A}(\boldsymbol{r})$ are obtained selfconsistently from the following equations

$$\Psi(\boldsymbol{r}) \ln \frac{T_c}{T} = 2\pi T \sum_{\omega_n > 0} \left[\frac{\Psi(\boldsymbol{r})}{\hbar\omega_n} - \langle \Omega(\phi,\theta) f \rangle \right], \tag{3}$$

$$\nabla \times \nabla \times \boldsymbol{A}(\boldsymbol{r}) = -\frac{4\pi^2 \hbar N_0 T}{\Phi_0} \operatorname{Im} \sum_{\omega_n > 0} \langle g \boldsymbol{v} \rangle . \tag{4}$$

Average over Fermi surface is denoted as $\langle \ldots \rangle$.

The other formalism of Bogoliubov-de Gennes equation (BdG) we employ is also useful to describe spatial varied superconducting order parameter, which is the case of FFLO. Namely, the Bogoliubov-de Gennes (BdG) equation for the quasi-particle wave functions $u_{\mathbf{q}}(\mathbf{r})$ and $v_{\mathbf{q}}(\mathbf{r})$ labeled by the quantum number $\mathbf{q}$:

$$\begin{bmatrix} \mathcal{K}_\uparrow - \mu_\uparrow & \Delta(\mathbf{r}) \\ \Delta^*(\mathbf{r}) & -\mathcal{K}_\downarrow^* + \mu_\downarrow \end{bmatrix} \begin{bmatrix} u_{\mathbf{q}}(\mathbf{r}) \\ v_{\mathbf{q}}(\mathbf{r}) \end{bmatrix} = \varepsilon_{\mathbf{q}} \begin{bmatrix} u_{\mathbf{q}}(\mathbf{r}) \\ v_{\mathbf{q}}(\mathbf{r}) \end{bmatrix}, \tag{5}$$

where $\mathcal{K}_\sigma = -\frac{\hbar^2}{2m}\nabla^2$ with the self-consistent equation

$$\Delta(\mathbf{r}) = g \sum_{|\varepsilon_q| \le \omega_D} u_{\mathbf{q}}(\mathbf{r}) v_{\mathbf{q}}^*(\mathbf{r}) f(\varepsilon_{\mathbf{q}}), \tag{6}$$

where $f(\varepsilon_{\mathbf{q}}) = 1/(e^{\varepsilon_{\mathbf{q}}/k_B T} + 1)$ is the Fermi-distribution function. The coupling constant g is negative and ω_D is the energy cutoff. Here the Pauli paramagnetic effect is included as the shifted chemical potentials μ_σ.

3. Orientational dependence of specific heat

The specific heat C at low T is one of fundamental thermodymamic quantities which characterize a superconductor of interest. The Sommerfeld coefficient $\gamma(H) = C/T$ at lower T is directly related to the zero-energy DOS $N(E = 0, H)$ obtained from the integration of LDOS over the vortex unit cell. When the external field H is rotated within the horizontal plane relative to the line node running vertically on the Fermi sphere by an angle θ, $N(H,\theta)/N_0$ (N_0 the DOS in the normal state) oscillates periodically. Its maxima (minima) are located at the anti-nodal (nodal) direction as seen from Fig.1(a). As H increases, the amplitude decreases which is

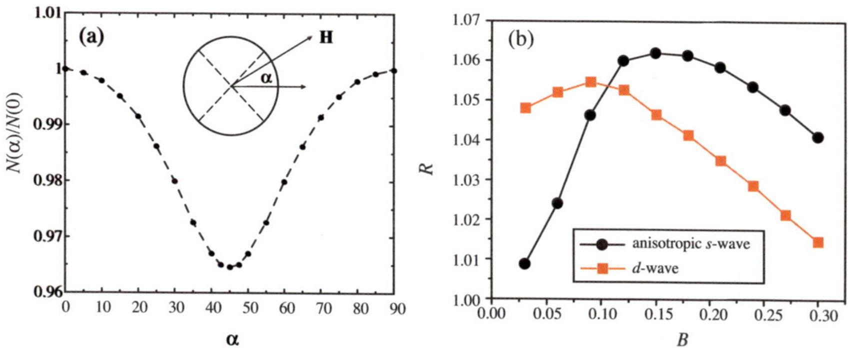

Figure 1. (a) Oscillation pattern for d-wave case where the line nodes (dashed lines) and angle α are shown in inset. (b) Field dependence of the oscillation amplitudes for two type gap structures.

displayed in Fig. 1(b). Towards $H \to 0$ the amplitude stays constant. The maximum amplitude of $N(H,\theta)/N_0$ is a few percent and the oscillation pattern is nearly sinusoidal as shown in Fig. 1(a). However, these characteristics are altered by the detailed gap structure near the node. Namely, the anisotropic s wave case also yields an oscillation in $N(H,\theta)$, but the amplitude decreases towards $H \to 0$ as seen from Fig. 1(b). Thus a careful specific heat measurement $\gamma(H,\theta)$ gives a powerful way to locate the nodal position or the gap anisotropy on the Fermi surface. Thermal conductivity $\kappa(H,\theta)$ at low T also yields similar information on the gap structure, but includes the scattering time effects, which sometimes hamper a clear conclusion.

It should be noticed that the oscillation in $\gamma(H,\theta)$ is also induced by the anisotropy of the Fermi velocity even when the gap is isotropic. In this case the oscillation amplitude tends to decrease towards $H \to 0$. Thus we can distinguish the origin of the oscillation in $\gamma(H,\theta)$ in these two cases.

4. Anisotropic diamagnetic response

The diamagnetic magnetization $M_{dia}(H,\theta)$ gives rise to valuable information on the gap structure when the external field is rotated. An advantage of $M_{dia}(H,\theta)$ over $\gamma(H,\theta)$ is that we can measure it not only along H, but also along T in H vs T plane. It is observed that in YNi_2B_2C that the oscillation pattern of the longitudinal component of $M_{dia}(H,\theta)$ changes its sign when H varies, keeping T fixed and vice versa[8].

We analyze this behavior in light of possible gap and Fermi velocity anisotropies to determine the nodal structure of this material[5]. Namely, the longitudinal diamagnetic magnetization is calculated by Eilenberger equation for various combinations of these two anisotropies. In fact we find a line in H vs T plane which signals the sign change of the oscillation pattern in $M_{dia}(H,T,\theta)$ for the nodal gap case with certain Fermi velocity anisotropy. The above line for sign change is favorably compared with the experimental data of YNi_2B_2C[8] which is known to have a large gap anisotropy.

We remark that compared with two quantities $\gamma(H,\theta)$ and $M_{dia}(H,T,\theta)$, the former is more directly connected to the anisotropic quasi-particle excitations than the latter. Thus the former is easier to extract the information on the gap topology, which is otherwise buried in the Fermi velocity anisotropy.

5. Field dependence of $\gamma(H)$

The field dependence of $\gamma(H)$, which is proportional to $N(E=0,H)/N_0$, reflects the low-lying excitations around a vortex core whose spatial structure is described by LDOS. It is known that $\gamma(H)\propto\sqrt{H}$ for nodal superconductors in $H\ll H_{c2}$ while $\gamma(H)\propto H$ for isotropic ones. We evaluate precisely the field dependence $\gamma(H)$ for the gap function $\Omega(\theta)=(1-\alpha\cos 6\theta)/\sqrt{1+\alpha^2/2}$, which interpolates between the isotropic gap ($\alpha=0$) and the nodal gap ($\alpha=1$) continuously. Here we notice that the Fermi velocity anisotropy is more or less integrated out in the spatially averaging process.

The DOS $N(H)/N_0$ vs H/H_{c2}, which is displayed in Fig. 2(a), is found as follows[4]:

(1) $N(H)/N_0$ is indeed linear in H/H_{c2} for $\alpha\neq 1$.

(2) The coefficient of this linear term increases as α increases.

(3) At $\alpha=1$, $N(H)/N_0\propto\sqrt{H}$.

(4) For $\alpha\neq 0,1$ $N(H)/N_0$ is linear in H/H_{c2} for $H<H^*$ and becomes non-linear for $H>H^*$.

(5) The cross-over H^* is a good indicator for the α value, that is, the gap anisotropy α is uniquely assigned by H^*.

(6) $H^*/H_{c2}\simeq$ for $\alpha=0$ and 0 for $\alpha=1$.

If there is a finite minimum energy gap, $N(H)/N_0$ or $\gamma(H)$ has a certain linear field region up to H^*, beyond which it becomes non-linear towards H_{c2}. Thus the precise determination of $\gamma(H)$ is important to know existence or non-existence of the minimum gap.

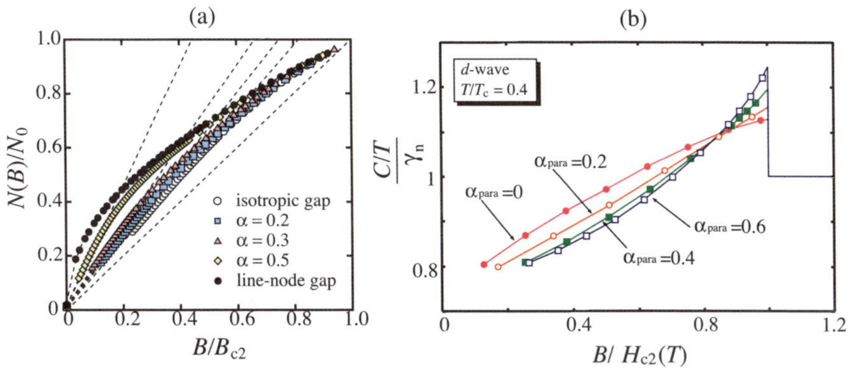

Figure 2. Field dependence of the low T specific heat. Results for various gap structures (a) and for the nodal case with the Pauli paramagnetic effect (b).

It has been known for sometime that in certain heavy Fermion superconductors, such as UBe_{13}[9], $CeCoIn_5$[10] and Sr_2RuO_4[11], the $\gamma(H)$ curve is concaved, showing a positive curvature. This is quite contrasting with the above categories where the $\gamma(H)$ curves have always a negative curvature. This is anomalous because as H increases vortices are entering into a system, resulting in accumulating the zero-energy DOS per vortex. Therefore we naively expected that $N(H)/N_0 \geq H/H_{c2}$. In order to resolve this paradox, we calculate the zero-energy DOS by taking account the Pauli paramagnetic effect into Eilenberger formalism[6].

The results, which are shown in Fig. 2(b), are summarized as
(1) The $C/\gamma_n T$ curves have indeed a positive curvature as the Maki parameter α_{para}, which indicates the relative strength of the paramagnetic effect, increases.
(2) This downward concave behavior occurs both for isotropic gap and nodal gap cases.
(3) Since α_{para} is approached to a critical value, the second order phase transition at H_{c2} changes into a first order one and simultaneously H_{c2} is suppressed. Thus the above downward concave curve in $N(H)/N_0$ is understood physically as non-linear mapping process from high field to low field.

According to a thermodynamic Maxwell relation

$$\frac{\partial \gamma(H)}{\partial H} = \frac{\partial^2 M(T)}{\partial^2 T} \tag{7}$$

at lower T, the total magnetization is directly related to $\gamma(H)$. Thus if the coefficient $\beta(H) = H^\delta$ in $M(T) = M_0 + \beta(H)T^2$ is an increasing function of H, then $\gamma(H) \propto H^{1+\delta}$ with $\delta > 0$. The isotropic (nodal) gap case without the Pauli effect corresponds to $\delta = 0$ ($\delta = -1/2$), meaning that $\gamma(H) \propto H$ ($\gamma(H) \propto \sqrt{H}$) and $\beta(H) \propto H^0$ ($\beta(H) \propto H^{-1/2}$). Superconductors with Pauli effect corresponds to $\delta \geq 1$. It should be also pointed out that among the total magnetization the paramagnetic component mainly responsible for this anomalous curvature in $\gamma(H)$[6].

6. Fulde-Ferrell-Larkin-Ovchinnikov state

When the Maki parameter α_{para} is further increased beyond a critical value where the Pauli paramagnetic effect dominates the orbital depairing effect and H_{c2} is limited by the former effect, the Fulde-Ferrell-Larkin-Ovchinnikov (FFLO) state becomes energetically more stable than the usual uniform BCS state. The spatially modulated FFLO accompanies the spatially varying moment, which is induced spontaneously. The interplay between FFLO and vortices is a quite interesting object to study[7].

It is believed on physical reasons that the modulation vector of FFLO is directed along the external field direction. Therefore a vortex along the z-direction intersects with the (x, y) plane at the origin. Thus on the (x, y) plane the order parameter vanishes, forming a nodal plane. The order parameter profile is shown in Fig.3(a). The nodal plane at $z = 0$ accommodates excess carriers with up-spin because the phase of the order parameter changes by π. This is also true for the vortex line where quasi-particles experience the same π shift when they across the vortex line with 2π phase winding. However, when they across the intersection point, namely, the origin with a finite angle to the nodal plane, the total phase shift they feel

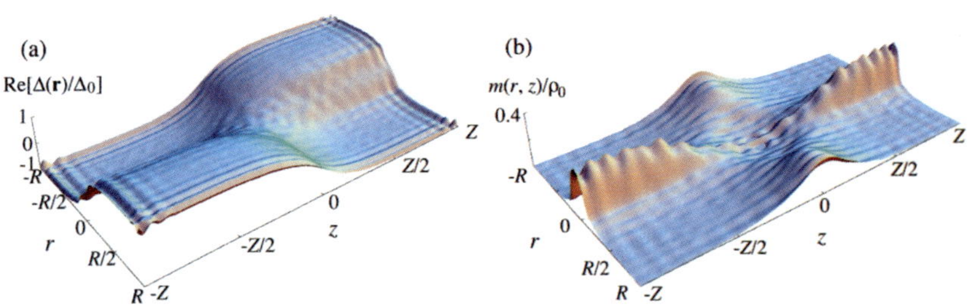

Figure 3. The order parameter (a) and the induced magnetization (b) profiles in FFLO state with single vortex.

is $\pi+\pi$, one coming from phase winding and the other from the nodal plane. Thus there is no π-shift for these quasi-particles, resulting in the void at the intersection point of the origin. The magnetization profile is shown in Fig.3(b). The spatial structure of the induced moment is peculiar, but topologically robust.

We can confirm this topological consideration by solving BdG equation in a single vortex case with isotropic gap, including the Pauli paramagnetic effect. This enable us to understand these peculiar state in terms of low-lying excitation spectrum microscopically[7].

7. Conclusion

We have given a unified view to help identify the pairing symmetry through vortex studies, emphasizing that the low-energy excitations around a core play a key role for this task.

References

1. K. Kakuyanagi, M. Saitoh, K. Kumagai, S. Takashima, M. Nohara, H. Takagi and Y. Matsuda, *Phys. Rev. Lett.* **94**, 047602 (2005) and references cited therein.
2. P. Miranović, N. Nakai, M. Ichioka and K. Machida, *Phys. Rev.* **B68**, 052501 (2003).
3. P. Miranović, M. Ichioka, K. Machida and N. Nakai, *J. Phys. Condens. Matter* **17**, 7971 (2005).
4. N. Nakai, P. Miranović, M. Ichioka and K. Machida, *Phys. Rev.* **70**, 100503 (2004).
5. H. Adachi, P. Miranović, M. Ichioka and K. Machida, *Phys. Rev. Lett.* **94**, 067007 (2005).
6. H. Adachi, M. Ichioka and K. Machida, *J. Phys. Soc. Jpn.* **74**, 2181 (2005).
7. T. Mizushima, M. Ichioka and K. Machida, *Phys. Rev. Lett.* **94**, 060404 (2005) and *Phys. Rev. Lett.* **95**, 117003 (2005)
8. L. Civale, A. V. Silhanek, J. R. Thompson, K. J. Song, C. V. Tomy and D. McK. Paul, *Phys. Rev. Lett.* **83**, 3920 (1999). V. G. Kogan, S. L. Bud'ko, P. C. Canfield and P. Miranović, *Phys. Rev.* **B60**, R12577 (1999).
9. A.P. Ramirez, C.M. Varma, Z. Fisk and J.L. Smith, *Phil. Mag.* **79**, 111 (1999).
10. H. Aoki, T. Sakakibara, H. Shishido, R. Settai, Y. Onuki, P. Miranović and, K. Machida, *J. Phys. Condens. Matter* **16**, L13 (2004).
11. K. Deguchi, Z.Q. Mao and Y. Maeno, *J. Phys. Soc. Jpn.* **75**, 1313 (2004).

MICROSCOPIC NEUTRON INVESTIGATION OF THE ABRIKOSOV STATE OF HIGH-TEMPERATURE SUPERCONDUCTORS*

J. MESOT

Laboratory for Neutron Scattering
ETH Zurich and Paul Scherrer Institute
CH-5232 Villigen-PSI, Switzerland
E-mail: joel.mesot@psi.ch

Using Small Angle Neutron Scattering we have been able to observe for the first time a well defined Vortex lattice (VL) structure both in the hole-doped LSCO and electron-doped NCCO superconductors. Our measurements on optimally doped LSCO reveal the existence of a magnetic field-induced phase transition from a hexagonal to a square coordination of the VL. Various scenarios to explain such phase transition are presented. In NCCO as well a clear square VL could be detected, which is unexpectedly kept down to the lowest measurable magnetic fields.

1. Introduction

Apart from the unusual electronic and magnetic behavior of the cuprate high-temperature superconductors (HTSC), experiments reveal a tremendously rich variety of mesoscopic phenomena associated with the flux vortices in the mixed state [1]. Because of their two-dimensional electronic structure, the HTSC are highly anisotropic. The anisotropy is characterized by the ratio $\gamma = \lambda_\perp/\lambda_\parallel$, where $\lambda_\perp, \lambda_\parallel$ are the superconducting penetration depths for currents flowing perpendicular and parallel to the two-dimensional CuO_2 planes. In $La_{2-x}Sr_xCuO_{4+\delta}$ (LSCO) the degree of anisotropy ($\gamma = 20$ for $x = 0.15$) lies between that of the $YBa_2Cu_3O_x$ (YBCO) and $Bi_2Sr_2CaCu_2O_{8+x}$ materials [2]. The cuprates are also extreme type-II superconductors, indicated by the high value of the Ginzburg-Landau parameter $\kappa = \lambda/\xi$, where ξ is the superconducting coherence

*This work is partially supported by the 21COE program on Topological Science and Technology from the Ministry of Education, Culture, Sport, Science and Technology of Japan, and by the Swiss National Science Foundation.

length. In HTSC the combination of high transition temperature T_c, high γ and high κ leads to exotic vortex behavior, such as the phenomenon of vortex lattice (VL) melting (see Figure 1 and Ref.[1]). For a conventional (isotropic) pairing mechanism such anisotropic conduction properties can lead to distortions of the vortex lattice as the applied field is tilted towards the CuO_2 planes, but the local coordination remains sixfold [3].

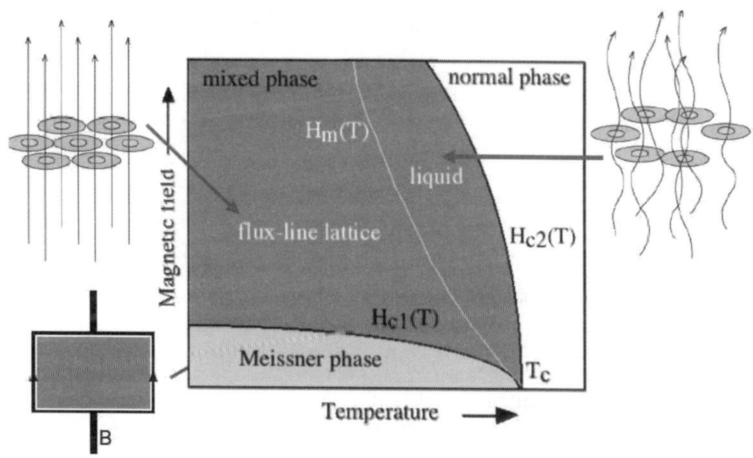

Figure 1. Schematic magnetic phase diagram of HTSC. For more details, see Ref.[1].

In the present paper, recent neutron results obtained in the Abrikosov phase of the hole-doped LSCO and electron-doped $Nd_{2-x}Ce_xCuO_4$ (NCCO) HTSC as a function of doping are presented. The very high-quality LSCO crystals have been grown by the Traveling Solvent Floating Zone (TSFZ) technique in the group of Profs N. Monomo, M. Oda and M. Ido, Hokkaido University, while the NCCO crystals were grown in the group of Prof. K. Yamada, Tohoku University.

2. Vortex Lattice in Optimally (Hole)-Doped LSCO

Using small angles neutron scattering (SANS) we have succeeded, for the first time, to measure a well-ordered vortex lattice (VL) structure at all doping regimes of LSCO. In the optimally to overdoped regime a field-induced transition from hexagonal to square coordination is reported around $H = 0.4$ T (see Fig.2 and Ref.[4]) with the square lattice oriented

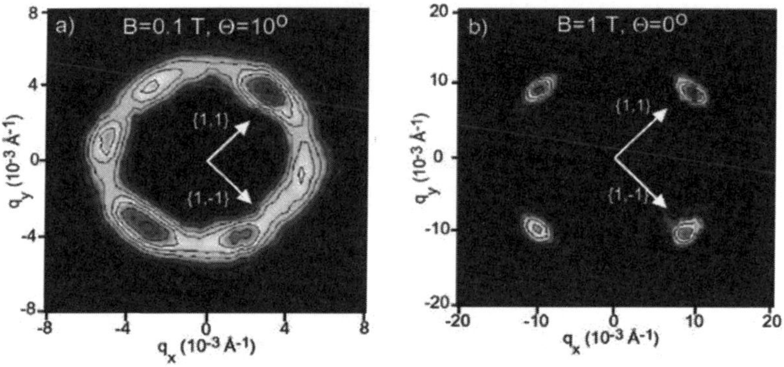

Figure 2. Vortex lattice measured in the superconducting phase of optimally doped LSCO a) at H=0.1 T (hexagonal structure) and b) H=1 T (square structure) [4].

along the anti-nodal direction of the d-wave superconducting gap[5]. In a recent experiment, Brown *et al.*[6] have observed a similar phase transition in the YBCO superconductor, however with two main differences: first, the critical field at which the transition occures is at least an order of magnitude higher than in LSCO; second, the square VL in YBCO is oriented along the nodal direction of the d-wave gap function.

Various scenarios are able to explain such a hexagonal-to-square field induced transition. It has been proposed that the importance of anisotropic vortex cores in a d-wave superconductor [7] would result in a square VL at high fields. Alternatively, coupling to other sources of anisotropy such as those provided by charge/stripe fluctuations or Fermi velocity anisotropies [8] should as well be considered. While the d-wave scenario favors a square VL aligned along the nodal direction (as observed in YBCO[6]), an anisotropy of the Fermi velocity would result in a VL aligned along the anti-nodal direction (as observed in LSCO[4]). These two competing scenarios are visualized in Figure 3.

In order to lift this apparent contradiction, Nakai *et al.*[8], based on photoemission data [9] suggested that the observed orientation of the square lattice in optimally doped LSCO above $0.4T$ is stabilized by the proximity of a van Hove singularity close to the Fermi level at the $(\pi, 0)$ anti-nodal point of the Brillouin zone. The extreme case of a van Hove singularity coinciding with the Fermi level, would result in a maximum anisotropy of the Fermi velocity along the nodal and anti-nodal directions. Such a strong anisotropy seems to be absent in YBCO, which would explain the different square lattice orientation. Nakai *et al.*[8] furthermore predict that

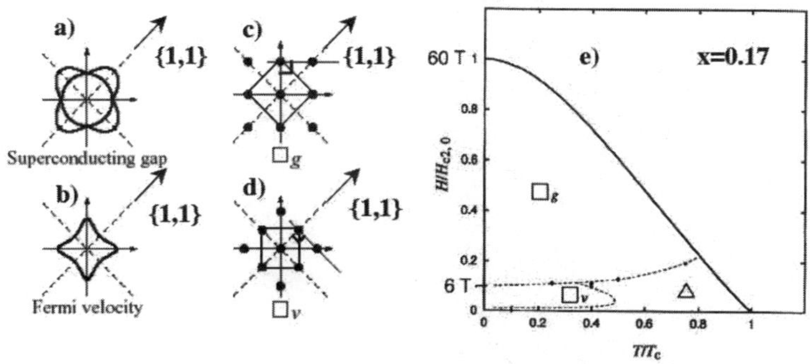

Figure 3. Both d-wave superconducting gap (panel-a) and Fermi velocities (panel-b) anisotropies will result in square lattices (panels c-d), but with different orientations with respect to the gap nodes. The resulting phase diagram as taken from Nakai *et al.* [8] for optimally doped LSCO is given in panel-e.

In LSCO, at high-enough fields ($H > 6T$), a second phase transitions into a square VL oriented along the nodal direction should be observed due to the increasing importance of the d-wave anisotropy at high fields. So far our experiments realised up to 10.5 Tesla [10] did not reveal such a transition, and experiments at even higher field-values are necessary.

3. Vortex Lattice in Underdoped-LSCO

The situation seems to be completely different in the underdoped regime of LSCO since a well ordered hexagonal VL could be observed only at very low fields, see Figure 4 [11]. By combining neutron scattering and muon spin rotation [11] it was concluded that the observed vanishing intensity with increasing field in the underdoped regime of LSCO is due to a transition into a so-called vortex-glass state [12]. Such a state, exemplified in Figure 5, is expected to occur when strong disorder is present.

4. Vortex Lattice in Electron-Doped NCCO

Recently, we performed the first successful measurements of a VL in the electron-doped NCCO system [13]. Here again a square lattice (see Fig. 6) was observed, however down to the lowest measurable magnetic fields. Several scenarios to explain our observation are currently examined.

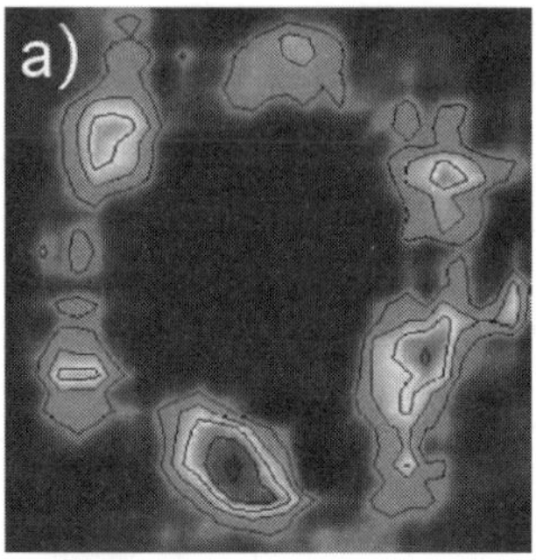

Figure 4. Small angle neutron diffraction pattern showing the existence of an ordered vortex phase at low fields (150 Oe) and temperature (6.2 K). The intensity of this pattern rapidly falls as the field is increased towards the vortex-glass phase [13].

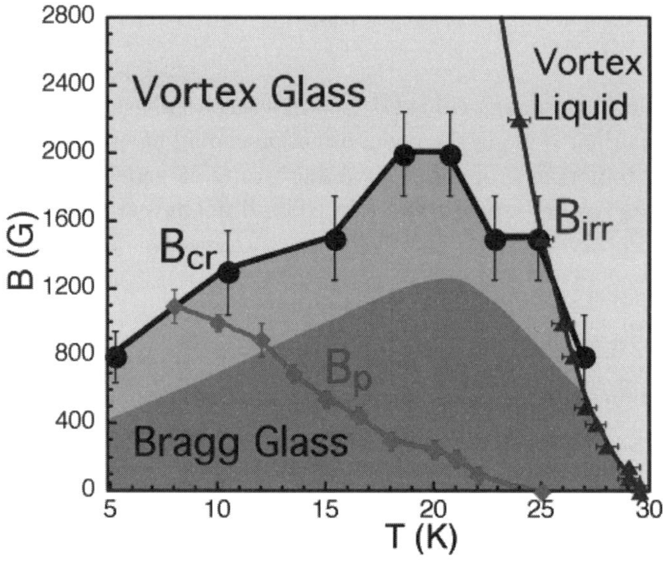

Figure 5. The magnetic phase diagram derived from the changes observed in the μSR field distributions. B_{cr} indicates the onset of the broadening at high field, which is significant only below 25 K, and should be considered as an upper limit for the Bragg-glass to vortex-glass transition. This uncertainty in the exact position of the transition is represented schematically by the shading below the line B_{cr}. B_{irr} indicates the irreversibility line as determined by bulk measurements of the field-cooledzero-field-cooled (FC-ZFC) magnetisation using a SQUID magnetometer. The position of the feature in the magnetisation curves B_{on} is also plotted [11].

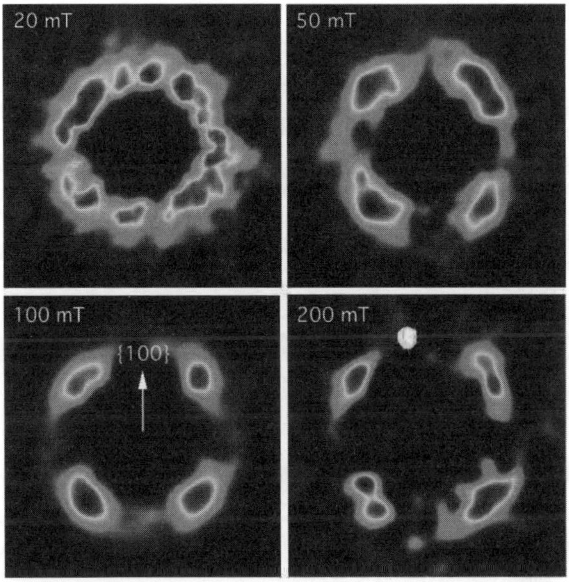

Figure 6. Vortex lattice in the electron doped HTSC NCCO at various magnetic fields between 20 and 200 mT [13].

Acknowledgments

This work would not have been possible without the intense collaboration of R. Gilardi, J. Kohlbrecher (ETH Zurich and Paul Scherrer Institute), U. Divakar, A. Drew, S. Lee (Uni. St-Andrews), S. Brown, E. M. Forgan (Uni. Birmingham), C. Dewhurst, B. Cubitt (Institute Laue-Langevin), N. Monomo, M. Oda and M. Ido (Uni. Hokkaido) and K. Yamada (Tohoku Uni.). The author also thanks K. Machida (Okayama Uni.) and K. Maki (Uni. S. California) for fruitful discussions.

References

1. For a review, see G. Blatter *et al., Rev. Mod. Phys.* **66**, 1125 (1994).
2. T. Sasagawa *et al., Phys. Rev. B* **61**, 1610 (2000).
3. S. L. Thiemann *et al., Phys. Rev. B* **39**, 11 406 (1989); M. Yethiraj *et al., Phys. Rev. Lett* **70**, 857 (1993); S. T. Johnson *et al., Phys. Rev. Lett.* **82**, 2792 (1999); M. Yethiraj *et al., Phys. Rev. Lett.* **71**, 3019 (1993);
4. R.Gilardi *et al., Phys. Rev. Lett.* **88**, 217003 (2002).
5. J. Mesot *et al., Phys. Rev. Lett.* **83**, 840 (1999).
6. S. P. Borwn *et al., Phys. Rev. Lett.* **96**, 067004 (2004).
7. A. J. Berlinsky *et al., Phys. Rev. Lett.* **75**, 2200 (1995); Ji-Hai Xu, Yong Ren,

and Ting Chin-Sen, *Phys. Rev. B* **53**, R2991 (1996); J. Shiraishi, M. Kohmoto, and K. Maki, *Phys. Rev. B* **59**, 4497 (1999); M. Ichioka, A. Hasegawa, and K. Machida, *Phys. Rev. B* **59**, 8902 (1999).
8. N. Nakai *et al., Phys. Rev. Lett.* **89**, 237004 (2002).
9. A.Ino *et al., Phys. Rev. B* **65**, 094504 (2002).
10. R.Gilardi *et al., Int. J. Mod. Phys.* **17**, 3411 (2003).
11. U. Divakar *et al., Phys. Rev. Lett.* **92**, 237004 (2004).
12. T. Giamarchi and P. Ledoussal *Phys. Rev. Lett.* **72**, 1530 (1994).
13. R.Gilardi *et al., Phys. Rev. Lett.* **93**, 217001 (2004).

ENERGY DISSIPATION AT NANO-SCALE TOPOLOGICAL DEFECTS OF HIGH-T_C SUPERCONDUCTORS: MICROWAVE STUDY

A. MAEDA

Department of Basic Sciences,
University of Tokyo, Tokyo 153-8902, Japan
E-mail: cmaeda@mail.ecc.u-tokyo.ac.jp

We investigated energy dissipation of QPs confined in the naturally prepared nano-scale topological defects; vortex core of high-temperature superconductors (HTSC), by using microwave surface impedance measurement techniques. We found a moderately clean nature of the core as a rather universal property of HTSC. This might be related to a novel dissipation mechanism proposed recently theoretically. We also found a sublinear flux flow resistivity in YNi_2B_2C as a function of magnetic field, which is believed to be common to anisotropic superconductors.

1. Introduction

In superconductors under magnetic field, there are many topological defects called as magnetic vortices, which are the quantized flux, $\Phi_0 \equiv hc/2e$ (h, c and e are Planck constant, speed of light and electron charge, respectively), accompanying circulating current. In the center of the vortex (core), whose dimension is about the coherence length of superconductor, $\xi \equiv 2\hbar v_F/\pi\Delta$ ($\hbar$, v_F, Δ are Planck constant, Fermi velocity, and superconducting gap magnitude, respectively), superconductivity is weakened, and quasiparticles (QPs) ($\simeq$ normal electrons) are confined in the core (Fig. 1). Since the coherence length, ξ, is inversely proportional to the critical temperature, T_c, the radius of vortex core of cuprate high-T_c superconductors (HTSCs) is in nano scale. Thus, the vortex core of HTSC is a naturally prepared nano scale topological defect immersed in the host superconductor. QPs confined in this nano-scale topological defect are important, since the energy dissipation of QPs in the core dominates the dynamics of the quantized vortex. This energy dissipation is closely related with the electronic structure of the QPs in the core, such as

$$\rho_f = \frac{B\Phi_0}{\pi\hbar n\Gamma}, \tag{1}$$

$$\Gamma \equiv \frac{\Delta E}{\delta E}. \tag{2}$$

which are the topics of much interest[1] (Figs. 2 and 3). Here, ρ_f, B, and n are flux-flow resistivity, magnetic field, and carrier concentration, respectively, and ΔE and δE are the spacing and the width of the quantized level of the QPs in the vortex core. Several STM experiments suggested that the core of HTSC was close to the quantum limit, where $k_F\xi$ (k_F is Fermi wave number and ξ is the coherence length) is close to unity. This criterion means that the dimension of the wave packet of the QPs in the core is almost equal to the core radius. Energy dissipation by the motion of such quantum core has not been investigated even theoretically, and we may expect new physics to show up.

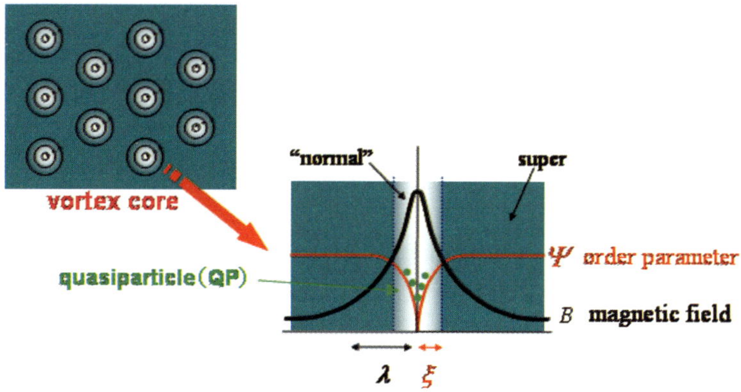

Figure 1. Schematic illustration of vortices of superconductor and the core of a vortex.

Based on these backgrounds, we investigated flux flow of vortices of HTSC at microwave frequencies in untwinned $YBa_2Cu_3O_y$ (YBCO)[2,3] and $Bi_2Sr_2CaCu_3O_y$ (BSCCO)[4], both of which were optimally doped. Also $(La,Sr)_2CuO_4$ (LSCO) with various carrier concentration were investigated[5], since one of the most essential features in physics of HTSC is a variation of physical properties as a function of carrier concentration. We found a moderately clean nature of the core as a rather universal property

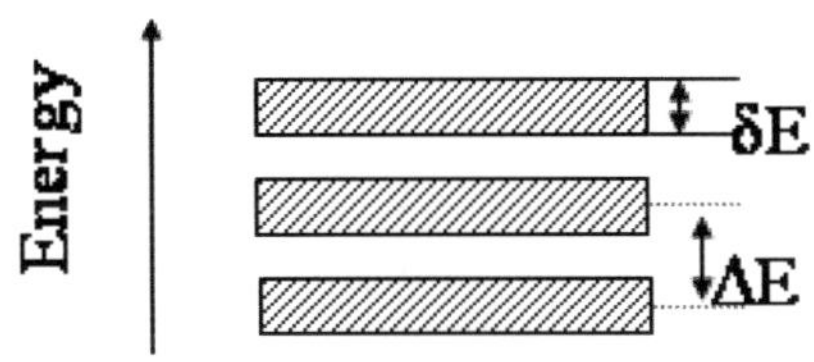

Figure 2. Electronic structure of QPs in the vortex core.

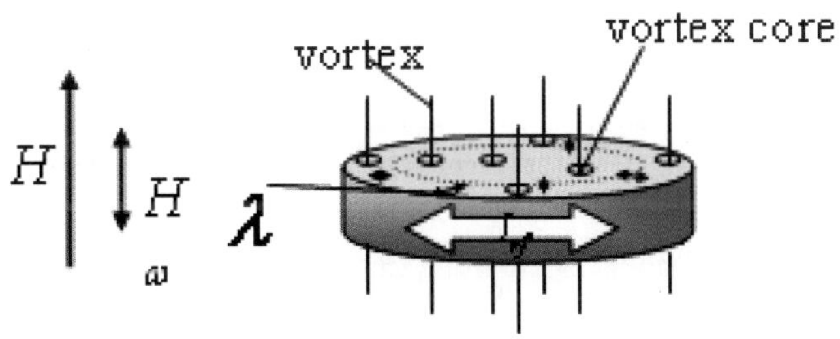

Figure 3. Schematic illustration of vortex shaking by ac electromagnetic field. λ represents the penetration depth.

of HTSC. This might be related to a novel dissipation mechanism proposed recently theoretically.

Another interest is to search common novel features in the flux flow of anisotropic superconductors, where the wave function of superconducting electrons are anisotropic, with nodes in some directions. Thus, we also investigated flux flow of other anisotropic superconductor, YNi_2B_2C[6], and found sublinear flux flow resistivity as a common property of anisotropic superconductors.

2. Experiments

To measure the flux flow (equivalently, viscosity) of a vortex, the most appropriate technique is the complex microwave surface impedance measurement. A small piece of a single crystal was put in the cavity resonator, and the resonance frequency, f, and the Q factor of the resonance were measured. From these raw data, surface impedance $Z_s \equiv R_s + iX_s$ (R_s and X_s are surface resistance and surface reactance, respectively) of the material was obtained as,

$$R_s = G\Delta(\frac{1}{2Q}), \tag{3}$$

$$X_s = G\Delta(\frac{f}{f_0}) + C, \tag{4}$$

where Δ represents the difference between the data with and without sample in the cavity, and f_0 denotes the resonance frequency without the sample, and G and C are constants determined geometrically. The complex conductivity, σ, was extracted from Z_s data as

$$Z_s = (\frac{i\omega\mu}{\sigma})^{1/2}, \tag{5}$$

where ω is the angular frequency and μ is the permeability of the material. With these procedures, the complex conductivity, σ, (or resistivity, $\rho \equiv 1/\sigma$) were obtained at the measured frequency. Detailes on the measurement techniques were described elsewhere[1,2,7]. Since vortices suffer from very strong pinning in HTSC, an appropriate general model[8] was necessary to extract viscosity of vortices[2]. According to Ref. 8, complex penetration depth, $\tilde{\lambda} \equiv Z_s/i\mu\omega$, in the low-temperature limit is expressed as

$$\frac{\tilde{\lambda}}{\lambda_L} = (1 + \frac{r}{1+r^2}b) + i(\frac{r}{1+r^2}b), \tag{6}$$

where λ_L is the London penetration depth, $r \equiv= \omega/\omega_p$ is the normalized frequency by the pinning frequency, ω_p, and $b = \Phi_0 B/\mu\omega\lambda_L^2\eta$ is a normalized magnetic field by some characteristic field including viscosity of vortex, η, as a parameter. Thus, we can obtain η and ω_p by this technique[1,2].

3. Experimental Results and Discussion

Figure 4 shows the surface impedance Z_s of an untwined YBCO, plotted in the real part (R_s) vs imaginary part (X_s). A curve with $r = \infty$ represents a free flux flow, where no pinning works. The data can be fitted with finite r values, which means that pinning is effective even at microwave frequencies. This is very much different from the situation in conventional superconductors. As was mentioned in the previous section, by using a mean-field model for vortiex motion[8], we can extract the vortex viscosity, η, without ambiguity, which is shown in Fig. 5, together with pinning constants, ω_p. This definitely shows that the vortex core of these materials was moderately clean ($k_F\ell \sim 1$ (ℓ is mean free path)). We found that the

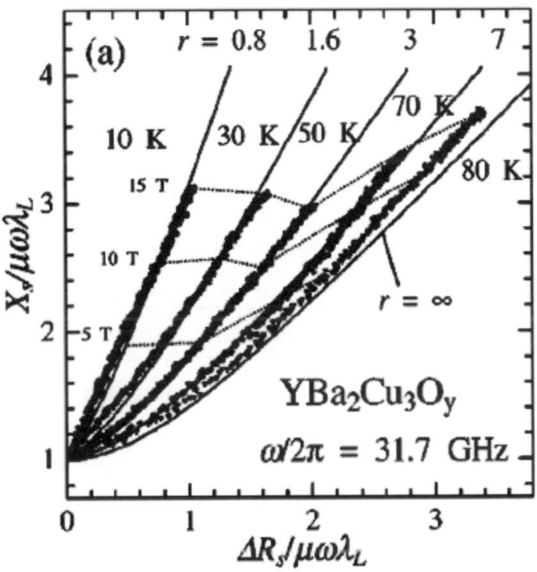

Figure 4. Surface impedance of YBCO as a function of magnetic field at various temperatures. See the text for details (Ref. 2).

moderately clean nature was rather robust when doping was changed, by investigating series of LSCO crystals with different carrier concentration[5]. On the other hand, STM studies suggested that almost no electronic states exist in the vortex core, which contradicts the moderately clean nature obtained by the microwave experiments.

To resolve this discrepancy, we prepared series of untwined YBCO crystals doped by Zn and Ni, and investigated the viscosity of a vortex core (flux flow resistivity)[3]. Since Ni and Zn act in very different manners for the creation of electronic states in the superconducting state, we expect big differences between the flux flow of Zn-doped and Ni-doped crystals if there are almost no QP states in the vortex core of pristine crystals, as was suggested by the STM experiments. However, we found that there was almost no difference in the flux flow of these samples (Fig. 6). We consider that this surprising result can be explained in terms of a recent theoretical suggestion[9] that an extra dissipation takes place in the vortex core of moderately clean superconductor because of the presence of a collective mode in the condensate. Alternatively, this result might suggest that a new physics

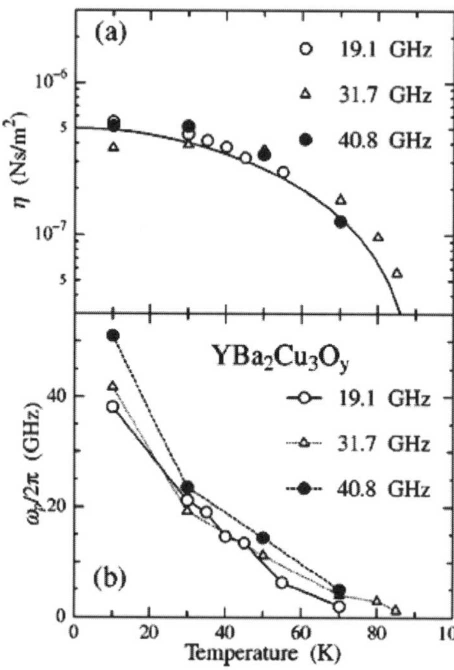

Figure 5. (a) viscosity and (b) pinning constant of vortices in YBCO (Ref. 2).

appears in the energy dissipation in the vortex core that is close to the quantum limit.

Figure 7 shows the flux flow resistivety of YNBC as a function of applied magnetic field[6]. A sublinear behaviour is clearly seen. Together with results in other anisotropic superconductors[10,11], we believe this to be a common, universal property of superconductors with node in the condensate wave function.

4. Conclusion

We investigated the energy dissipation of QPs confined in the naturally prepared nano-scale topological defects; vortex core of HTSC, by using microwave surface impedance measurement techniques. We found a moderately clean nature of the core as a rather universal property of HTSC. This might be related to a novel dissipation mechanism proposed recently theoretically. We also found a sublinear flux flow resistivity in YNBC,

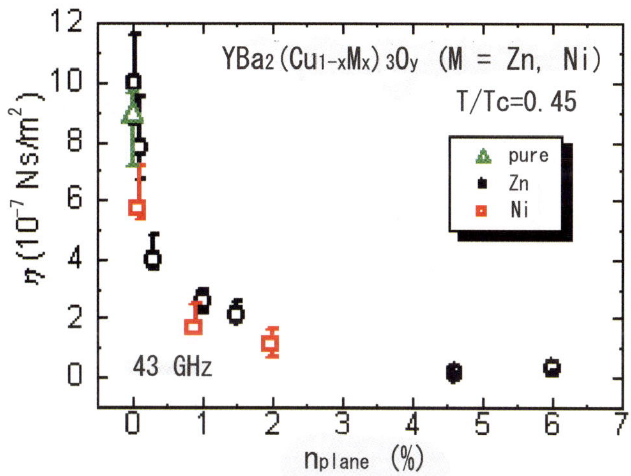

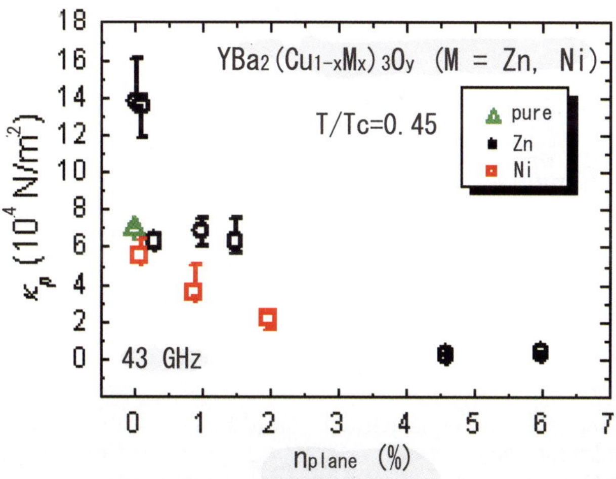

Figure 6. η (upper panel) and ω_p (lower panel) of impurity doped YBCO as a function of impurity concentration, n_{plane}, in a CuO_2 plane. Note that $n \simeq (1/3)n_{\rm nominal}$ for Ni doped crystals.

which are believed to be common to anisotropic superconductors.

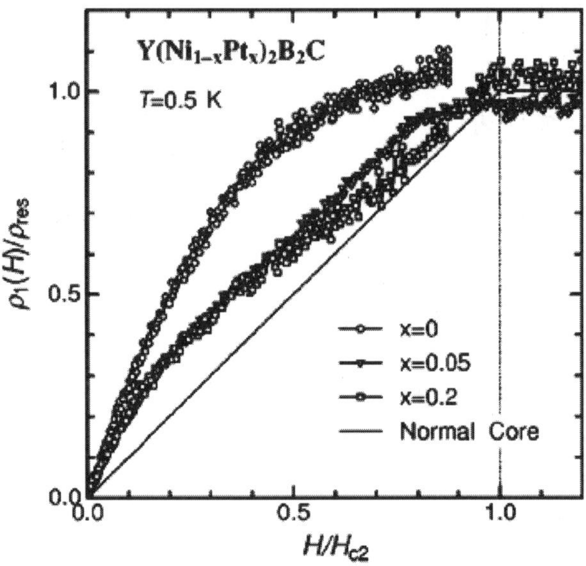

Figure 7. Flux flow resistivity of YNBC (Ref. 6).

Acknowledgments

This work was in collaboration with T. Hanaguri, H. Kitano, Y. Tsuchiya, T. Umetsu, K. Takaki, T. Nishizaki, N. Kobayashi, H. Takagi, and M. Nohara.

References

1. For a review, A. Maeda, H. Kitano, and R. Inoue : *J. Phys. Condens. Matter.* **17** R143-R185 (2005).
2. Y. Tsuchiya *et al.*, *Phys. Rev.* **B63**, 184517 (2001).
3. K. Kinoshita *et al.*, *Physica* **C412-414**, 530 (2004), and *submitted.*
4. A. Maeda *et al.*, *Physica* **C362**, 127 (1991)
5. T. Umetsu *et al.*, *unpublished.*
6. K. Takaki *et al.*, *Phys. Rev.* **B66**, 184511 (2002).
7. T. Hanaguri *et al.*, *Rev. Sci. Inst.* **74**, 4436 (2003).
8. M. W. Coffey and J. R. Clem, *Phys. Rev. Lett.* **67**, 386 (1991).
9. M. Eschrig, J. A. Sauls and D. Rainer, *Phys. Rev.* **B60**, 10447 (1999).
10. S. Kambe *et al.*, *Phys. Rev. Lett.* **83**, 1842 (1999).
11. Y. Matsuda *et al.*, *Phys. Rev.* **B66**, 014527 (2002).

PRESSURE INDUCED TOPOLOGICAL PHASE TRANSITION IN THE HEAVY FERMION COMPOUND CEAL$_2$

H. MIYAGAWA, M. OHASHI, G. OOMI

Department of Physics, Kyushu University,
4-2-1 Ropponmatsu, Chuo-ku, Fukuoka 810-8560, Japan
E-mail: ohashi@rc.kyushu-u.ac.jp

I. SATOH, T. KOMATSUBARA

IMR, Tohoku University,
Katahira, Aoba-ku, Sendai, 980-8577, Japan

N. MIYAJIMA, T. YAGI

ISSP, University of Tokyo,
5-1-5 Kashiwanoha, Kashiwa, 277-8581, Japan

The X-ray diffraction measurement has been measured on $CeAl_2$ under pressure up to 30 GPa. Although there was no pressure induced structure transition in the pressure range, An anomaly was observed on the compression curve. Detailed comparison was made with a band structure calculations, and investigated that the Lifshitz nature of the topological transition is responsible for the anomaly in the compression curve

1. Introduction

$CeAl_2$ has been well known as a prototype of heavy fermion Kondo compound, which has a large coefficient of the specific heat, $\gamma = 140$ mJ/mol K^2 and $logT$ temperature dependence of the electrical resistivity [1]. These characteristic properties are mainly originated from the existence of unstable $4f$ electrons and the hybridization between local f electrons and conduction band electrons. Since the magnitude of hybridization is affected strongly by an application of external pressure and a lot of interesting phase transition is expected to induce, the investigation of these compounds under application of external forces like high pressure has been carried out extensively by many authors[2]. In the present work, we report the occurrence of the pressure induced phase transition in $CeAl_2$ by the changes occurring electronic structure.

2. Experimental

A single crystal of $CeAl_2$ was grown by using Czochralski pulling method. The sample was found to be of single phase and the lattice parameter of cubic Laves phase in $CeAl_2$ was determined to be $a = 8.055 \pm 0.002$ Åby the X-ray diffraction analysis at ambient pressure. The powdered sample was loaded in a 250 μm hole of preindented stainless steel gasket of the diamond anvil cell. A 4:1 methanol: ethanol mixture or helium gas was used as the pressure-transmitting medium. Ruby fluorescence was used for the estimation of pressure. X-ray diffraction measurement was made at room temperature using Mo Kα radiation from a 5.4 kW Rigaku rotating anode generator equipped with a 100 μm collimator. An image plate (IP) was used as the detector. The d-values were obtained by recording the diffraction.

3. Results and discussion

Figure 1 shows the diffraction patterns at several pressures by using a 4:1 methanol: ethanol mixture as the pressure-transmitting medium. At high pressure all the diffraction peaks shifted to higher 2θ. The dependence of the d-spacings on pressure could be followed up to 30 GPa. There is no pressure induced structure transition in the pressure range, and the cubic structure is stable up to 30 GPa at room temperature.

The half width increases as increasing pressure for all the diffraction peaks, indicating that it comes from uniaxial stress by freezing of the pressure-transmitting medium. In order to show the effect of the hydrostaticity of the pressure- transmitting medium, A high-pressure powder X-ray diffraction experiment has been carried out with He pressure-transmitting medium which achieves the best hydrostatic conditions.

Figure 2 shows the diffraction patterns by using helium gas as the pressure-transmitting medium. It is noted that all of the diffraction peaks are sharp compared with Fig. 1, and that the (331) and (422) reflection is clearly observed even at the highest pressure of 23.3 GPa. It means that good hydrostaticity is retained compared with that by using a 4:1 methanol: ethanol mixture.

The lattice parameter a is determined mainly from the reflections (111), (200) and (311). Figure 3 shows a as a function of pressure. Although sev-

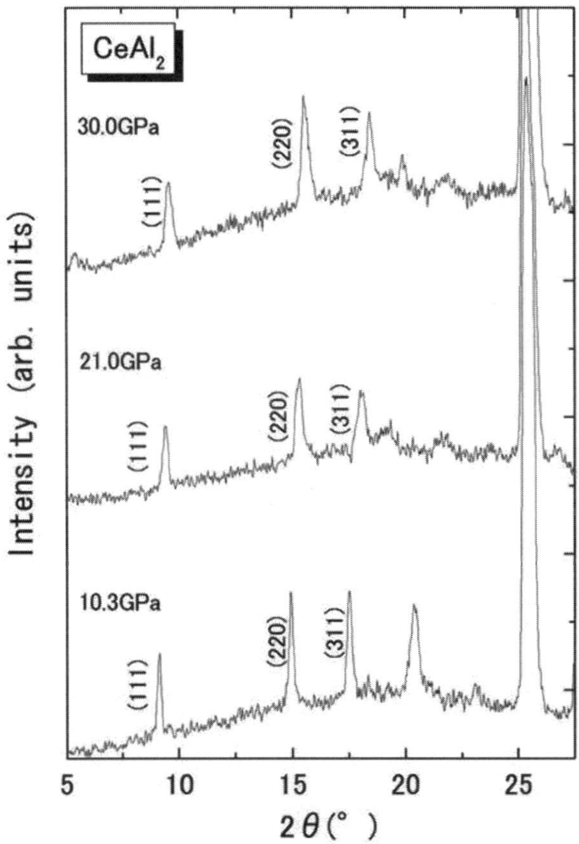

Figure 1. X-ray diffraction pattern of $CeAl_2$ at room temperature at different pressures by using a 4:1 methanol: ethanol mixture as the pressure- transmitting medium. Indices of some of the prominent peaks are also shown. A sharp peak is observed at $2\theta \sim 26°$ the diffraction patterns at every pressures. It comes from the diffraction of the diamond.

eral independent groups studied on the compression curve of $CeAl_2$, two different results have been reported, namely there is a discontinuous change in [3] but no discontinuous change is observed [4,5]. In the present work, a decreases monotonically with increasing pressure and no discontinuous change is observed within experimental error in support of the recent reports [4,5].

The new result is that an anomaly exists near 17 GPa on the compression curve. Takemura investigated in Zn that the anomaly on the compression curve is induced by the nonhydrostaticity associated with the solidi-

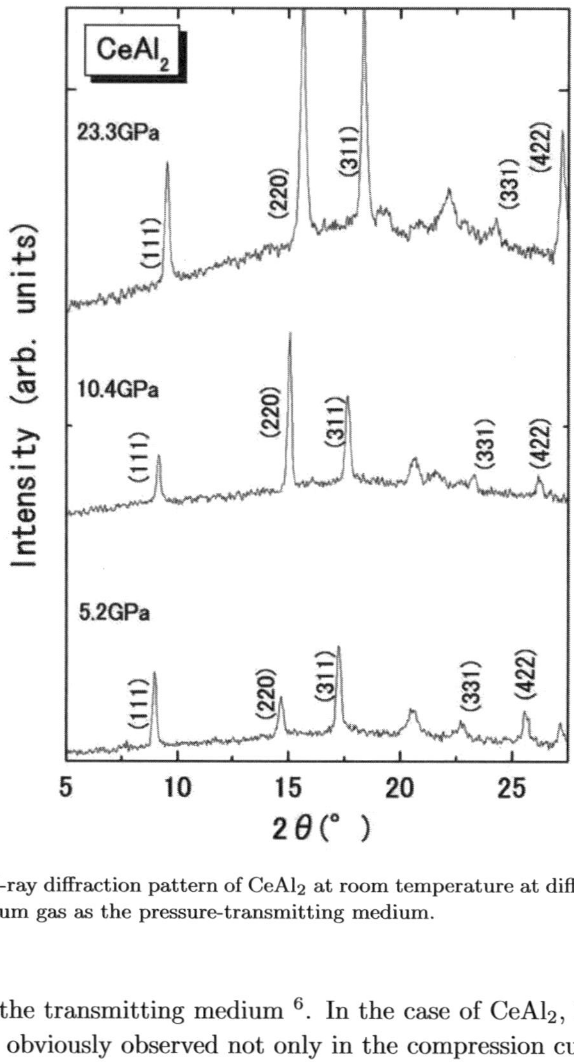

Figure 2. X-ray diffraction pattern of $CeAl_2$ at room temperature at different pressures by using helium gas as the pressure-transmitting medium.

fication of the transmitting medium [6]. In the case of $CeAl_2$, however, the anomaly is obviously observed not only in the compression curve by using a 4:1 methanol: ethanol mixture, but also by using helium as the best hydrostatic conditions. It means that the anomaly does not come from the nonhydrostaticity but from the pressure induced phase transition.

It is well known that if the variation of an external parameter like pressure causes changes in the Fermi surface topology, it is likely to influence the otherwise monotonic variation of density of states [7]. Chandra Shekar *et al.*, performed band structure calculations, and investigated that the electronic transition seen in this pressure range could be the cause for the

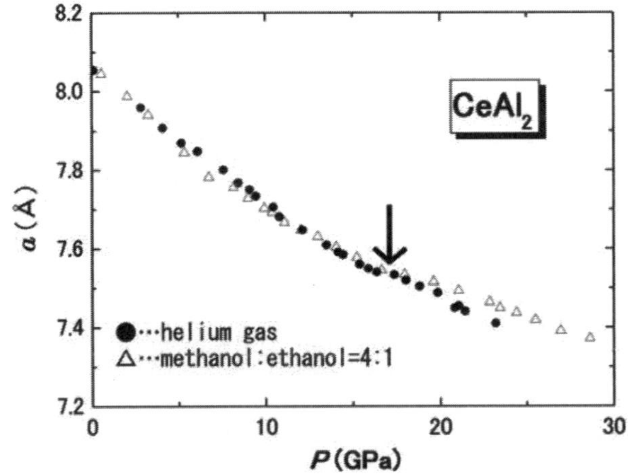

Figure 3. The lattice parameter a of $CeAl_2$ as a function of pressure.

anomaly seen in the compression curve [5].

Moreover, an anomaly is also observed in the pressure dependence of the electrical resistivity near 17 GPa [8]. It indicates the possibility of some significant electronic changes near the pressure-induced phase transition. Indeed, in the case of $AuIn_2$, the Lifshitz nature of the topological transition is responsible for the anomaly in the high-pressure electrical resistivity[9].

References

1. Y. Onuki, Y. Furukawa and T. Komatsubara, *J. Phys. Soc. Jpn* **53**, (1984) 2734.
2. M. Ohashi, G. Oomi, S. Koiwai, M. Hedo, Y. Uwatoko, *Phys. Rev. B***68**, (2003), 144428., M.Ohashi, N. Takeshita, H. Mitamura, T. Matsumura, T. Suzuki, T. Mori, T. Goto and N. Môri, *J. Magn. Magn. Mater.* **226**, (2001) 158.
3. M. Croft, A. Jayaraman, *Solid State Commun.* **29** (1979) 9.
4. B. Barbara, J. Beille, B. Cheiato, J. M. Laurant, M. F. Rossignol, A. Waintal, S. Zemirli, *Physics Letters A* **113** (1986) 381.
5. V. N. V. Chandra Shekar, P. C. Sahu, M. Yousuf, K. G. Rajan, M. Rajagopalan, *Solid State Commun.* **111** (1999) 529.
6. K. Takemura, *Phys. Rev. B* **60** (1999) 6171.
7. I. M. Lifshitz, Soviet *Physics-JETP* **11** (1960) 1130.
8. H. Miyagawa et al., private communication.
9. B. K. Godval, A. Jayaraman, S. Meenakshi, R. S. Rao, S. K. Sikka and V. Vijayakumar, *Phys. Rev. B* **57** (1998) 773.

EXPLANATION FOR THE UNUSUAL ORIENTATION OF LSCO SQUARE VORTEX LATTICE IN TERMS OF NODAL SUPERCONDUCTIVITY

MIGAKU ODA

Department of Physics, Hokkaido University, Sapporo 060-0810, Japan.

In this paper, we provide a possible explanation for the unusual orientation of square vortex lattice in a high-T_c cuprate $La_{2-x}Sr_xCuO_4$ (LSCO) , which was recently demonstrated by the small angle neutron scattering technique to be oriented along the anti-nodal directions of d-wave superconducting order parameter. Such an orientation of the square vortex lattice might be consistent with " nodal superconductivity" that the nodal Fermi arc around $(\pi/2, \pi/2)$ plays a crucial role in the coherence of superconductivity or the coherent part of energy gap, that is, the effective superconducting gap formed by coherent pairing develops only on the nodal Fermi arc and is suppressed on the anti-nodal Fermi surface.

1. Introduction

Recently, it has been demonstrated in small angle neutron scattering (SANS) experiments on $La_{2-x}Sr_xCuO_4$ (LSCO) by Gilardi *et al.* that the vortex lattice undergoes a gradual transition from triangular to square coordination with the increase of magnetic field (Fig. 1).[1] As is well known, in the case of d-wave superconductors such as high-T_c cuprates, where the superconducting (SC) order parameter or the SC gap takes a maximum along the $(\pi, 0)$ and $(0, \pi)$ directions and has nodes along the (π, π) directions, the spatial variation of a vortex has a fourfold symmetry; the low-energy quasiparticle excitations around a vortex core occur along the node directions, and the supercurrent circulating around a vortex core extends more largely towards the anti-nodal directions.[2] Therefore, under high magnetic fields, where the vortex spacing is small, the vortices are expected to form a square lattice so that the strongly anisotropic, low-energy quasiparticle excited regions can overlap with each other between neighboring vortices and their repulsive interactions due to the circulating supercurrent can be reduced; that is, the square vortex lattice will be oriented along the node directions. On the other hand, under low magnetic fields, where the vortex spacing is large, the vortices form a triangular lattice with the hexagonal closest-packed structure so as to reduce only their repulsive interactions most effectively. Thus, the field-induced vortex lattice transition from triangular to square in LSCO could be understood in terms of the anisotropic d-wave SC gap. However, the orientation of square vortex lattice in this system is 45° different from the node or (π, π) directions; the square vortex lattice is oriented along the anti-nodal $(\pi, 0)$ and $(0, \pi)$ directions, where the SC gap is expected to be largest.[1] This fact implies that the SC gap in LSCO might be more complicated than a simple d-wave one. As is well known, the vortex is a typical topological defect and, in general, the local properties of topological defects induced in an ordered system reflect the grobal nature of its order parameter. Therefore, studies for the local properties of vortices in LSCO give us an important clue to understand the spatial structure of the SC order parameter in details.

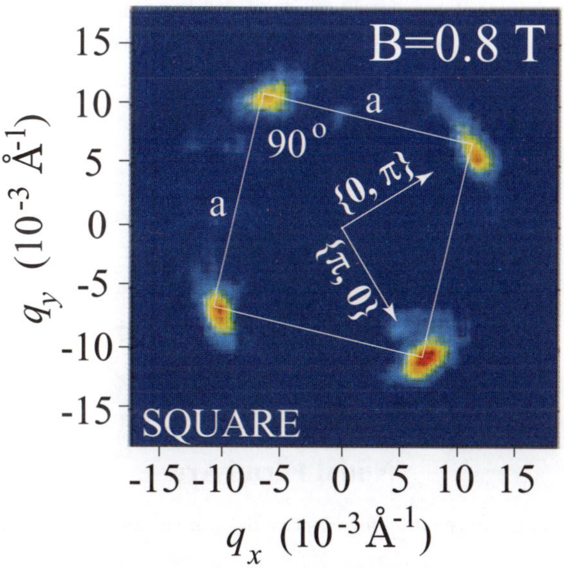

Figure 1. Square vortex lattice observed in SANS experiments on a LSCO crystal with $x = 0.17$ by Gilardi *et al.*[1] This shows that the square lattice is oriented along the anti-nodal $(\pi, 0)$ and $(0, \pi)$ directions or the Cu-O bond directions.

2. Results and Discussion

According to spectroscopic studies on the electronic states of high-T_c cuprates, such as angle-resolved photoemission spectroscopy (ARPES), scanning tunneling microscopy/spectroscopy (STM/STS) and electronic Raman scattering experiments,[3–13] it seems that the low-energy electronic states around the Fermi level (E_F) are strongly inhomogeneous in the k-space, and the Fermi surface (FS) is qualitatively classified into two regions. One is the FS around $(\pi, 0)$ and $(0, \pi)$, the so-called "anti-nodal FS", where the quasiparticles are expected to have a short lifetime because they are strongly scattered through their couplings with antiferromagnetic excitations; namely, the quasiparticles on the anti-nodal FS will be incoherent. On the anti-nodal FS, furthermore, a pseudogap opens up at temperatures lower than $\sim T^*$ well above T_c. The other one is the FS near $(\pi/2, \pi/2)$, including the gap node point, the so-called " nodal Fermi arc ", where the quasiparticles have a much higher mobility compared with those on the anti-nodal FS. On the nodal Fermi arc, furthermore, the SC gap opens below $\sim T_c$. Since mobile pairs play a crucial role in the development of phase coherence in their collective motion, the coherence will be of long-range as the mobile quasiparticles on the nodal Fermi arc near $(\pi/2, \pi/2)$ start to participate in pairing at $\sim T_c$.[10]

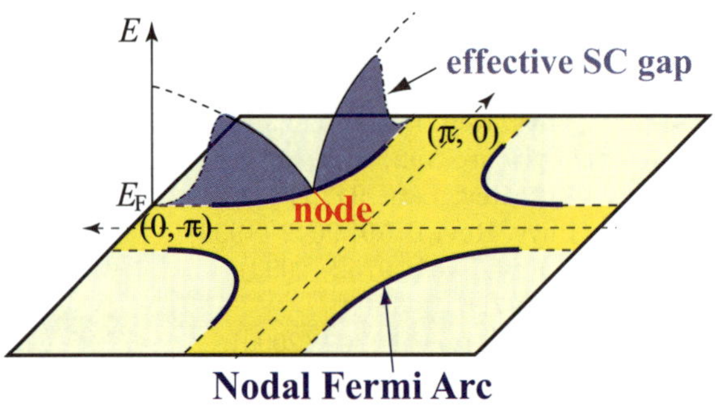

Figure 2. Schematic illustration for effective SC gap on the nodal Fermi arc in the first Brillouin zone.

Indeed, it has been suggested in ARPES, electronic Raman scattering and STM/STS experiments on high-T_c cuprates that the energy gap formed by coherent paring will appear only around the nodal Fermi arc; the coherence seems to be very weak for pairs around the anti-nodal FS even at sufficiently low temperatures ($T << T_c$).[5–10,12,13] Therefore, the effective SC gap Δ_c or the coherent part of energy gap has been considered to develop around the nodal Fermi arc, as schematically shown in Fig. 2, and such a superconductivity has been called " nodal superconductivity ". In this case, the low-energy quasiparticle excitations around vortex cores, whose shape reflects the spatial dependence of SC coherence length, would be enhanced largely towards the anti-nodal $(\pi, 0)$ and $(0, \pi)$ directions or the Cu-O bond directions, leading to a square vortex lattice oriented along these directions. Thus, the unusual orientation of square vortex lattice in LSCO, which was observed in SANS experiments,[1] could be understand in terms of the nodal superconductivity.

The author would like to thank Prof. M. Ido, Prof. S. Tanda, Dr. N. Momono, Dr. K. Inagaki, Dr. Y. Asano, Dr. M. Takigawa and Dr. J. Mesot for valuable discussions. This work was partially supported by the 21st century COE program on " Topological Science and Technology " and Grant in Aids for Scientific Research from the Ministry of Education, Culture, Sports, Science and Technology of Japan.

References

1. R. Gilardi *et al.*, Phys. Rev. Lett. **88**, 217003 (2002).
2. M. Ichioka *et al.*, Phys. Rev.B **53**, 15316 (1996).
3. J. M. Harris *et al.*, Phys. Rev. B **54**, 15665 (1996).
4. M. R. Norman *et al.*, Nature **392**, 157 (1998).
5. T. Yoshida *et al.*, Phys. Rev. Lett. **91**, 27001 (2003).
6. J. G. Naeini *et al.*, Phys. Rev. B **59**, 9642 (1999).
7. M. Opel *et al.*, Phys. Rev. B **61**, 9752 (2000).
8. G. Deutscher, Nature **397**, 410 (1999).
9. N. Momono *et al.*, Physica C **341**, 909 (2000).
10. M. Oda *et al.*, J. Phys. Soc. Jpn. **69**, 983 (2000).
11. R. M. Dipasupil *et al.*, J. Phys. Soc. Jpn. **71**, 1535 (2002).
12. N. Momono *et al.*, J. Phys. Soc. Jpn. **71**, 2832 (2002).
13. K. McEloy *et al.*, cond-mat/0404005.

LOCAL ELECTRONIC STATES IN $Bi_2Sr_2CaCu_2O_{8+\delta}$

A. HASHIMOTO, Y. KOBATAKE, Y. ICHIKAWA, S. SUGITA,
N. MOMONO, M. ODA AND M. IDO
Department of Physics, Hokkaido University, Sapporo, 060-0810, Japan.

In this work, we have examined the spatial variation of low-temperature ($T \ll T_c$) tunneling spectra in a slightly underdoped $Bi_2Sr_2CaCu_2O_{8+\delta}$ (BSCCO) sample using the STM/STS technique, to clarify whether the nano-scale granular type superconductivity is intrinsic or not. We report that the superconductivity takes place rather homogeneously at least in the slightly underdoped sample.

1. Introduction

Recently, Lang *et al.* have reported in STM/STS experiments on BSCCO that the electronic excitation spectra of underdoped (UD) samples are heterogeneous at $T \ll T_c$ and classified into two different types. One is a V-shaped gap with sharp coherence peaks at the edges, which is consistent with a d-wave superconducting (dSC) gap. The other is a much wider gap without coherence peaks and has a subgap structure at low energies, which is called a zero-temperature pseudogap (ZTPG).[1,2] They claimed that the ZTPG region is non-SC and the SC regions, which are ~ 3 nm in size, are separated by ZTPG, leading to the superconductivity of a granular type in UD BSCCO.

In STM/STS experiments on long term annealed BSCCO samples, on the other hand, it was reported that the dSC gap appears homogeneously at least over the distance of ~ 30 nm for doping levels p above the optimal-doping p_o ($p > p_o$), and the heterogeneity could be the matter of sample quality.[3,4] Very recently, McElroy *et al.* reported that the fractional area where ZTPG is observed markedly changes with doping; it is $55 \sim 60$ % for $p \sim 0.1$ and reduces to ~ 2 % for $p \sim p_o$.[2] This result suggests that the contrasting STM/STS results could mesh with each other by taking into account the marked reduction of ZTPG regions for $p > p_o$. However, the heterogeneous SC state in UD BSCCO is still controversial, and strongly affects phenomena such as vortices, which are related to the properties of its order parameter. It is therefore interesting to clarify whether the heterogeneity is intrinsic or not to the UD electronic states. In the present study, to clarify this issue, the spatial variation of low-T local electronic states was examined on a slightly UD BSCCO sample by STM/STS.

2. Experiment

BSCCO single crystals ($p \sim 0.12$, $T_c \sim 74$ K) were grown by the traveling solvent floating zone method. These crystals are cleaved at 8 K under ultrahigh vacuum in STM/STS experiments. STS data were taken using a conventional ac lock-in technique.

3. Results and Discussion

Figure 1(a) shows the spatial variation of STS spectra taken at 8 K over the distance of $\sim$ 18 nm. The two bright line edges along the center, at sample bias voltage of $\sim \pm 50$ mV, corresponds the coherence peaks. We can see that the gap size Δ_0 is very homogeneous throughout this region.

Every spectrum shows a V-shaped gap with a small zero-bias conductance and clear coherence peaks at the gap edges, typical dSC gap spectrum is showed in Fig. 1(b), which is distinct sign of the dSC gap. Furthermore, there is no subgap structure, although it was reported that the subgap appears not only in ZTPG but also in SC gap measured near the ZTPG region on the sample with $p = 0.14$, close to the present doping level $p \sim 0.12$.[1,2] These results qualitatively indicate that the present gaps observed over $\sim$ 18 nm are all typical dSC gaps and show no sign of the existence of ZTPG.

STM image of Fig. 1(c) was taken at sample bias voltage of 0.6 mV and set point current of 0.3 mA just before STS measurements. This STM image exhibits atomic resolution, which guarantees that spatial resolution in the present STS measurements reaches to the scale of atom, though there are a few defects of atoms. The spacing between neighboring STS spectra ($\sim$ 0.9 nm) is smaller than the mean size of the ZTPG region reported in Ref. 1. Thus, ZTPG is detectable in the present STM/STS experiments if it exists.

Shown in Fig. 2 is the histogram for the spatial distribution of the gap size Δ_0, where Δ_0 is defined as half energy-difference between the coherence peaks at positive and negative bias voltages. The mean value of Δ_0 is 47 meV and the full width at half maximum (FWHM) is $\sim$ 8 meV, indicating that Δ_0 has small spatial distribution. As mentioned above, Lang *et al.*

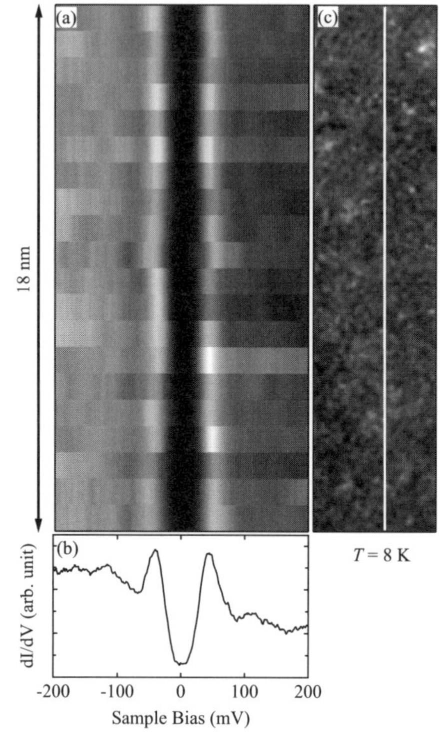

Figure 1. Spatial variation of STS spectra. (a) 2D image plot of STS spectra taken along the line of (b) over $\sim$ 18 nm. (b) STM image measured at 8K. (c) A typical dSC gap spectrum of (a).

already reported the distributions of Δ_0 for both dSC and ZTPG of the UD BSCCO sample with $p \sim 0.14$; FWHM is $\sim$ 10 meV for the distribution of dSC gap and $\sim$ 20 meV for the total distribution including both the dSC gap and the ZTPG.[1] The FWHM value obtained in the present study is close to the former besides much smaller than the latter (Fig. 2). Moreover, they recently reported that the gap size of ZTPG (dSC gap) is always larger (smaller) than $\sim$ 65 meV ($= \Delta_{\rm ZTPG}$) regardless of doping level; that is, the threshold value of gap size between ZTPG and dSC is $\sim$ 65 meV.[2] The result that every gap is not ZTPG but dSC gap is consistent with that report, because the gap size maximum in the present study ($\sim$ 53 meV) is obviously smaller than $\Delta_{\rm ZTPG}$.

In summary, only typical dSC gaps were observed at least over the

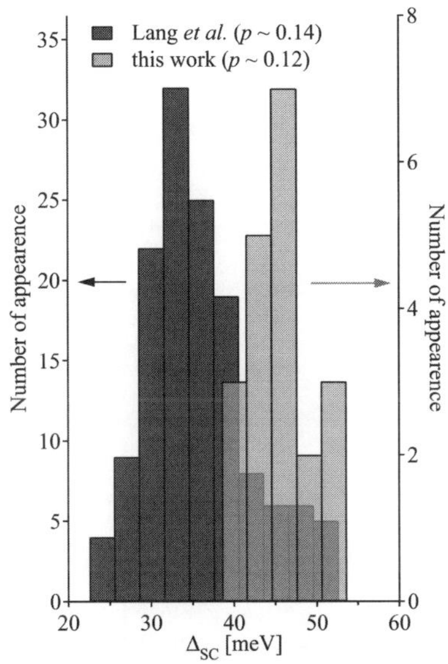

Figure 2. Histogram of the superconducting gap size Δ_0. The black bars represent statistical distribution of Δ_0 reported in Ref. 1 and the grey bars represent the Δ_0 in this work.

distance of $\sim$ 18 nm in the present STM/STS experiments on a slightly UD BSCCO sample ($p \sim 0.12$). The heterogeneous electronic excitation spectra may become remarkable at doping levels less than $p \sim 0.12$.

This work was supported in part by Grants-in-Aid for Scientific Research and the 21st century COE program "Topological Science and Technology" from the Ministry of Education, Culture, Sports, Science and Technology of Japan.

References

1. K. M. Lang, V. Madhavan, J. E. Hoffman, E. W. Hudson, H. Eisaki, S. Uchida, and J. C. Davis, *Nature* **415**, 412 (2002).
2. K. McEloy, D.-H. Lee, J. E. Hoffman, K. M. Lang, J. Lee, E. W. Hudson, H. Eisaki, S. Uchida, J. Lee, and J. C. Davis, *Phys. Rev. Lett.* **94**, 197005 (2005).
3. C. Renner and Ø. Fischer, *Phys. Rev. B* **51**, 9208 (1995).
4. B. W. Hoogenboom, K. Kadowaki, B. Revaz, and Ø. Fischer, *Physica C* **391**, 376 (2003).

V

Topology in Quantum Phenomena

TOPOLOGICAL VORTEX FORMATION IN A BOSE-EINSTEIN CONDENSATE OF ALKALI-METAL ATOMS*

MIKIO NAKAHARA

Department of Physics
Kinki University
Higashi-Osaka, 577-8502, Japan
E-mail: nakahara@math.kindai.ac.jp

BEC of alkali-metal atoms is different from conventional BEC of ^{4}He in several aspects. One of the remarkable properties of this newly discovered BEC is its internal degrees of freedom called the hyperfine spin. This spin is easily controllable by an external magnetic field. We proposed to utilise the hyperfine spin to create a vortex in the condensate by manipulating the external magnetic field, which was subsequently demonstrated experimentally at MIT and Kyoto University. In this contribution, we show that this vortex formation is understood in terms of the Berry phase and is topological in nature. Finally it is shown that the gravitational field introduces a subtle difference between topological vortex formation with light Na atoms (MIT) and with heavy Rb atoms (Kyoto).

1. Introduction

Since its discovery in 1995[1,2], it was expected that the Bose-Einstein condensate (BEC) of alkali-metal atoms had coherent properties and bahaved as a superfluid analogous to superfluid ^{4}He[3,4,5,6]. The existence of a quantised vortex is the direct consequence of these properties and, therefore, attempts to create vortices in this system started right after its discovery.

BEC of alkali-metal atoms is different form superfluid ^{4}He in that (a) the order parameter has hyperfine spin degrees of freedom, which can be controlled by external magnetic fields and laser beams, (b) the coupling constants and the confining potential are easily controllable, (c) it is a gas for which the dilute gas approximation is applicable with good precision

*Partially supported by the Grant-in-Aid for Scientific Research on Priority Areas, Project No. 13135215

and (d) the order parameter profile is easily visualised by miscellaneous techniques.

The above properties are fully utilised in vortex formation scenarios, namely (1) dynamical phase imprinting, (2) stiring by an "optical spoon", (3) rotation of asymmetric trap and (4) topological phase imprinting. I will concentrate on the last scenario in my contribution[7,8,9,10].

2. Topological Phase Imprinting

A Ioffe-Pritchard (IP) trap is made of four current carrying bars (Ioffe bars), which produce quadrupole magentic field at the centre of the trap, and two loop coils for confinement along the axis, see Fig. 1.

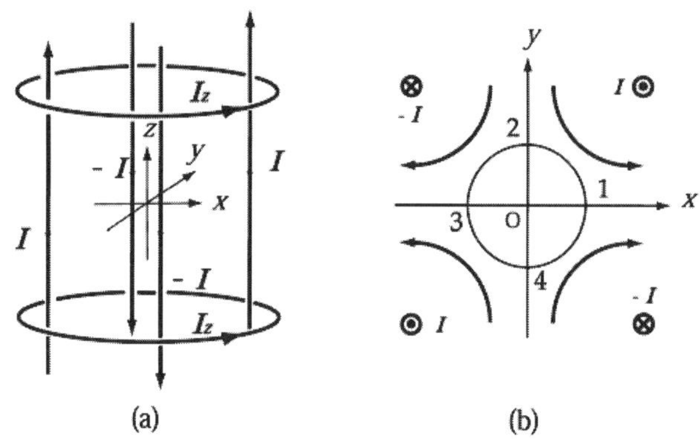

Figure 1. (a) Ioffe-Pritchard trap with four Ioffe bars, each carrying current I. A field produced by the loops at top and bottom prevents Majorana flips, which would take place along the axis. It also confines the condensate along the axial direction. (b) Quadrupole field in the xy-plane, produced by the Ioffe bars, radially confines the condensate. See Fig. 2 for the numbers $1 \sim 4$.

Consider BEC with hyperfine spin $F = 1$ in an IP trap. The condensate is subject to a linear confining potential $V(r) = \gamma_\mu B'r$, where γ_μ is the gyromagnetic ratio while $B'r$ is the magnitude of the quadrupole magentic field $\boldsymbol{B}_\perp$. Confinement is weak along the z-axis perpendicular to $\boldsymbol{B}_\perp$ and therefore BEC is assumed to be cigar shaped with its axis along the z-axis. Suppose a uniform axial field $\boldsymbol{B}_0$ is applied to the condensate. The confining potential then takes the form $\gamma_\mu B = \gamma_\mu \sqrt{B'^2 r^2 + B_0^2}$. The weak-field-seeking state of the condensate is confined by this potential while the other components escape and disappear from the trap, see Fig. 2. It is

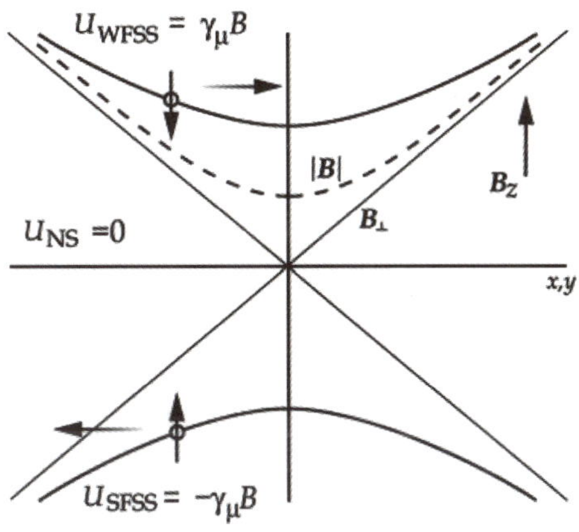

Figure 2. Potential profiles of of the weak-field seeking state (U_{WFSS}), the neutral state (U_{NS}) and the strong-field seeking state (U_{SFSS}). The magnitude of the field is $B = \sqrt{B'^2 r^2 + B_0^2}$.

important to notice that the hyperfine spin $\boldsymbol{F}$ of the weak-field-seeking state is always antiparallel to the magentic field so long as the field changes adiabatically.

Suppose that the axial field $\boldsymbol{B}_0$ reverses its sign as

$$\boldsymbol{B}_0(t) = \begin{cases} B_z(0)(1 - 2t/T_{\mathrm{rev}})\hat{\boldsymbol{z}} & 0 \leq t \leq T_{\mathrm{rev}} \\ -B_z(0)\hat{\boldsymbol{z}} & t > T_{\mathrm{rev}} \end{cases}$$

It is assumed that $B_z(0) \gg |\boldsymbol{B}_\perp|$ in the domain of interest.

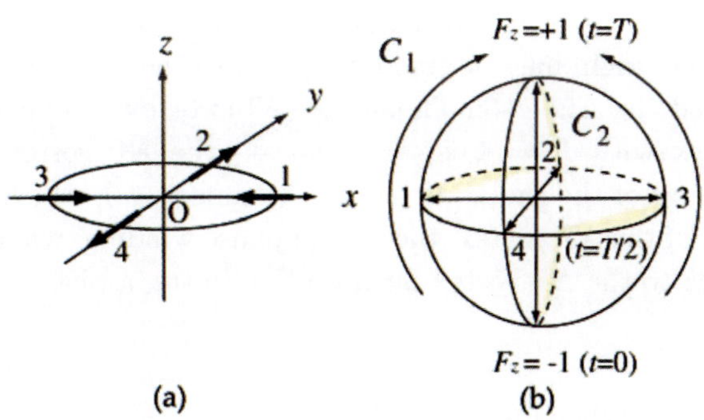

Figure 3. (a) Hyperfine spins at $t = T_{\mathrm{rev}}/2$ on a circle surrounding a vortex along the z-axis. (b) The paths spins follow while $\boldsymbol{B}_0$ is reversed.

Let us see what happens to the $\boldsymbol{F}$ spin of the atoms in the condensate. All the $\boldsymbol{F}$ spins point down ($F_z = -1$) at $t = 0$ when $\boldsymbol{B} \simeq \boldsymbol{B}_0$. As $\boldsymbol{B}_0$ is reversed, the spin at position 1 in Fig. 3 (a) follows the path C_1 in Fig. 3 (b), which ends up with $F_z = 1$ at $t = T_{\rm rev}$ while the spin at 2 follows the path C_2 and ends up with the same point. Note that the two paths surround area π on the unit sphere in the $\boldsymbol{F}$ space. Berry's theorem [11] then claims that an atom following the path C_2 acquires a relative phase $e^{i\pi}$ at time $t = T_{\rm rev}$ with respect to an atom following the path C_1. Repeating the same argument for spins at 3 and 4, we conclude that a formation of a vortex with 4π phase winding (2 units of circulation) takes place at the centre $r = 0$ at $t = T_{\rm rev}$. If, instead, BEC with $F = 2$ is employed, the vortex has the phase winding of 8π (4 units of circulation).

The above observation has been verified also by solving the relevant Gross-Pitaevskii equation numerically[8,9,10].

Leanhardt *et al.* experimentally observed, with $F = 1$ and $F = 2$ states of ^{23}Na atom, the formation of higher winding number vortices according to our scenario[12].

Let us comment on the similarity between a topologically created vortex in BEC and a coreless vortex in superfluid ^{3}He before we close this section. It was pointed out in the seventies that the A-phase of superfluid ^{3}He is able to support a vortex without a sigular core[13,14]. The orbital part of the A-phase order parameter takes its value in SO(3). A well known formula of the fundamental group[15]

$$\pi_1(\mathrm{SO}(3)) = \mathbb{Z}_2 \tag{1}$$

claims that two singly quantised vortices are merged to yield a doubly quantised vortex without a singular core. This is a physical interpretation of the algebra $1 + 1 = 0$ in $\mathbb{Z}_2$. Therefore a vortex with the phase winding 4π far from the origin may be smoothly glued to the centre without a core, which is called the Anderson-Thoulouse (AT) vortex[14]. Our topologically created vortex with $F = 1$ is, so to speak, the AT vortex in the time domain, instead of the radial direction. At time $t = 0$, which corresponds to the centre of the AT vortex, there is no phase winding, while at $t = T_{\rm rev}$, corresponding to the AT vortex far from the centre, a phase winding of 4π is imprinted.

3. Effect of Gravitational Field

Recently Hirotani *et al.* conducted similar experiments with BEC of ^{87}Rb atoms and found that fine-tuning of $T_{\rm rev}$ is necessary to achieve stable vortex formation[16]. Figure 4 shows the CCD image of their data, in which B_0 is reduced from 0.4 G to -0.2 G with $T_{\rm rev} = 2.2$ ms. They observed vortex fragmentation or a condensate without a vortex for other choices of $T_{\rm rev}$. We have analysed the topological phase imprinting under gravi-

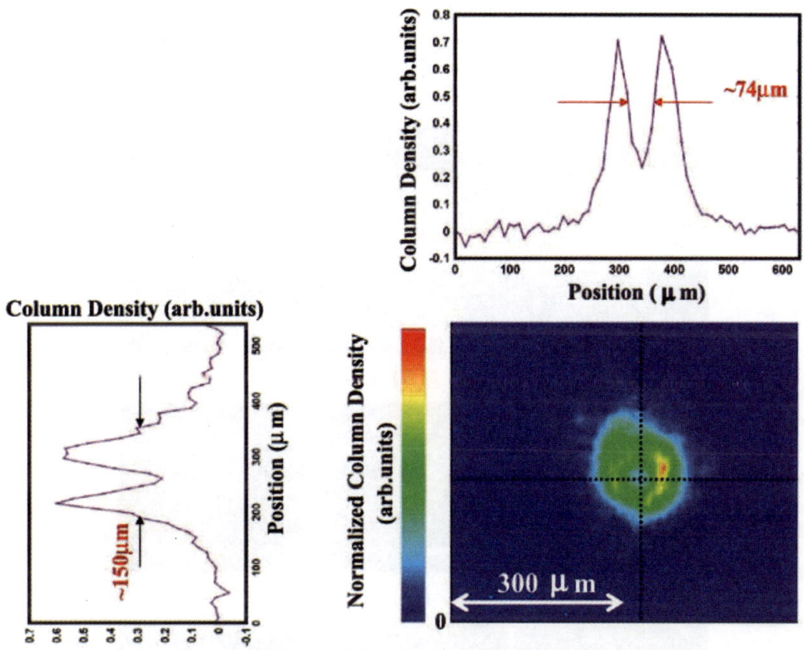

Figure 4. The CCD image of a topologically created vortex in BEC of ^{87}Rb atoms obtained after a TOF of 13 ms. The dimple at the centre of the density profile shows that vortex formation took place along the axis. The number of atoms in the condensate is $\sim 2 \times 10^5$ and the radial extension is ~ 150 μm, while the core size of the vortex is ~ 74 μm. Image courtesy of Yoshiro Takahashi.

tational field[17]. The gravitational effect is more prominent for BEC with heavy atomic mass. This is seen from that fact that the condensate with the hyperspin component m_B along the local magnetic field is subject to the trapping potential

$$V(x, y, t) = \frac{1}{2} m_B \gamma_\mu \sqrt{B'^2(x^2 + y^2) + B_z(t)^2} - Mgx, \qquad (2)$$

where the gravitational force acts along the x-axis, and that this potential has a minimum at

$$(x_0, y_0) \equiv \left(\frac{2Mg|B_z(t)|}{B'\sqrt{B'^2 m_B^2 \gamma_\mu^2 - 4M^2 g^2}}, 0 \right) \tag{3}$$

where we have assumed that $B' m_B \gamma_\mu > 2Mg$. Note that the vortex formation takes place at the centre of the quadrupole potential $(0,0)$. Equation (3) shows that the effect of the gravitational field is stronger for large M.

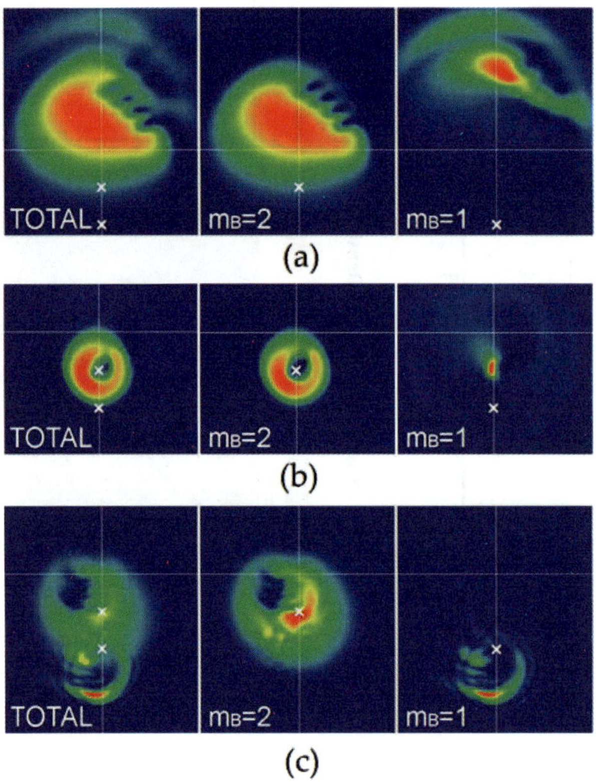

Figure 5. Total density profile of $F = 2$ BEC (left) and individual profiles for hyperfine states with $m_B = 2$ (centre) and 1 (right) at $t = T_{\rm rev}$, when the field inversion has been completed. We have taken $B_z(0) = 0.5$ G and $T_{\rm rev} =$ (a) 1.5 ms, (b) 2.0 ms and (c) 3.0 ms. The white lines are the symmetry axes of the quadrupole field while the cross ($\times$) shows the minimum of the confining potential.

Note also that the confining potential is dependent upon the eigenvalue

m_B. The positions of two weak-field-seeking states with $m_B = 1$ and 2 deviate from each other for BEC with $F = 2$. We solved the time-dependent Gross-Pitaevskii equation[17] and found two decay processes for $F = 2$ BEC: (i) the vortex is created off centre of the condensate and splits into singly quantised vortices (Fig. 5 (a)), and (ii) two hyperfine states interact in a very complicated way leading to vortex fragmentation (Fig. 5 (c)). It is found that the stable vortex is obtained by fine-tuning T_{rev} (Fig. 5 (b)). Our analysis quantitatively accounts for the results of Hirotani *et al.*

Acknowledgement

The author would like to thank the organisers of the symposium TOP 2005 for giving him the opportunity to report our works. He thanks Tomoya Isoshima, Yuki Kawaguchi, Kazushige Machida, Naoki Matsumoto, Mikko Möttönen, Shin-ichiro Ogawa, Tetsuo Ohmi, and Hisanori Shimada for collaboration and Takashi Hirotani, Takeshi Fukuhara, Jun Kobayashi, Mitsutaka Kumakura, Yoshiro Takahashi and Tsutomu Yabuzaki for kindly allowing him for using their experimental result (Fig. 4) prior to publication.

References

1. M. H. Anderson, J. R. Ensher, M. R. Matthews, C. E. Wieman, and E. A. Cornell, Science **269**, 198 (1995).
2. K. B. Davis, M.-O. Mews, M. R. Andrews, N. J. van Druten, D. S. Durfee, D. M. Kurn, and W. Ketterle, Phys. Rev. Lett. **75**, 3969 (1995).
3. M. Inguscio, S. Stringari, and C. E. Wieman (eds.), *Bose-Einstein Condensation in Atomic Gases*, (IOS Press, Amsterdam, 1999).
4. S. Martellucci, A. N. Chester, A. Aspect, and M. Inguscio (eds.), *Bose-Einstein Condensates and Atomic Lasers*, (Kluwer Academic/ Plenum Publishers, New York, 2000).
5. C. J. Pethick and H. Smith, *Bose-Einstein Condensation in Dilute Gases*, (Cambridge University Press, Cambridge, 2002).
6. L. P. Pitaevskii and S. Stringari, *Bose-Einstein Condensation*, (Oxford University Press, Oxford, 2003).
7. M. Nakahara, T. Isoshima, K. Machida, S.-i. Ogawa, and T. Ohmi, Physica B **284-288**, 17 (2000).
8. T. Isoshima, M. Nakahara, T. Ohmi, and K. Machida, Phys. Rev. A **61**, 063610 (2000).
9. S.-I. Ogawa, M. Möttönen, M. Nakahara, T. Ohmi and H. Shimada, Phys. Rev. **66**, 013617 (2002).
10. M. Möttönen, N. Matsumoto, M. Nakahara and T. Ohmi, J. Phys.: Condens. Matter **14**, 29 (2002).

11. M. V. Berry, Proc. R. Soc. Lond. A **392**, 45 (1984).
12. A. E. Leanhardt, A. Görlitz, A. P. Chikkatur, D. Kielpinski, Y. Shin, D. E. Pritchard, and W. Ketterle, Phys. Rev. Lett. **89**, 190403 (2002).
13. N. D. Mermin and T.-L. Ho, Phys. Rev. Lett. **36**, 594 (1976).
14. P. W. Anderson and G. Toulouse G, Phys. Rev. Lett. **38**, 408 (1977).
15. M. Nakahara, *Geometry, Topology and Physics* (2nd ed.), (IOP Publishing, Bristol, 2003).
16. T. Hirotani, T. Fukuhara, J. Kobayashi, M. Kumakura, Y. Takahashi and T. Yabuzaki, private communication.
17. Y. Kawaguchi, M. Nakahara, and T. Ohmi, Phys. Rev. A **70**, 043605 (2004).

QUANTUM PHASE TRANSITION OF ^{4}HE CONFINED IN NANO - POROUS MEDIA*

KEIYA SHIRAHAMA, KEI-ICHI YAMAMOTO AND YOSHIYUKI SHIBAYAMA

Department of Physics, Keio University,
Hiyoshi 3-14-1, Kohoku - ku,
Yokohama 223-8522, Japan
E-mail: keiya@rk.phys.keio.ac.jp

Superfluid ^{4}He confined in nano - porous media provides us with novel topological matters, in which the system topology can be controlled by size, dimensionality and interconnectivity of the porous structures. We study superfluidity and liquid - solid phase transition of ^{4}He confined in a porous glass which has nanopores of 2.5 nm in diameter. The pressure - temperature phase diagram is quite unprecedented: The superfluid transition temperature approaches 0 K at 3.4 MPa, and the freezing pressure shifts up about 1 MPa from the bulk one. The features indicate that the confined ^{4}He undergoes a superfluid - nonsuperfluid - solid quantum phase transition at 0 K. The nonsuperfluid phase may be a localized Bose - condensed state, in which global phase coherence is destroyed by strong correlation between ^{4}He atoms or by random potential (i.e. the Bose glass).

1. Introduction

Vortex in superfluid ^{4}He is one of the simplest topological objects in condensed matter physics[1]. Various types of vortices have been found and studied: vortex rings, lattices, and tangles in bulk ^{4}He and vortex pairs in ^{4}He films. The vortices and the topological structures of ^{4}He can be controlled by confining ^{4}He in nano - porous media. Thanks to recent development of material systhesis, a great variety of nano - porous materials have been available. Such porous materials can introduce novel topological structures to superfluid ^{4}He. Moreover, ^{4}He in porous media is a model system of strongly correlated Bose liquid and solid under confinement potential. Bosons under potential have recently been of great interest. A

*This work is supported by Grant - in - Aid for Scientific Research (A) of Ministry of Science, Education and Sports, Japan.

number of exotic ground states, such as Mott insulator, Bose glass, and supersolid, have been theoretically predicted[2]. Experimental studies have been done employing cold alkali gas atoms with optical lattices[3], superconducting Josephson junction arrays[4], and granular superconducting films[5]. In porous materials, the density of ^{4}He atoms can be varied in a very wide range, from nearly ideal Bose gas regime[6] to strongly correlated solid or liquid ones[7]. This changeability is quite unique and is never achieved in the other experimental systems.

In order to search for new quantum phenomena and topological structures, we study superfluidity of ^{4}He confined in a nano - porous glass. Our present work was motivated by past studies of ^{4}He confined in porous Vycor glass, which have three - dimansionally interconnected nanopores of 7 nm in diameter. It was found that, the freezing presure is elevated about 1.5 MPa from bulk freezing curve, and the superfluid transition temperature T_c shifts down about 0.2 K from the bulk λ line[8,9,10]. From these experiments, we have got an idea to reduce the pore size down to a few nanometer. Confinement into smaller pores may lead to further suppression of superfluidity, in particular at high pressures. Here we report on studies of superfluidity and solidification of ^{4}He confined in a nano - porous Gelsil glass, whose nominal pore diameter is 2.5 nm[7].

2. Superfluidity and Solidification of ^{4}He in Nanopores

We have explored superfluidity and liquid - solid (L - S) transition, by torsional oscillator technique and an isochoric pressure measurement, respectively. The torsional oscillator consists of a brass cell in which a porous glass disk is mounted, and a hollow Be - Cu torsion rod, which acts as the ^{4}He filling line. The cell is oscillated electrostatically at its resonance frequency $f \sim 1956$ Hz, with a very high stability, 10 μHz. We measure the shift in frequency caused by superfluidity of ^{4}He as a function temperature. The liquid ^{4}He introduced to the oscillator cell is pressurized by a room temperature gas handling system. The pressure in the cell is monitored by the pressure gauge located at room temperature. At high pressures where the filling line is blocked by solid ^{4}He, we estimate the pressure in the cell from the oscillator frequency[7].

The isochoric pressure measurement was done separately. Three porous glass disks are installed in the Be - Cu cylindrical cell, in which one of the flat cell wall acts as a pressure sensitive diaphragm.

Typical data of the torsional oscillator experiments are shown in Fig. 1.

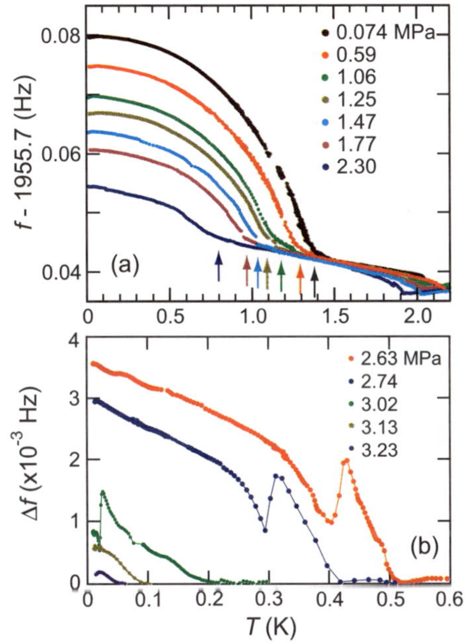

Figure 1. Temperature dependence of the torsional oscillator frequency. (a): Data at $P < 2.53$ MPa (bulk freezing pressure). To clalify the superfluid transitions, the ordinates of the data are shifted so as to collapse onto a single curve at $T \sim 1.5$ K. Small upturns seen around 2 K are due to the superfluid transition of bulk liquid in the interspace between the porous glass sample and the cell wall. Arrows indicate the superfluid transitions inside the nanopores. (b): The frequency shift data above the bulk freezing. In this case the bulk ^{4}He solidifies, and the frequency shift is determined employing the frequency of the empty cell.

In Fig. 1(a), aside from the small frequency shift due to the superfluid transition of bulk ^{4}He inside the oscillator cell, we observe clearly the superfluid transition of ^{4}He confined in the nanopores. At low pressures the superfluid transition temperature T_c is about 1.4 K, which is already much lower than the bulk λ point, 2.17 K. Both T_c and the frequency shift decrease as the pressure increases; the superfluidity is suppressed by pressurization.

Above the bulk freezing pressure $P_f(T < 0.8 \text{ K}) \sim 2.53$ MPa, the cell pressure is estimated from the oscillating frequency[7]. The frequency shifts are clearly observed; That is, liquid ^{4}He is supercooled down to 0 K and still shows superfluidity. T_c and the frequency shift are further suppressed with increasing pressure. The lowest T_c we have observed so far is 38 mK at $P = 3.33$ MPa.

We determine the superfluid phase boundary by plotting the T_c data in the $P - T$ phase diagram shown in Fig. 3. It is clear that T_c approaches 0 K, at a "critical" pressure $P_c \sim 3.4$ MPa. The frequency shift, which is proportional to the superfluid density ρ_s, also decreases continuously to zero at P_c. These continuous suppressions of T_c and ρ_s to zero are quite unprecedented for ^{4}He in other confined geometries.

As the torsional oscillator technique is not sensitive to solidification of ^{4}He inside the pores, we have investigated the L - S transition by measuring the isochoric pressure directly at low temperatures.

We show in Fig. 2 typical pressure data sets. The pressure is measured in the isochoric condition is realized by the capillary blocking method. We firstly prepare a high pressure liquid ^{4}He at $T \geq 2.5$ K and at $5 < P < 7$ MPa, and then cool the whole cell by slowly operating the dilution refrigerator. The liquid ^{4}He solidifies firstly somewhere in the inlet line located between the 4K vacuum flange and the still of the refrigerator, and the cell pressure drops when the bulk ^{4}He inside the cell starts to freeze. The bulk ^{4}He in the cell is in the L - S coexistence, and follows the bulk L - S boundary as the temperature is lowered. After the bulk freezing is completed, the system becomes apart from the L - S boundary (at Point A in Fig. 2(a)). Apart from the bulk L - S boundary, the pressure decreases gently with cooling due to the density change of both bulk solid and supercooled liquid inside the pores.

At high starting pressures (Fig. 2 (a)), the pressure abruptly drops at Point B. This pressure drop indicates that the freezing takes place inside the pores[8]. We denote the Point B as (P_F, T_F). The freezing proceeds in the finite temperature range from Point B to C. After passing C the pressure becomes nearly independent of temperature down to 20 mK. In the warming process from 20 mK, the pressure shows a hysteretic behavior. The melting may be completed at Point D, above which the warming data coincide with the cooling one. The Point D is denoted as (P_M, T_M). All the features seen in Fig. 2(a) are similar to the isochores of ^{4}He in Vycor studied by Adams *et al.*[8]. We therefore conclude that ^{4}He clearly undergoes a L - S transition in the 2.5 nm pores.

As the starting pressure decreases, the pressure drop becomes less prominent, and eventually disappears. Figure 2(b) shows such data. The data exhibit only a broad minimum at about 1.7 K. We have observed so far the pressure drop which indicates freezing down to 3.62 MPa. We plot the data sets, (P_F, T_F) and (P_M, T_M), in the phase diagram Fig. 3. Both the freezing and melting curve shift from the bulk L - S curve to lower

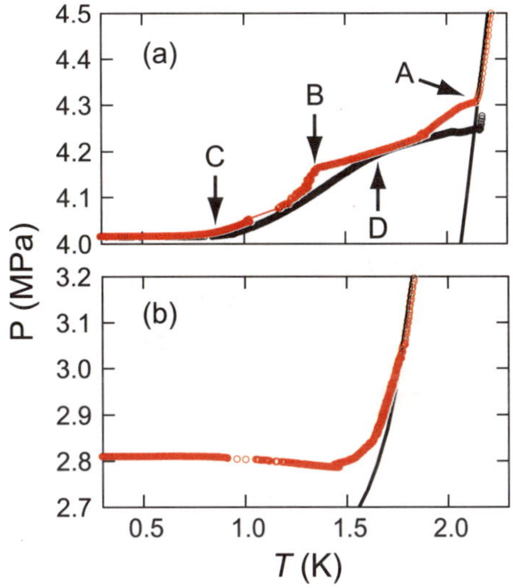

Figure 2. Pressure isochores for typical two starting pressures. Solid lines denote the bulk L - S boundary. (a): High pressure data. The red and black data are taken during cooling and warming, respectively. In the cooling run, freezing starts at Point B $(P_\mathrm{F}, T_\mathrm{F})$ and is completed at Point C. Melting is completed at Point D $(P_\mathrm{M}, T_\mathrm{M})$ in the warming process. (b): Low pressure data. The data are taken during the cooling. No abrupt pressure drop is observed. Instead the pressure shows a minimum at 1.45 K.

temperatures and higher pressures. Below 1.2 K, P_F becomes independent of temperature. In the present sample cell, in which the cell pressure is controlled only by the starting pressure, no data were available below 0.8 K. However, we can determine the "lower limit" of the L -S phase boundary to be about 3.4 MPa, below which no freezing was observed. We determine the possible phase boundary as shown in Fig. 3.

It is quite remarkable that the overall shape of the L - S boundary is similar to that of bulk ^{4}He. In particular, below 1 K, the freezing pressure is nearly independent of temperature. The Clapeyron - Clausius relation for the L - S boundary is given by $dP_\mathrm{F,M}/dT = (S_\mathrm{L} - S_\mathrm{S})\,/\,(V_\mathrm{L} - V_\mathrm{S})$, where S and V denote the molar entropy and molar volume, respectively, and the subscripts $_\mathrm{L}$ and $_\mathrm{S}$ indicate the liquid (nonsuperfluid) and solid states, respectively. As the solid ^{4}He has very small entropy below 1 K, the observed flat L - S boundary indicates that the liquid phase has also small entropy, although it is not superfluid.

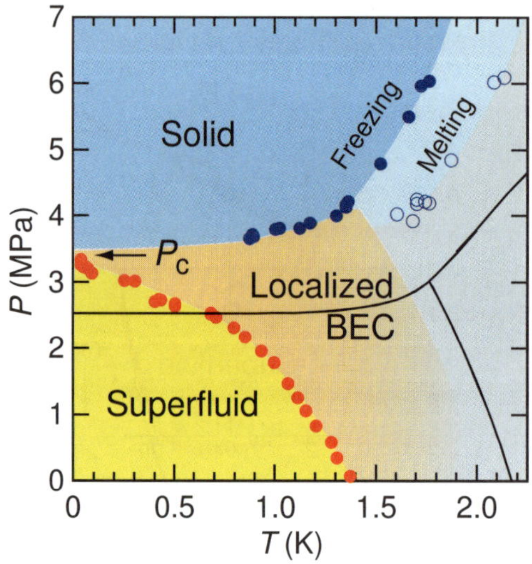

Figure 3. The $P - T$ phase diagram. The solid line shows the bulk phase diagram.

3. Quantum Phase Transition and Localized BEC

The revealed phase diagram is quite unique and has interesting aspects. The continuous suppressions of T_c and ρ_s near P_c strongly suggest that ^{4}He confined in the nanopores undergoes a continuous quantum phase transition (QPT) at P_c and at 0 K, in the sense that the superfluid - nonsuperfluid - solid transition, which is not first order phase transition, is driven at 0 K by changing pressure as an externally controllable parameter[11].

The most intriguing question is the nature of the "nonsuperfluid, but low - entropy" phase adjacent to the superfluid and solid phases. As shown in Fig. 3, this phase may be a localized Bose - Einstein condensation (BEC) state. The idea of the localized BEC was first proposed by Glyde *et al.*[12], in their interpretation of inelastic neutron scattering studies for liquid ^{4}He confined in Vycor and Gelsil glasses near saturated vapor pressure. They observed clear roton peaks, which indicates BEC or superfluidity, at temperatures higher than superfluid transition temperature determined by torsional oscillator or ultrasound studies. Their interpretation is the following: Since the pore size in the porous glasses is not spatially uniform, the superfluid (or BEC) transition temperature has also spatial distribution. When the temperature is lowered from the bulk T_λ, the Bose condensates may grow from large pores. The size of the BEC's is roughly limited to

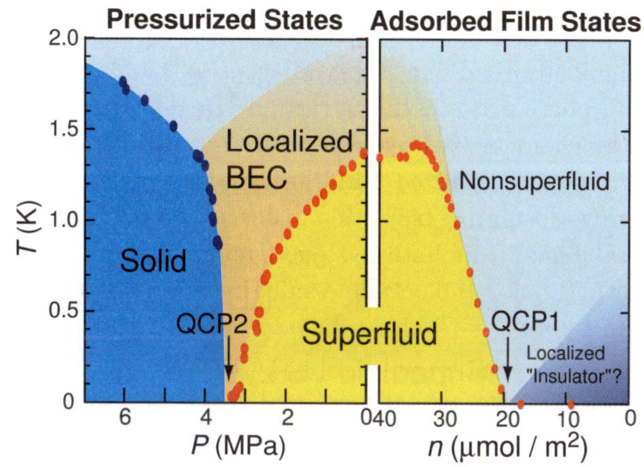

Figure 4. The "global" phase diagram, covering from the thin film states to the compressed solid. The phase diagram can be characterized by two quantum critical points.

the pore size, so the BEC's exist locally. In this situation the system has no global phase coherence and does not exhibit superfluidity which would be detected by macroscopic measurement. As temperature is lowered, the localized BEC's grow continuously. Macroscopic superfluidity is realized when most of the BEC's coalesce.

In our experiment with 2.5 nm pores, macroscopic superfluidity is greatly suppressed by pressurization. As the pressure increases, the spatial exchanges of ^{4}He atoms, which are indispensable for phase coherence, can be suppressed at the thin pores, because of the hard - core nature of He atoms. The interruption of atom exchanges prevents the condensate growth, and hence results in the suppression of T_c.

In the above scenario, only the smallness of the pore size is essential for the superfluid suppression. Disorder or randomness in the porous structure and the pore walls may also lead to the localization of BEC's. That is, the nonsuperfluid phase is a Bose glass state, which has been predicted theoretically in the Bose - Hubbard model[2].

4. Similarity to Other Systems

The possibility of the quantum phase transition in ^{4}He was firstly proposed by Crowell *et al.* for ^{4}He films adsorbed on Vycor glass[13]. They attributed the nonsuperfluid ^{4}He layer adjacent to the pore walls to localization of ^{4}He by substrate randomness. This may be also the case for our ^{4}He - nanopore system. In Fig. 4, we show the "glabal" phase diagram by displaying the phase diagram of adsorbed film states adjacent to the $P-T$ phase diagram. The superfluid phase exists within a certain ^{4}He density range. The phase diagram can be characterized by two quantum critical points.

The above idea of localized BEC has analogies with a model for the superconducting - insulating transition in Josephson junction arrays[4,11] and granular metal films[5]. In both systems suppression of the Cooper pair tunneling by the Coulomb blockade results in phase fluctuation. A similar phase diagram to our $P-T$ phase diagram has been presented.

The global phase diagram in Fig. 4 has a striking similarity to the phase diagram of high - T_c cuprates. Emery and Kivelson proposed that, in the pseudo gap state in the underdoped regime, superconductivity is destroyed by phase fluctuation, although the amplitude of the order parameter has a finite value[14]. Their proposal is quite similar to our picture for the superfluid suppression, and the phase diagrams resemble each other. Although this similarity might be apparent, it gives us a novel aspect for the ^{4}He - nanopore system.

Our findings open a new avenue for studying strongly correlated bosons under confinement potential.

References

1. R. J. Donnelly, *Quantized Vortices in Helium II*, Cambridge University Press (1991).
2. M. P. A. Fisher *et al.*, *Phys. Rev.* **B40**, 546 (1989).
3. M. Greiner *et al.*, *Nature* **415**, 39 (2002).
4. L. J. Geerligs *et al.*, *Phys. Rev. Lett.* **63**, 326 (1989).
5. L. Merchant *et al.*, *Phys. Rev.* **B63**, 134508 (2001).
6. J. D. Reppy *et al.*, *Phys. Rev. Lett.* **84**, 2060 (2000).
7. K. Yamamoto *et al.*, *Phys. Rev. Lett.* **93**, 075302 (2004).
8. E. D. Adams *et al.*, *Phys. Rev. Lett.* **52**, 2249 (1984).
9. J. R. Beamish *et al.*, *Phys. Rev. Lett.* **50**, 425 (1983).
10. C. Lie-Zhao *et al.*, *Phys. Rev.* **B33**, 106 (1986).
11. S. L. Sondhi *et al.*, *Rev. Mod. Phys.* **69**, 315 (1997).
12. H. R. Glyde *et al.*, *Phys. Rev. Lett.* **84**, 2646 (2000); O. Plantevin *et al.*, *Phys. Rev.* **B65**, 224505 (2002).
13. P. A. Crowell *et al.*, *Phys. Rev. Lett.* **75**, 1106 (1995).
14. V. J. Emery and S. A. Kivelson, *Nature* **374**, 434 (1995).

A NEW MEAN-FIELD THEORY FOR BOSE-EINSTEIN CONDENSATES

T. KITA

Division of Physics, Hokkaido University,
Sapporo 060-0810, Japan
E-mail: kita@phys.sci.hokudai.ac.jp

We formulate a conserving gapless mean-field theory for Bose-Einstein condensates based on a Luttinger-Ward thermodynamic functional. It is applied to a weakly interacting uniform gas with density n and s-wave scattering length a to clarify its basic thermodynamic properties. It is found that the condensation here occurs as a first-order transition. The shift of the transition temperature from the ideal-gas result T_0 is positive and given to the leading order by $\Delta T_c = 2.33an^{1/3}T_0$, in agreement with a couple of previous estimates.

1. Introduction

Much effort has been devoted recently to calculate the leading-order correction $\Delta T_c \equiv T_c - T_0$ to the transition temperature of a weakly interacting uniform Bose gas; see Refs. 1 and 2 for an overview. However, the results differ substantially among different approaches, with ΔT_c ranging from negative to positive. Unlike the quantities at $T=0$, perturbational calculations are difficult for ΔT_c. The situation also highlights the absence of an established mean-field theory for Bose-Einstein condensates (BEC), corresponding to the Bardeen-Cooper-Schirieffer theory of superconductivity, with which one could estimate ΔT_c easily. Thus, we still do not have a quantitative description of BEC at finite temperatures, especially near T_c.

As was reported by Hohenberg and Martin in 1965 [3] and emphasized recently by Griffin [4], there are a couple of key words that can be used in formulating a mean-field theory for BEC: "conserving" and "gapless." Here, application of the conventional Wick-factorization procedure to the Hamiltonian of interacting Bose particles, known as the Hartree-Fock-Bogoliubov (HFB) theory,[4,5] fails to fulfill the Hugenholtz-Pines theorem[6] giving rise to an unphysical energy gap in the excitation spectrum. Then it has become customary to introduce a further approximation, now called "Popov" approximation,[4] of completely ignoring the anomalous quasiparticle pair correlation to recover a gapless excitation. However, when the same approximation is adopted for dynamics, it does not satisfy various conservation laws as required. Also, it is not clear whether it is permissible to neglect the pair correlation which has played an essential role in the pioneering perturbation theory of Bogoliubov.[7]

With these observations, we formulate a new mean-field theory for BEC with the desired conserving gapless characters. To this end, we do not follow the previous approaches but ask the question: Can we construct a Luttinger-Ward thermodynamic functional[8] for BEC within a mean-field approximation which satisfies both the conservation laws and the Hugenholtz-Pines theorem? This is indeed possible as will be shown below. The predictions of the resulting mean-field theory for the uniform system will be presented later with an expression for ΔT_c. We put $\hbar = k_B = 1$.

2. Formulation

As usual, we express the field operator $\psi(\mathbf{r})$ as a sum of the condensate wave function $\Psi(\mathbf{r})$ and the quasiparticle field $\phi(\mathbf{r})$. We also adopt Beliaev's Green's function approach[9] and define our Matsubara Green's function in the Nambu space as

$$\hat{G}(\mathbf{r},\mathbf{r}',\tau) \equiv -\hat{\tau}_3 \langle T_\tau \boldsymbol{\phi}(\mathbf{r},\tau)\boldsymbol{\phi}^\dagger(\mathbf{r}')\rangle, \quad \boldsymbol{\phi} = \begin{bmatrix} \phi \\ \phi^\dagger \end{bmatrix}, \quad \boldsymbol{\phi}^\dagger = [\phi^\dagger \ \phi], \tag{1}$$

where $\hat{\tau}_3$ is the third Pauli matrix; it is usually absent[3,4] but essential for our purpose. Using the Fourier transform $\hat{G}(\mathbf{r},\mathbf{r}',\omega_n)$ with ω_n the Matsubara frequency, we write our Luttinger-Ward functional $\Omega = \Omega(\Psi, \Psi^*, \hat{G})$ as

$$\Omega = \int \Psi^*(\mathbf{r})(H_0 - \mu)\Psi(\mathbf{r})\, d\mathbf{r} + \frac{1}{2\beta}\sum_{n=-\infty}^{\infty} \mathrm{Tr}\left[\ln(\hat{H}_0 + \hat{\Sigma} - \mu\hat{\tau}_3 - i\omega_n\hat{1}) + \hat{G}\hat{\Sigma}\right]\hat{1}(\omega_n) + \Phi\,. \tag{2}$$

Here H_0 denotes the kinetic-energy operator plus the external potential, μ is the chemical potential, $\beta \equiv T^{-1}$, $\hat{1}$ is the unit matrix, and

$$\hat{H}_0 \equiv \begin{bmatrix} H_0 & 0 \\ 0 & -H_0^* \end{bmatrix}, \quad \hat{1}(\omega_n) \equiv \begin{bmatrix} e^{i\omega_n 0_+} & 0 \\ 0 & e^{-i\omega_n 0_+} \end{bmatrix}, \tag{3}$$

with 0_+ an infinitesimal positive constant. The symbol Tr in Eq. (2) also includes an integration over space variables with multiplications of $\hat{H}_0$, $\hat{\tau}_3$, and $\hat{1}$ by the unit matrix $\delta(\mathbf{r}-\mathbf{r}')$ implied. Finally, $\hat{\Sigma}$ is the irreducible self-energy obtained from the functional $\Phi = \Phi(\Psi, \Psi^*, \hat{G})$ by

$$\hat{\Sigma}(\mathbf{r}, \mathbf{r}', \omega_n) = -2\beta\, \delta\Phi/\delta\hat{G}(\mathbf{r}', \mathbf{r}, \omega_n) \,. \tag{4}$$

With this relation, Ω becomes stationary with respect to a variation in $\hat{G}$ satisfying Dyson's equation:

$$\hat{G}^{-1} = i\omega_n \hat{1} - \hat{H}_0 - \hat{\Sigma} + \mu\hat{\tau}_3 \,. \tag{5}$$

A key quantity in Eq. (2) is Φ. We choose it so that the conservation laws and the Hugenholtz-Pines theorem are simultaneously satisfied as

$$\begin{aligned}\Phi = \int d\mathbf{r} \int d\mathbf{r}'\, U(\mathbf{r}-\mathbf{r}') \Bigg\{ & \frac{1}{2} |\Psi(\mathbf{r})|^2 |\Psi(\mathbf{r}')|^2 \\ & -\frac{1}{\beta}\sum_n \Bigg[|\Psi(\mathbf{r})|^2 \frac{1}{2} \mathrm{Tr}\, \hat{\tau}_3 \hat{G}(\mathbf{r}', \mathbf{r}', \omega_n)\hat{1}(\omega_n) \\ & \qquad + \frac{1}{2}\mathrm{Tr}\, \hat{\tau}_3 \boldsymbol{\Psi}(\mathbf{r})\boldsymbol{\Psi}^\dagger(\mathbf{r}')\hat{G}(\mathbf{r}', \mathbf{r}, \omega_n)\hat{1}(\omega_n) \Bigg] \\ & + \frac{1}{2\beta^2}\sum_{n,n'} \Bigg[\frac{1}{2}\mathrm{Tr}\, \hat{\tau}_3 \hat{G}(\mathbf{r}, \mathbf{r}, \omega_n)\hat{1}(\omega_n)\, \frac{1}{2}\mathrm{Tr}\, \hat{\tau}_3 \hat{G}(\mathbf{r}', \mathbf{r}', \omega_{n'})\hat{1}(\omega_{n'}) \\ & \qquad + \frac{1}{2}\mathrm{Tr}\, \hat{G}(\mathbf{r}, \mathbf{r}', \omega_n)\hat{1}(\omega_n)\hat{G}(\mathbf{r}', \mathbf{r}, \omega_{n'})\hat{1}(\omega_{n'}) \Bigg] \,, \end{aligned} \tag{6}$$

where U denotes the interaction potential and $\boldsymbol{\Psi}$ and $\boldsymbol{\Psi}^\dagger$ are defined in the same way as $\boldsymbol{\phi}$ and $\boldsymbol{\phi}^\dagger$ in Eq. (1), respectively. The first and second terms in the two square brackets of Eq. (6) are the Hartree and the Fock terms in the Nambu space, respectively. The difference from the HFB theory lies in the Fock terms. Indeed, the HFB theory[3,4] is recovered from Eq. (6) by replacing $\hat{G}$ and $\hat{\tau}_3\boldsymbol{\Psi}(\mathbf{r})\boldsymbol{\Psi}^\dagger(\mathbf{r}')$ in the two Fock terms by $\hat{\tau}_3\hat{G}$ and $\boldsymbol{\Psi}(\mathbf{r})\boldsymbol{\Psi}^\dagger(\mathbf{r}')$, respectively.

Now, the self-energy can be calculated explicitly by Eq. (4), yielding the $(1,1)$ and $(1,2)$ components as

$$\begin{aligned}\Sigma(\mathbf{r}, \mathbf{r}') = \delta(\mathbf{r}-\mathbf{r}') \int d\mathbf{r}''\, U(\mathbf{r}-\mathbf{r}'') \left[|\Psi(\mathbf{r}'')|^2 + \langle \phi^\dagger(\mathbf{r}'')\phi(\mathbf{r}'')\rangle \right] \\ + U(\mathbf{r}-\mathbf{r}') \left[\Psi(\mathbf{r})\Psi^*(\mathbf{r}') + \langle \phi^\dagger(\mathbf{r}')\phi(\mathbf{r})\rangle \right] ,\end{aligned} \tag{7a}$$

$$\Delta(\mathbf{r}, \mathbf{r}') = U(\mathbf{r}-\mathbf{r}') \left[\Psi(\mathbf{r})\Psi(\mathbf{r}') + \langle \phi(\mathbf{r})\phi(\mathbf{r}')\rangle \right] , \tag{7b}$$

respectively. The $(2,1)$ and $(2,2)$ components are given by $-\Delta^*(\mathbf{r}, \mathbf{r}')$ and $-\Sigma^*(\mathbf{r}, \mathbf{r}')$, respectively. We can also derive the equation for $\Psi(\mathbf{r})$ from

$\delta\Omega/\delta\Psi^*(\mathbf{r})=0$. Noting $\delta\Omega/\delta\hat{G}=\hat{0}$, we only need to consider the explicit Ψ^* dependences to obtain

$$(H_0-\mu)\Psi(\mathbf{r})+\int[\Sigma(\mathbf{r},\mathbf{r}')\Psi(\mathbf{r}')-\Delta(\mathbf{r},\mathbf{r}')\Psi^*(\mathbf{r}')]\,d\mathbf{r}'=0\,. \tag{8}$$

In the uniform case $\Psi=$ constant with no external potential, Eq. (8) reduces to the Hugenholtz-Pines relation $\mu=\Sigma_{\mathbf{k}=\mathbf{0}}-\Delta_{\mathbf{k}=\mathbf{0}}$, as desired. The expression for the particle number $N=-\partial\Omega/\partial\mu$ is obtained similarly as

$$N=\int d\mathbf{r}\left[|\Psi(\mathbf{r})|^2+\langle\phi^\dagger(\mathbf{r})\phi(\mathbf{r})\rangle\right]. \tag{9}$$

In order to calculate thermodynamic quantities, it is convenient to diagonalize the Green's function of Eq. (5). To this end, consider the following eigenvalue problem:

$$\int\hat{H}(\mathbf{r},\mathbf{r}')\begin{bmatrix}u_\nu(\mathbf{r}')\\-v_\nu^*(\mathbf{r}')\end{bmatrix}d\mathbf{r}'=E_\nu\begin{bmatrix}u_\nu(\mathbf{r})\\-v_\nu^*(\mathbf{r})\end{bmatrix}, \tag{10a}$$

where $\hat{H}\equiv\hat{H}_0+\hat{\Sigma}-\mu\hat{1}$, the subscript ν specifies the eigenstate, and (u_ν,v_ν) should be normalized as

$$\int\left[|u_\nu(\mathbf{r})|^2-|v_\nu(\mathbf{r})|^2\right]d\mathbf{r}=1\,. \tag{10b}$$

One can show that E_ν is real when Eq. (10b) can be satisfied.[11] It also follows from Eq. (8) that, without the condition (10b), Eq. (10a) has a special solution $E=0$ for $u=v=\Psi$. We hence expect $E_\nu>0$ under the condition (10b) if the system is stable; an appearance of a negative eigenvalue marks an instability of the assumed $\Psi(\mathbf{r})$. Finally, one can show by using the symmetry $\hat{H}=-\hat{\tau}_1\hat{H}^*\hat{\tau}_1$ that the eigenstate of $\hat{H}$ corresponding to $-E_\nu$ is given by $(-v_\nu,u_\nu^*)^{\mathrm{T}}$. Now, we can provide explicit expressions for $\langle\phi^\dagger(\mathbf{r}')\phi(\mathbf{r})\rangle=-G_{11}(\mathbf{r},\mathbf{r}',\tau=-0_+)$ and $\langle\phi(\mathbf{r})\phi(\mathbf{r}')\rangle=-G_{12}(\mathbf{r},\mathbf{r}',0)$ as

$$\langle\phi^\dagger(\mathbf{r}')\phi(\mathbf{r})\rangle=\sum_\nu\left[u_\nu(\mathbf{r})u_\nu^*(\mathbf{r}')f_\nu+v_\nu(\mathbf{r})v_\nu^*(\mathbf{r}')(1+f_\nu)\right], \tag{11a}$$

$$\langle\phi(\mathbf{r})\phi(\mathbf{r}')\rangle=\sum_\nu\left[u_\nu(\mathbf{r})v_\nu(\mathbf{r}')+v_\nu(\mathbf{r})u_\nu(\mathbf{r}')\right](0.5+f_\nu), \tag{11b}$$

where $f_\nu\equiv(\mathrm{e}^{E_\nu/T}-1)^{-1}$ is the Bose distribution function. In deriving Eq. (11b), use has been made of the identity $\sum_\nu u_\nu(\mathbf{r})v_\nu(\mathbf{r}')=\sum_\nu v_\nu(\mathbf{r})u_\nu(\mathbf{r}')$ as can be proved from the completeness of the eigenfunctions from Eq. (10).

Equations (7)–(11) form a closed set of self-consistent equations satisfying both the Hugenholtz-Pines theorem and various conservation laws. Also, the pair correlation $\langle\phi\phi\rangle$ is adequately included in Eqs. (8) and (10); neglecting this contribution yields the HFB-Popov theory.

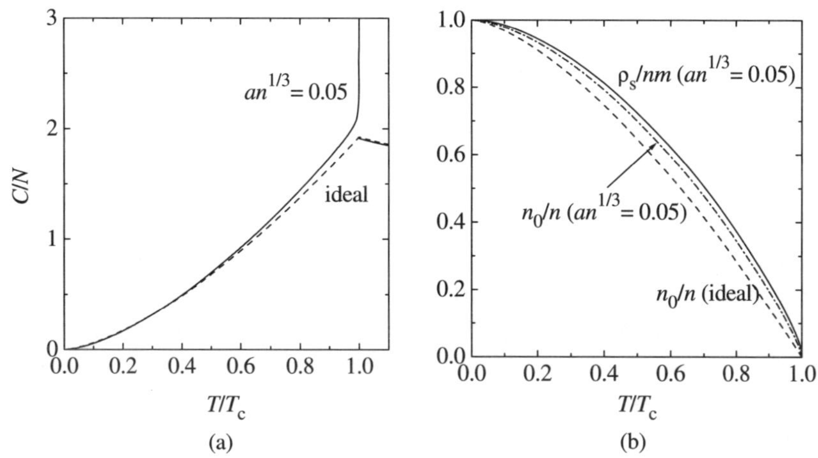

Figure 1. (a) Specific heat C/N as a function of T/T_c for $an^{1/3}=0.05$ in comparison with the ideal gas result. (b) Normalized condensate density n_0/n and superfluid density ρ_s/nm as a function of T/T_c for $an^{1/3}=0.05$ in comparison with the ideal gas result for $n_0/n\,(=\rho_s/nm)$

3. Weakly Interacting Uniform Gas

We have applied Eqs. (7)-(11) to a uniform system with the δ-function interaction: $U(\mathbf{r}-\mathbf{r}')=(4\pi a/m)\delta(\mathbf{r}-\mathbf{r}')$, where m is the particle mass and a is the s-wave scattering length. Thermodynamic quantities near T_c have been calculated analytically in Ref. 15. It is shown that the transition temperature T_c is enhanced over T_0 as

$$\frac{T_c}{T_0} = 1 + 2.33an^{1/3}\,, \tag{12}$$

with n the particle density. Our expression for $\Delta T_c/T_0 = T_c/T_0 - 1$ agrees with the analytic result by Baym *et al.*[13] as well as the numerical one by Holzmann and Krauth.[14] It is also shown that the condensate density n_0 grows continuously from T_c, but the superfluid density ρ_s and the chemical potential μ are discontinuous at T_c. Thus, the transition is weakly first order. Note the differences from the HFB-Popov theory where $T_c = T_0$ and n_0 is discontinuous at T_c. The specific heat $C = T(\partial S/\partial T)$ just below T_c is divergent as $C/N \sim 13.0(an^{1/3})^{3/2}(1-T/T_c)^{-1/2}$. Similar calculations at $T=0$ lead to the expressions of the Bogoliubov theory, as it should.[11]

Figure 1 shows overall temperature dependences of C/N, n_0/n, and ρ_s/nm as a function of T/T_c obtained numerically for a weakly interacting Bose gas of $an^{1/3}=0.05$. For comparison, the corresponding results for the ideal gas ($a=0; T_c=T_0$) are also plotted. The curves of $an^{1/3}=0.05$ satisfy the limiting behaviors of $T \to T_c$ and 0 mentioned above, although $\Delta\rho_s(T_c)$ and $1-n_0(0)/n$ are too small in the present scale to be seen clearly.

We comment on our result $\Delta T_c = 2.33an^{1/3}T_0$ in connection with previous numerical estimates. Among them, the path-integral approach by Holzmann and Krauth[14] seems most reliable, since they calculated ΔT_c in terms of a correlation function in the non-interacting Bose gas derived by an expansion from $a = 0$. They thereby obtained the result $\Delta T_c \sim 2.3an^{1/3}T_0$ compatible with our analytical one. On the other hand, a different value $\Delta T_c \sim 1.3an^{1/3}T_0$ has been obtained from Monte Carlo calculations by Arnold and Moore[1] and by Kashurnikov *et al.*[17] However, they both adopted as a starting point an effective action with a finite interaction, where the effects from the pair correlation $\langle\phi\phi\rangle$ may not have been included appropriately. Now, one may ask: Are the results on the transition compatible with the second-order transition in superfluid ^{4}He with $T_c < T_0$? It is expected that, as the interaction between particles is increased from zero, T_c initially increases towards a maximum, decreasing eventually below T_0, as rationalized by Grüter *et al.*[16]. Also, the transition will change its character during the course from first- to second-order.

References

1. P. Arnold and G. Moore, Phys. Rev. Lett. **87**, 120401 (2001).
2. V.I. Yukalov, cond-mat/0408599.
3. P.C. Hohenberg and P.C. Martin, Ann. Phys. (N.Y.) **34**, 291 (1965).
4. A. Griffin, Phys. Rev. B **53**, 9341 (1996).
5. M. Girardeau and R. Arnowitt, Phys. Rev. **113**, 755 (1959).
6. N.M. Hugenholtz and D. Pines, Phys. Rev. **116**, 489 (1959).
7. N.N. Bogoliubov, J. Phys. (USSR) **11**, 23 (1947).
8. J.M. Luttinger and J.C. Ward, Phys. Rev. **118**, 1417 (1960).
9. S.T. Beliaev Zh. Eksp. Teor. Fiz. **34**, 417 (1958) [Sov. Phys. JETP **7**, 289 (1958)].
10. T. Kita, J. Phys. Soc. Jpn. **65** 1355, (1996).
11. A.L. Fetter, Ann. Phys. (N.Y.) **70**, 67 (1972).
12. See, e.g., C.J. Pethick and H. Smith, *Bose-Einstein Condensation in Dilute Gases* (Cambridge University Press, Cambridge, 2002) Chap. 5.
13. G. Baym, J.-P. Blaizot, and J. Zinn-Justin, Europhys. Lett. **49**, 150 (2000).
14. M. Holzmann and W. Krauth, Phys. Rev. Lett. **83**, 2687 (1999).
15. T. Kita, J. Phys. Soc. Jpn. **74**, No. 7 (2005); cond-mat/0411296.
16. P. Grüter, D. Ceperley, and F. Laloë, Phys. Rev. Lett. **79**, 3549 (1997).
17. V.A. Kashurnikov, N.V. Prokof'ev, and B.V. Svistunov, Phys. Rev. Lett. **87**, 120402 (2001).

SPIN CURRENT IN TOPOLOGICAL CRISTALS

YASUHIRO ASANO
Applied Physics, Hokkaido University, Sapporo 060-8628, Japan
E-mail: asano@eng.hokudai.ac.jp

A formula of the spin current in mesoscopic superconductors is derived from the mean-field theory of superconductivity. The pair potentials in spin-triplet superconductivity is characterized by $\vec{d}$ vectors. The spatial fluctuations of $\vec{d}$ generate the spin current in equilibrium. We apply the obtained formula to the spin current in Josephson junctions and isolated rings of p-wave superconductors.

1. Introduction

The generation and control of the spin polarized current have been the main topics in spintronics [1,2,3]. The spin devices have been originally proposed on ferromagnets or half-metals because the electric current in such materials basically carries the spin at the same time. In a recent study, a possibility of spin current has been discussed in nonmagnetic semiconductors [4]. In contrast to ferromagnetic metals and semiconductors, superconductors themselves are not major materials in spintronics. This might be because Cooper pairs do not carry the spin in conventional s-wave superconductors. Cooper pairs in p-wave superconductors, however, have the spin degree of freedom. Thus spin active transport can be expected in spin-triplet superconductors [5].

The spin states of Cooper pairs are described by three components of $\vec{d}$. Throughout this paper, vectors in spin space are denoted by $\vec{\cdots}$. The properties of $\vec{d}$ are similar to those of magnetic moments $\vec{m}$ in ferromagnets because both of them characterize the spin polarization. There are, however, mainly three important differences between $\vec{d}$ and $\vec{m}$. Firstly the time-reversal symmetry (TRS) is broken in ferromagnets, whereas it is preserved in unitary superconducting states. Secondly $\vec{m}$ is an observable in ferromagnets, whereas $\vec{d}$ itself is nonobservable. This is because $\vec{d}$ is a part of the wave function of Cooper pairs. Finally $\vec{m}$ is parallel to the spin polarization while $\vec{d}$ points a direction perpendicular to the spin of Cooper pairs.

In this paper, we derive a formula of the spin current based on the mean-field theory of superconductivity. In Josephson junctions of spin-triplet superconductors, the spin current is represented by the Andreev reflection coefficients, which implies that Cooper pairs carry the spin. The spin flows in equilibrium is possible when the TRS is broken by the spatial fluctuations of $\vec{d}$. We also discuss the spin current in topological crystals [6] of p-wave superconductors.

2. Spin current formula

In the mean-field theory of superconductivity, electronic states in superconductors are described by the Bogoliubov-de Gennes equation,

$$\int d\boldsymbol{r}' \begin{bmatrix} h(\boldsymbol{r},\boldsymbol{r}')\hat{\sigma}_0 & \hat{\Delta}(\boldsymbol{r},\boldsymbol{r}') \\ -\hat{\Delta}^*(\boldsymbol{r},\boldsymbol{r}') & -h^*(\boldsymbol{r},\boldsymbol{r}')\hat{\sigma}_0 \end{bmatrix} \begin{bmatrix} \hat{u}(\boldsymbol{r}') \\ \hat{v}(\boldsymbol{r}') \end{bmatrix} = E \begin{bmatrix} \hat{u}(\boldsymbol{r}) \\ \hat{v}(\boldsymbol{r}) \end{bmatrix}, \tag{1}$$

$$h(\boldsymbol{r},\boldsymbol{r}') = \delta(\boldsymbol{r}-\boldsymbol{r}') \left\{ -\frac{\hbar^2 \boldsymbol{D}_{\boldsymbol{r}'}}{2m} + V(\boldsymbol{r}') - \mu \right\}, \tag{2}$$

where $\boldsymbol{D}_{\boldsymbol{r}} = \boldsymbol{\nabla}_{\boldsymbol{r}} - (ie/\hbar c)\boldsymbol{A}_{\boldsymbol{r}}$, $\cdot\hat{\cdot}\cdot$ indicates 2×2 matrix describing the spin space, $\hat{\sigma}_0$ is the unit matrix and μ is the Fermi energy. In uniform superconductors, the pair potentials are given by

$$\hat{\Delta}\left(\boldsymbol{r},\boldsymbol{r}'\right) = \frac{1}{V_{vol}} \sum_{\boldsymbol{k}} \hat{\Delta}_{\boldsymbol{k}} e^{i\boldsymbol{k}\cdot(\boldsymbol{r}-\boldsymbol{r}')}, \tag{3}$$

$$\hat{\Delta}_{\boldsymbol{k}} = \begin{cases} i\,\vec{d}(\boldsymbol{k}) \cdot \hat{\vec{\sigma}}\hat{\sigma}_2 e^{i\varphi} & : \text{triplet} \\ i\,d_0(\boldsymbol{k})\,\hat{\sigma}_2 e^{i\varphi} & : \text{singlet,} \end{cases} \tag{4}$$

where $\hat{\sigma}_i$ with $i = 1, 2$ and 3 are the Pauli matrices, and φ is a macroscopic phase. Using the conservation low of the spin density, we derive a formula of the spin current,

$$\vec{\boldsymbol{J}}_s = \frac{\hbar}{4im} \lim_{\boldsymbol{r}'\to\boldsymbol{r}} k_B T \sum_{\omega_n} \mathrm{Tr}$$

$$\times \begin{bmatrix} \boldsymbol{D}_{\boldsymbol{r}} - \boldsymbol{D}^*_{\boldsymbol{r}'} & 0 \\ 0 & \boldsymbol{D}^*_{\boldsymbol{r}} - \boldsymbol{D}_{\boldsymbol{r}'} \end{bmatrix} \check{\mathfrak{G}}_{\omega_n}(\boldsymbol{r},\boldsymbol{r}') \frac{\hbar}{2} \begin{bmatrix} \hat{\vec{\sigma}} & \hat{0} \\ \hat{0} & \hat{\vec{\sigma}}^* \end{bmatrix}, \tag{5}$$

where $\check{\mathfrak{G}}_{\omega_n}(\boldsymbol{r},\boldsymbol{r}')$ is the Matsubara-Green function of Eq. (1) in the 4×4 matrix form, and Tr is carried out over the Nambu $\times$ spin space. When $\hbar\hat{\vec{\sigma}}/2$ is replaced by $-e\hat{\sigma}_0$ in Eq. (5), we obtain the formula of the Josephson electric current ($\boldsymbol{J}_e$) [7].

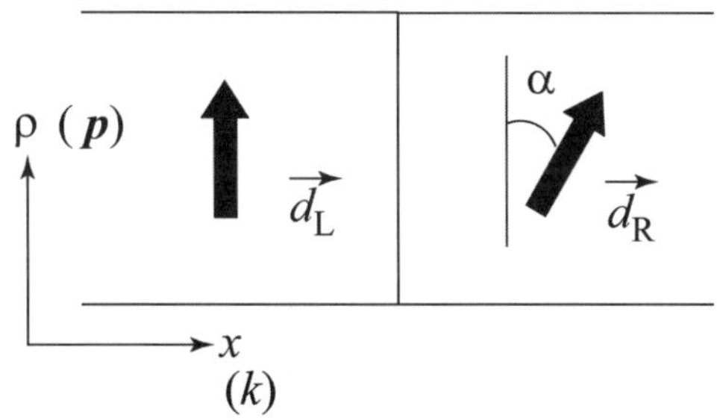

Figure 1. The Josephson junction under consideration.

We first apply the formula to a p-wave superconductor / p-wave superconductor junction in the clean limit as shown in Fig. 1, where x points the normal of the interface and $\boldsymbol{\rho}$ denotes a vector in transverse directions to x. When triplet-superconductors are in the unitary states, the spin current is calculated from Eq. (5) and the Green function of the junction [7],

$$\vec{\boldsymbol{J}}_s = -\sum_{\boldsymbol{p}} \frac{k_B T}{4\hbar} \sum_{\omega_n} \mathrm{Tr}\left[\frac{1}{\Omega_+}\left\{\hat{\Delta}_+ \hat{a}_1 \frac{\hbar\hat{\vec{\sigma}}}{2} + \hat{a}_1 \hat{\Delta}_+ \frac{\hbar\hat{\vec{\sigma}}^*}{2}\right\} - \frac{1}{\Omega_-}\left\{\hat{a}_2 \hat{\Delta}_-^\dagger \frac{\hbar\hat{\vec{\sigma}}}{2} + \hat{\Delta}_-^\dagger \hat{a}_2 \frac{\hbar\hat{\vec{\sigma}}^*}{2}\right\}\right], \tag{6}$$

where $\Omega_\pm = \sqrt{\hbar^2\omega_n^2 + |\vec{d}_\pm|^2}$, $\hat{\Delta}_\pm = i\ \vec{d}_\pm \cdot \hat{\vec{\sigma}}\hat{\sigma}_2$, $\vec{d}_\pm = \vec{d}(\pm k, \boldsymbol{p})$ are the vectors in the left superconductor. In the momentum space, k and $\boldsymbol{p}$ are the wave numbers on the Fermi surface in the x and $\boldsymbol{\rho}$ directions, respectively. The electric Josephson current is also given by $\hbar\hat{\vec{\sigma}}/2 \to -e\hat{\sigma}_0$ in Eq. (6). In particular in the case of spin-singlet s-wave symmetry, the expression reduces to so called Furusaki-Tsukada formula [8] for the electric Josephson current. The spin current is represented by the two Andreev reflection coefficients of a quasiparticle incident from the superconductor on the left hand side, (i.e., $\hat{a}_1$ and $\hat{a}_2$). Thus Cooper pairs are possible to carry the spin.

In Fig. 1(a), we assume that amplitudes of $\vec{d}$ in the two superconductors are identical to each other and that $\vec{d}$ is real vectors satisfying $\vec{d}_+ = \vec{d}_-$. The directions of the $\vec{d}$ on the right hand side is oriented by α from the $\vec{d}$ on the left hand side. In the absence of the potential barrier between the two superconductors, the Andreev reflection coefficients in Eq. (6) result in

$$\hat{a}_1 = \frac{i\hat{\Delta}_L^\dagger}{2\hbar\omega_n g^*}\left[u^2\chi^*\frac{\hat{\Delta}_L\hat{\Delta}_R^\dagger}{|\vec{d}|^2} - v^2\chi\right]\left[\chi^*\frac{\hat{\Delta}_R\hat{\Delta}_L^\dagger}{|\vec{d}|^2} - \chi\right], \tag{7}$$

$$\hat{a}_2 = \left[u^2\chi - \frac{\hat{\Delta}_L\hat{\Delta}_R^\dagger}{|\vec{d}|^2}v^2\chi*\right]\left[\chi - \frac{\hat{\Delta}_R\hat{\Delta}_L^\dagger}{|\vec{d}|^2}\chi^*\right]\frac{i\hat{\Delta}_L}{2\hbar\omega_n g}, \tag{8}$$

$$g = \left(1 + \frac{|\vec{d}|^2}{2(\hbar\omega_n)^2}\right)\cos\varphi + \frac{|\vec{d}|^2}{2(\hbar\omega_n)^2}\cos\alpha + i\left(\frac{\Omega}{\hbar\omega_n}\right)\sin\varphi, \tag{9}$$

where $\chi = e^{i\varphi/2}$ with $\varphi = \varphi_L - \varphi_R$ and u^2 $(v^2) = (\omega_n + (-)\Omega)/(2\omega_n)$. The pair potentials are defined as $\hat{\Delta}_{L,R} = i\vec{d}_{L,R}(\boldsymbol{k}) \cdot \vec{\sigma}\ \sigma_2$ with $|\vec{d}_L| = |\vec{d}_R| = |\vec{d}|$. The spin currents of junctions in Fig. 1(a) are calculated at the zero temperature,

$$J_e = e\sum_{\boldsymbol{p}} \frac{|\vec{d}|}{4\hbar}\left[F_+ + F_-\right], \tag{10}$$

$$\vec{J}_s = \frac{\hbar}{2}\vec{n}\sum_{\boldsymbol{p}} \frac{|\vec{d}|}{4\hbar}\left[F_+ - F_-\right], \tag{11}$$

$$F_\pm = \frac{\sin\left(\frac{\varphi\pm\alpha}{2}\right)\cos\left(\frac{\varphi\pm\alpha}{2}\right)}{\left|\cos\left(\frac{\varphi\pm\alpha}{2}\right)\right|}, \tag{12}$$

where $\vec{n} = \vec{d}_R \times \vec{d}_L/|\vec{d}_R \times \vec{d}_L|$. The current-phase relation of spin is shown for several choices of α in Fig. 2(a), where we fix temperature at at zero. The electric current is usually the odd function of φ, whereas the spin current is always the even function of φ [9]. The spin flow is allowed even at $\varphi = 0$. At $\varphi = 0$, the spin current is proportional to $\sin(\alpha/2)$ as shown in Fig. 2(b). The spin flow is possible when $\vec{d}$ in the two superconductors are not aligned in parallel. Noting that the spin current vanishes at $\alpha = 0$ and π irrespective of φ. The results in Fig. 2 indicate that the spin current is caused by the spatial fluctuations of $\vec{d}$.

Generally speaking, the direction of $\vec{d}$ is fixed to the underlying lattice of superconductors because it is determined by the spin anisotropy due to the spin-orbit coupling. Thus the spatial fluctuations of $\vec{d}$ are not expected in usual bulk samples and thin films. The situation, however, would be changed in topological crystals [6] shown in Fig. 3. In Fig. 3(a), no spin current is expected because there is no spatial fluctuations in $\vec{d}$. In (b), however, it is easily found that $\vec{d}$ is rotated by 2π while it circles the ring.

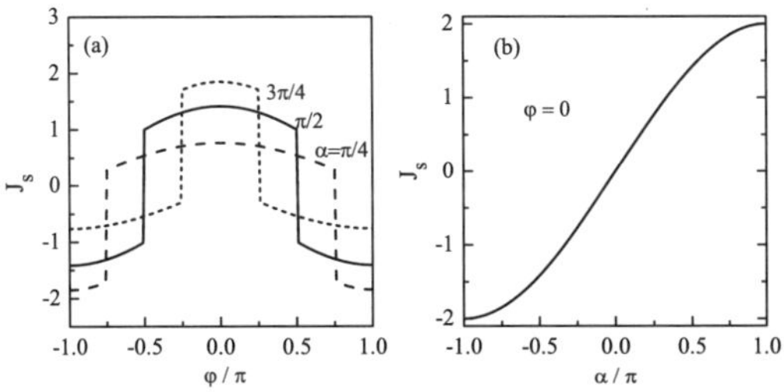

Figure 2. Relations between the spin current and the phase difference across junctions are plotted for several choices of α in (a). The spin current disappears at $\alpha = 0$ and π. The spin current is shown as a function of α for $\varphi = 0$ in (b). Temperature is fixed at zero and the vertical axes are normalized by appropriate values.

The pair potential in Fig. 1(c) can be described by

$$\hat{\Delta}_{\bar{\theta}}(k, \boldsymbol{p}) = i\vec{e}_r(\bar{\theta}) \cdot \hat{\vec{\sigma}}\hat{\sigma}_2 d(k, \boldsymbol{p}), \tag{13}$$

where $\bar{\theta} = (\theta + \theta')/2$, $d(k, \boldsymbol{p})$ is the spatial part of the pair potential, $\vec{e}_r$ is the unit vector in the $\vec{r}$ direction. As shown in Fig. 3(c), the wave numbers in θ and (r, z) directions are denoted by k and $\boldsymbol{p}$, respectively. To estimate the spin current, we first analytically calculate the Green function of the two-dimensional ring as shown in Fig. 3(b) with Eq. (13). By substituting the results into Eq. (5), the spin current of the ring in Fig. 3 (b) becomes

$$\vec{\boldsymbol{J}}_s \simeq -\frac{\hbar}{2}\vec{e}_z \frac{\epsilon_0}{\hbar} \boldsymbol{n}_\theta N_c, \tag{14}$$

where $\vec{e}_z$ is the unit vector in the z direction, $\boldsymbol{n}_\theta$ is the unit vector in real space in the θ direction, N_c is the number of propagating channels on the Fermi surface, and $\epsilon_0 = \hbar^2/4mr^2$. The current expression in Eq. (14) depends only on the shape of rings such as the diameter and the height. The spin polarized in the $+z$ direction flows in the clockwise. In other words, the spin polarized in the $-z$ direction flows in the anti-clockwise. Under the spatial fluctuations of $\vec{d}$, the TRS is broken due to the non-commutativity of the spin algebra. As a consequence, the circulating spin current flows in equilibrium in the p-wave superconducting rings. In small ferromagnetic rings, it has been pointed out that the non-commutativity of the spin algebra is a source of the persistent electric current [10].

In summary, we have derived the formula of the spin current in spin-triplet superconducting systems. The spin flow in equilibrium which is described by the Andreev reflection is possible when the time-reversal symmetry is broken by the spatial fluctuations of $\vec{d}$. We have also discussed the

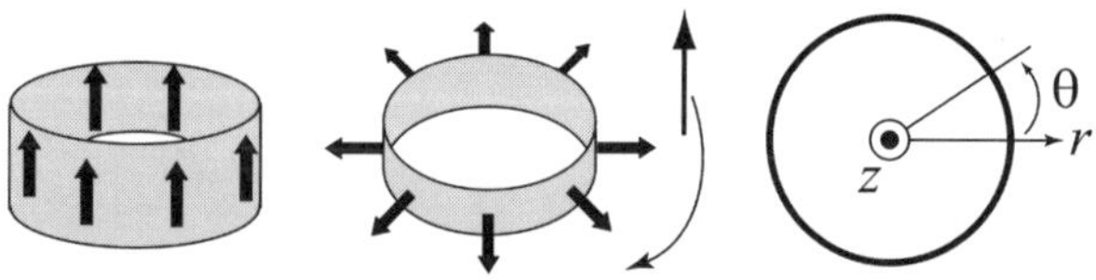

(a) (b) (c)

Figure 3. Schematic figures of p-wave ring shaped crystals. Thick arrows on the strips indicate directions of $\vec{d}$.

circulating spin current on topological crystals of triplet superconductors. The spin-up and -down symmetry in Cooper pair is violated with respect to their directions of flow; this is a new feature of superconductors. It is possible to connect the direction of the spin current with winding numbers characterizing the spatial configurations of $\vec{d}$. Details will be given elsewhere [11].

Acknowledgments

The author thanks Y. Maeno, Y. Tanaka, G. Tatara and S. Tanda for useful discussion. This work has been supported by Grant-in-Aid for the 21st Century COE "Topological Science and Technology" Scientific Research from the Ministry of Education, Culture, Sport, Science and Technology of Japan.

References

1. G. A. Printz, Science, **282**, 1660 (1998).
2. S. A. Wolf, D. D. Awchalom, R. A. Buhrman, J. M. Daughton, S. von Molnar, M. L. Roukes, A. Y. Chtchelkanova, and, D. M. Treger, Science, **294**, 1488 (2001).
3. G. Tatara, An article in this Book.
4. S. Murakami, N. Nagaosa, and S. C. Zhang, Science, **301**, 1348 (2003).
5. Y. Maeno, H. Hashimoto, K. Yoshida, S. Nishizaki, T. Fujita, J. G. Bednorz, and F. Lichtenberg, Nature **372**, 532 (1994).
6. S. Tanda, T. Tsumeta, Y. Okajima, K. Inagaki, K. Yamaya, and N. Hatakenaka, Nature **417**, 397 (2002).
7. Y. Asano, Phys. Rev. B **64**, 224515 (2001).
8. A. Furusaki and M. Tsukada, Solid State Commun. **78**, 299 (1991).
9. G. Rashedi and Yu. A. Kolesnichenko, cond-mat/0501211.
10. G. Tatara and H. Kohno, Phys. Rev. B **67**, 113316 (2003).
11. Y. Asano, Phys. Rev. B **72**, 092508 (2005).

ANTIFERROMAGNETIC DEFECTS IN NON-MAGNETIC HIDDEN ORDER OF THE HEAVY-ELECTRON SYSTEM URU_2SI_2

H. AMITSUKA, K. TENYA

Graduate School of Science, Hokkaido University,
N10W8, Sapporo 060-0810, Japan
E-mail: amiami@phys.sci.hokudai.ac.jp

M. YOKOYAMA

Faculty of Science, Ibaraki University,
Mito 310-8512, Japan
E-mail: makotti@mx.ibaraki.ac.jp

We discuss unusual competition between antiferromagnetic order and nonmagnetic hidden order in the heavy-electron compound URu_2Si_2, on the basis of neutron scattering, ^{29}Si-NMR and μSR measurements performed under hydrostatic pressure applied up to $\sim$ 1.5 GPa. The experimental results suggenst the presence of a bicritical point in the P-T phase diagram, around which the antiferromagnetic order might surve as topological defects of the underlying hidden order.

Since the discovery of coexistence between weak antiferromagnetism (WAF) and non-BCS superconductivity (SC) in UPt_3 and URu_2Si_2, the nature of small moments and their role playing in the normal and superconducting properties of heavy-electron systems have been intensively investigated. In this contribution, we review our recent investigation on the microscopic characterization of WAF for URu_2Si_2 based on neutron scattering (NS), ^{29}Si-NMR and μSR measurements performed under high pressure. The experimental results obtained strongly suggest that the WAF of this system is not ascribed to the small moments but results from inhomogeneous antiferromagnetic (AF) order developing in some non-magnetic ordered phase.

The WAF of URu_2Si_2 (I4/mmm; a = 4.13 Å, c = 9.57 Å) develops below about 17.5 K ($\equiv T_o$), and coexists with SC below 1.2 K ($\equiv T_c$).[1] The size of the ordered moment μ_{ord} is estimated to be $\sim$ 0.02–0.04 μ_B/U

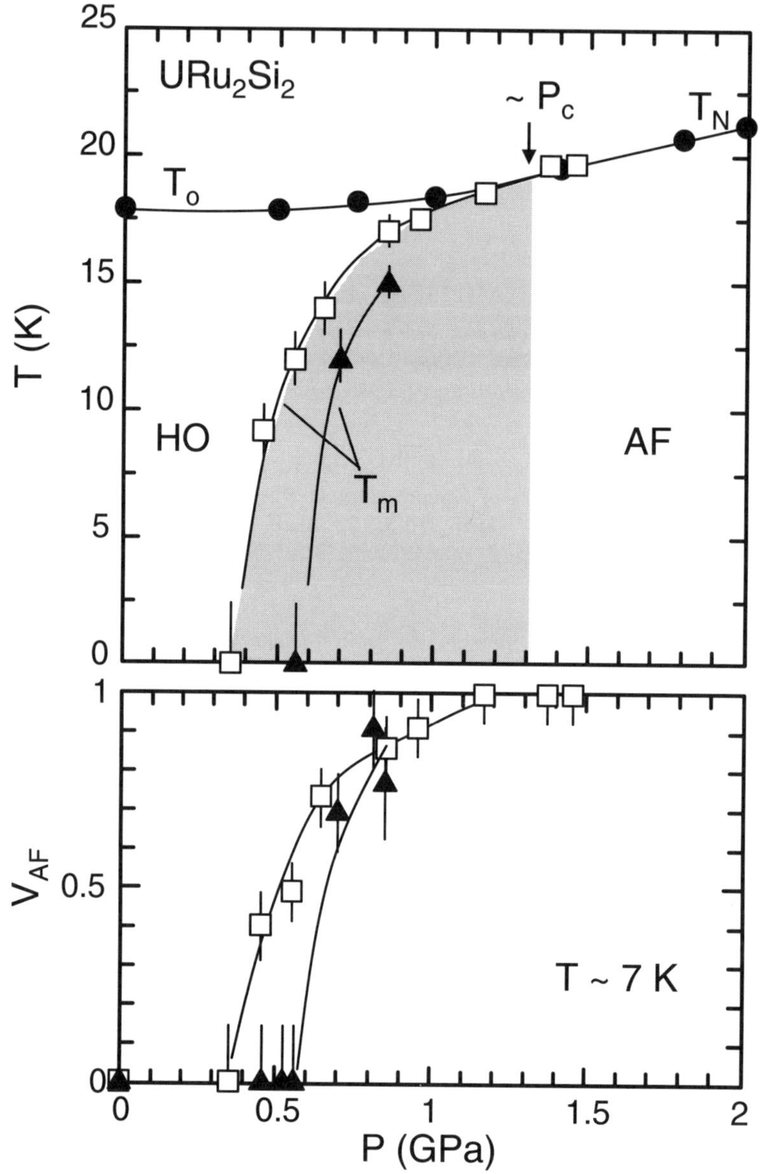

Figure 1. Magnetic entropy $S_{\rm mag}$ versus ordered moment $\mu_{\rm ord}$ for several heavy-electron antiferromagnets.[3]

from the NS measurements.[2] Such a reduction of the ordered moment is now accepted as one of the general aspects in heavy-electron physics, but URu_2Si_2 has another unique feature that the tiny moment is concomitant with large macroscopic anomalies. The situation is illustrated in Fig. 1, where the magnetic entropy S_{mag} for typical HE antiferromagnets is plotted as a function of μ_{ord}.[3] It is apparent that $S_{mag}(\sim 0.2R\ln 2)$ of URu_2Si_2 is disproportionately large compared to its μ_{ord} value, deviating from the behavior of the other systems.

This contradiction has aroused a variety of new ideas for the order parameters, such as d-type spin density waves,[4,5] uranium dimers,[6] bond currents,[7] quadrupoles,[8] and so on. Topological order parameters have also been discussed in very recent years.[9,10] Despite a large number of extensive studies, none of the proposed order parameters have directly been detected yet, and thus this transition is referred to as "hidden order" (HO).

To clarify the origin of WAF and its relation to the bulk anomalies, we first performed the NS measurements under hydrostatic pressure.[11] We found that the AF Bragg scattering intensity I_{AF} is enlarged with P by hundred times, which corresponds to an increase of the staggered moment μ_{ord} from $\simeq 0.02\ \mu_B/U$ ($P \simeq 0$) to $\simeq 0.25\ \mu_B/U$ ($\simeq 1$ GPa) if the order is uniform. In parallel, ^{29}Si-NMR measurements have revealed that the central (paramagnetic) line is 'partially' split into two symmetrically located lines below T_o, where the P-variable quantity was found not to be the resonant frequency but to be the intensity of the satellite lines.[12] The P evolution of I_{AF} is thus ascribed to an unusual P variation of the AF volume fraction $V_{AF}(P)$. Simple extrapolation using previous NS data may give the estimation, $V_{AF} \simeq 0.5$ %, at ambient pressure. This $V_{AF}(P = 0)$ value falls below the detectable limit of NMR, leading to the most reasonable explanation for the longstanding mystery that the staggered fields due to WAF have never been detected by NMR.

μSR measurements have further proven the above phanomena to be of a bulk property.[13,14] The main results are summarized in Fig. 2. We found a spontaneous muon-spin oscillation developing as P exceeds ~ 0.5 GPa for both as-grown and well-annealed (1000 °C, 10 d.) single-crystalline samples. For weak pressures, the asymmetry of the oscillatory component gradually increases below T_m, which is lower than T_o, whereas the frequency ν occurs discontinuously. This confirms that the phase transition between AF and HO is of first order. This is also obvious from the observation that the system exhibits a two-phase separation phenomenon in a wide P-T range (the shaded area in Fig. 2). Upon further compression, T_m

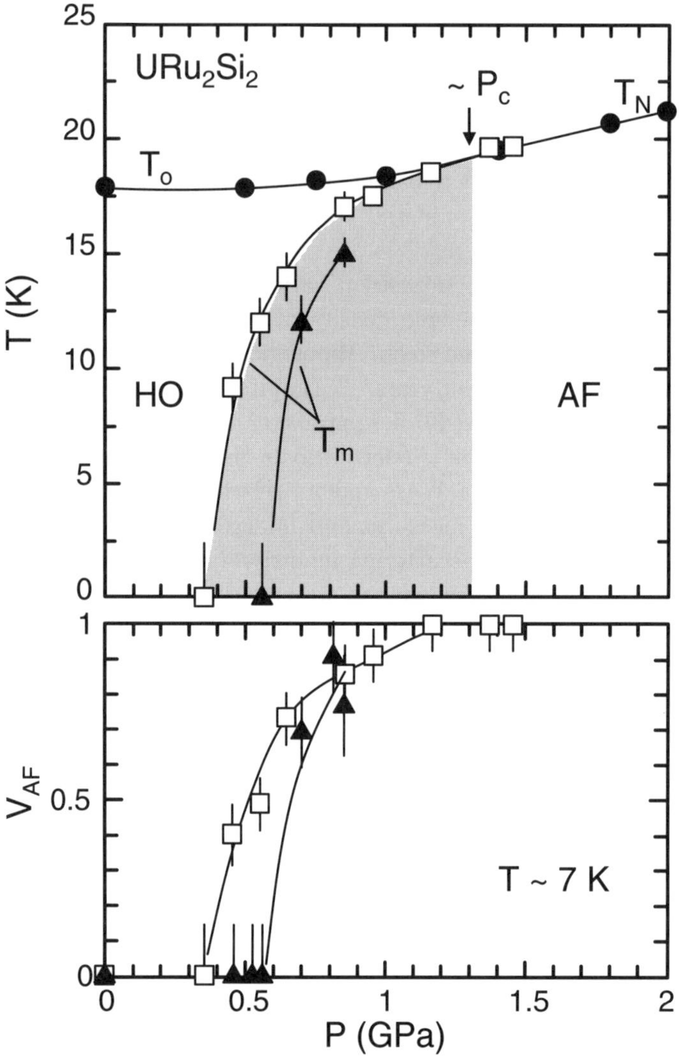

Figure 2. (a) Pressure variations of the transition temperature T_o (or T_N) obtained by electrical resistivity (circles) and the onset temperature of the AF phase T_m obtained from the μSR measurements for annealed (squares) and as-grown (triangles) samples. The shaded area shows the region of two-phase separation. (b) Pressure variations of V_{AF} observed at 7 K.

merges with T_o at $P_c \sim 1.3$ GPa, above which ν behaves as the usual order parameter with $V_{AF} = 1$. We thus conclude that there is a bicritical point on the P-T phase diagram at around P_c, where T_o meets the Néel point T_N. We also observed that the AF state is not affected by the SC transition at weak pressures.

The presence of the bicritical point is relevant to the possible HO scenarios, because the phase diagram of this type is derived when the free energy of the system includes the leading cross-term in the form of $\propto \psi^2 m^2$. In this situation, it is not necessary for the hidden order parameter ψ to break time reversal symmetry and possess the same ordering Q vector as the AF moment m. Such possibilities have acctually been pointed out by the models that assume ψ to be uranium dimers,[6] quadrupoles[8] and bond currents[7]. Interestingly, in these models the local magnetic moments can be induced by rotation or tilting of the hidden order parameters, and thus may surve as defects or domain boundaries of the HO. This would be consistent with the experimental fact that the inhomogeneous magnetism is unusually stable, covering a wide region on the P-T phase diagram. Further work to characterize spatial patterns and low-energy excitations of this mixed phase are in progress.

Acknowledgments

We wish to acknowledge experimental supports and useful discussions with N. Metoki, M. Sato, S. Kawarazaki, K. Kuwahara, K. Matsuda, Y. Kohori, T. Kohara, D. Andreika, A. Schenck, F.N. Gygax, A. Amato, Huang Ying Kai, and J.A. Mydosh.

References

1. T.T.M. Palstra *et al.*, *Phys. Rev. Lett.* **55**, 2727 (1985).
2. C. Broholm *et al.*, *Phys. Rev. Lett.* **58**, 1467 (1987).
3. H. Amitsuka*et al.*, *Physica B* **312-313**, 390 (2002).
4. H. Ikeda and Y. Ohashi, *Phys. Rev. Lett.* **81**, 3723 (1998).
5. A. Virosztek *et al.*, *Int. J. Mod. Phys. B* **16**, 1667 (2002).
6. T. Kasuya, *J. Phys. Soc. Jpn.* **66**, 3348 (1997).
7. P. Chandra *et al.*, *Nature (London)* **417**, 831 (2002).
8. F.J. Ohkawa and H. Shimizu, *J. Phys.: Condens. Matter* **11**, L519 (1999).
9. T. Senthil *et al.*, *Phys. Rev. B* **69**, 035111 (2004).
10. C.M. Varma and Lijun Zhu, *cond-mat/0502344* (2005).
11. H. Amitsuka *et al.*, *Phys. Rev. Lett.* **83**, 5114 (1999).
12. K. Matsuda *et al.*, *Phys. Rev. Lett.* **87**, 87203 (2001).
13. H. Amitsuka *et al.*, *Physica B* **326**, 418 (2003).
14. A. Amato *et al.*, *J. Phys.: Condens. Matter* **16**, S4403 (2004).

MAGNETIC-FIELD DEPENDENCES OF THERMODYNAMIC QUANTITIES IN THE VORTEX STATE OF TYPE-II SUPERCONDUCTORS

KOICHI WATANABE

Division of Physics, Hokkaido University, Sapporo 060-0810, Japan

TAKAFUMI KITA

Division of Physics, Hokkaido University, Sapporo 060-0810, Japan

MASAO ARAI

National Institute for Materials Science, Namiki 1-1, Tsukuba, Ibaraki 305-0044, Japan

We develop an alternative method to solve the Eilenberger equations numerically for the vortex-lattice states of type-II superconductors. Using it, we clarify the magnetic-field and impurity-concentration dependences of the magnetization, the entropy, the Pauli paramagnetism, and the mixing of higher Landau levels in the pair potential for two-dimensional s- and $d_{x^2-y^2}$-wave superconductors with a cylindrical Fermi surface.

1. Introduction

Recent experiments[1,2,3,4,5] have shown that magnetic-field dependences of thermodynamic quantities in the vortex state of type-II superconductors provide unique information on the pairing symmetry and gap anisotropy. On the theoretical side, however, calculations of those quantities still remain a fairly difficult task to perform. The quasiclassical equations derived by Eilenberger[6] provide a convenient starting point for this purpose.

With these backgrounds, we here develop an alternative method to solve the Eilenberger equations for the vortex-lattice states. A key point lies in expanding the pair potential and the quasiclassical f function in the basis functions of the vortex-lattice states, thereby transforming the differential equations into algebraic equations.

This method is applied here to calculate magnetic-field dependences of the magnetization, the entropy, the Pauli paramagnetism, and the pair potential at various temperatures for the two-dimensional s- and $d_{x^2-y^2}$-wave superconductors in the clean and dirty limits. These quantities have been obtained near H_{c2} for the s-wave pairing.[7,8] Our purpose here is to clarify the overall field dependence of those quantities.

This paper is organized as follows. Section II gives the formulation. Section III presents numerical results. Section IV summarizes the paper. We use $k_{\rm B}=1$ throughout.

2. Formulation

2.1. *Eilenberger equations*

We take the external magnetic field $\mathbf{H}$ along the z axis and express the vector potential as

$$\mathbf{A}(\mathbf{r}) = Bx\hat{\mathbf{y}} + \tilde{\mathbf{A}}(\mathbf{r})\,. \tag{1}$$

Here B is the average flux density produced jointly by the external current and the internal supercurrent, and $\tilde{\mathbf{A}}$ is the spatially varying part of the magnetic field satisfying $\int \boldsymbol{\nabla}\times\tilde{\mathbf{A}}\,d\mathbf{r}=\mathbf{0}$. We choose the gauge such that $\boldsymbol{\nabla}\cdot\tilde{\mathbf{A}}=\mathbf{0}$.

The Eilenberger equation for the even-parity pairing without Pauli paramagnetism is given by[6,9]

$$\left(\varepsilon_n+\frac{\hbar}{2\tau}\langle g\rangle+\frac{1}{2}\hbar\mathbf{v}_{\rm F}\cdot\boldsymbol{\partial}\right)f=\left(\Delta\phi+\frac{\hbar}{2\tau}\langle f\rangle\right)g\,. \tag{2a}$$

Here $\varepsilon_n=(2n+1)\pi T$ ($n=0,\pm1,\pm2,\cdots$) is the Matsubara energy with T the temperature, τ is the relaxation time by nonmagnetic impurity scattering, and $\langle\cdots\rangle$ denotes the Fermi-surface average:

$$\langle g\rangle\equiv\int dS_{\rm F}\frac{g(\varepsilon_n,\mathbf{k}_{\rm F},\mathbf{r})}{(2\pi)^3N(0)|\mathbf{v}_{\rm F}|}\,,$$

with $dS_{\rm F}$ an infinitesimal area on the Fermi surface, $N(0)$ the density of states per spin and per unit volume at the Fermi energy in the normal state, and $\mathbf{v}_{\rm F}$ the Fermi velocity. The operator $\boldsymbol{\partial}$ in Eq. (2a) is defined by

$$\boldsymbol{\partial}\equiv\boldsymbol{\nabla}-i\frac{2\pi}{\Phi_0}\mathbf{A}\,,$$

with $\Phi_0\equiv hc/2e$ the flux quantum, $\Delta(\mathbf{r})$ is the pair potential, and $\phi(\mathbf{k}_{\rm F})$ specifies the gap anisotropy satisfying $\langle\phi(\mathbf{k}_{\rm F})\rangle=1$. Finally, the quasiclassical Green's functions f and g for $\varepsilon_n>0$ are connected by $g=(1-ff^\dagger)^{1/2}$ with $f^\dagger(\varepsilon_n,\mathbf{k}_{\rm F},\mathbf{r})=f^*(\varepsilon_n,-\mathbf{k}_{\rm F},\mathbf{r})$.

Equation (2a) has to be solved simultaneously with the self-consistency equation for the pair potential and the Maxwell equation for $\tilde{\mathbf{A}}$, which are given respectively by

$$\Delta(\mathbf{r})\ln\frac{T_{c0}}{T} = 2\pi T\sum_{n=0}^{\infty}\left[\frac{\Delta(\mathbf{r})}{\varepsilon_n} - \langle\phi(\mathbf{k}_\mathrm{F})f(\varepsilon_n,\mathbf{k}_\mathrm{F},\mathbf{r})\rangle\right], \tag{2b}$$

$$-\nabla^2\tilde{\mathbf{A}}(\mathbf{r}) = -i\frac{16\pi^2 eN(0)T}{c}\sum_{n=0}^{\infty}\langle\mathbf{v}_\mathrm{F}g(\varepsilon_n,\mathbf{k}_\mathrm{F},\mathbf{r})\rangle, \tag{2c}$$

with T_{c0} the transition temperature for $\tau=\infty$.

2.2. *Algebraic Eilenberger equations*

By using a set of basis function to describe arbitarary vortex-lattice structures[10], we now expand Δ, f, and $\tilde{A}$ in the basis functions of the vortex lattice as

$$\Delta(\mathbf{r}) = \sqrt{V}\sum_{N=0}^{\infty}\Delta_N\,\psi_{N\mathbf{q}}(\mathbf{r}), \tag{3a}$$

$$f(\varepsilon_n,\mathbf{k}_\mathrm{F},\mathbf{r}) = \sqrt{V}\sum_{N=0}^{\infty}f_N(\varepsilon_n,\mathbf{k}_\mathrm{F})\,\psi_{N\mathbf{q}}(\mathbf{r}), \tag{3b}$$

$$\tilde{A}(\mathbf{r}) = \sum_{\mathbf{K}\neq\mathbf{0}}\tilde{A}_\mathbf{K}\mathrm{e}^{i\mathbf{K}\cdot\mathbf{r}}, \tag{3c}$$

where $\mathbf{K}$ is the reciprocal-lattice vector.[10] Substituting Eq. (3) into Eq. (2) and using the orthogonality of the basis functions, Eq. (2) is transformed into a set of algebraic equations for $\{f_N\}$, $\{\Delta_N\}$, and $\{\tilde{A}_\mathbf{K}\}$ as

$$\begin{aligned}&\varepsilon_n f_N + \bar{\beta}^*\sqrt{N+1}f_{N+1} - \bar{\beta}\sqrt{N}f_{N-1}\\ &= \frac{1}{\sqrt{V}}\int\psi^*_{N\mathbf{q}}\left(\Delta\phi g + \hbar\frac{\langle f\rangle g - \langle g\rangle f}{2\tau} - \bar{\beta}^*\tilde{A} + \bar{\beta}\tilde{A}^*\right)d\mathbf{r},\end{aligned} \tag{4a}$$

$$\Delta_N\ln\frac{T_{c0}}{T} = 2\pi T\sum_{n=0}^{\infty}\left[\frac{\Delta_N}{\varepsilon_n} - \langle\phi(\mathbf{k}_\mathrm{F})f_N(\varepsilon_n,\mathbf{k}_\mathrm{F})\rangle\right], \tag{4b}$$

$$\tilde{A}_\mathbf{K} = -\frac{16\pi^2 N(0)T}{(Kl_\mathrm{c}B)^2 V}\sum_{n=0}^{\infty}\int\langle\beta\,g(\varepsilon_n,\mathbf{k}_\mathrm{F},\mathbf{r})\rangle\mathrm{e}^{-i\mathbf{K}\cdot\mathbf{r}}d\mathbf{r}, \tag{4c}$$

with

$$\bar{\beta} \equiv \frac{\hbar(c_2 v_{\mathrm{F}x} + i c_1 v_{\mathrm{F}y})}{2\sqrt{2} l_{\mathrm{c}}}, \quad \beta \equiv \frac{\hbar(c_1 v_{\mathrm{F}x} + i c_2 v_{\mathrm{F}y})}{2\sqrt{2} l_{\mathrm{c}}}. \tag{5}$$

Together with the equation to determine H_{c2} derived recently,[9] the above coupled equations form a basis for efficient numerical calculations of the Eilenberger equations for vortex-lattice states with arbitrary Fermi-surface structures.

For a given vortex-lattice structure specified by the basis vectors, the coupled equation (4) may be solved iteratively in order of Eqs. (4a), (4b), and (4c) by adopting a standard technique to solve nonlinear equations.[11] Now, the integrations can be performed by the trapezoidal rule; its convergence is excellent for periodic functions.

Once Δ, f, and $\tilde{\mathbf{A}}$ are determined as above, we can calculate thermodynamic quantities of the vortex-lattice state. The free-energy functional corresponding to Eq. (2) is given by the equation used in Ref.7 The entropy are obtained by deriving from the free-enegy functional. Also, when it is much smaller than the diamagnetism by supercurrent, the magnetization due to Pauli paramagnetism is given by the equation used in Ref.8

3. Results

We now present numerical results for two-dimensional systems with the cylindrical Fermi surface which is placed in the xy plane perpendicular to $\mathbf{H}$. We have considered the energy gap of s-wave pairing with and without the impurity effect. The vortex-lattice structure has been fixed as hexagonal so that finite contributions in the expansion (3a) come only from $N = 0, 6, 12, \cdots$ Landau levels.

Figures 1 plot the entropy S_{s} and the magnetization M_{sP} due to Pauli paramagnetism, respectively, as a function of B/H_{c2} in the clean limit at several temperatures. To see the field dependence clearly, the entropy is normalized by using $S_{\mathrm{s}0} \equiv S_{\mathrm{s}}(B=0)$ and $S_{\mathrm{n}} \equiv S_{\mathrm{s}}(B=H_{c2})$ as $(S_{\mathrm{s}} - S_{\mathrm{s}0})/(S_{\mathrm{n}} - S_{\mathrm{s}0})$; it varies from 1 to 0 for $H_{c2} \geq B \geq 0$. The same normalization is adopted for M_{sP}. All curves deviate upwards from the linear behavior $\propto B/H_{c2}$ and become more and more convex upward as $T \to 0$.

Figure 2 shows the field dependences of S_{s} and M_{sP} for $\tau = 0.01\hbar/\Delta(0)$ at various temperatures. Compared with Figs. 3(a) and 4(a), the curves are more monotonic with the almost linear behavior $\propto B/H_{c2}$. Looking at the temperature dependence more closely, however, we observe a change from a convex-upward behavior at high temperatures to a convex-downward

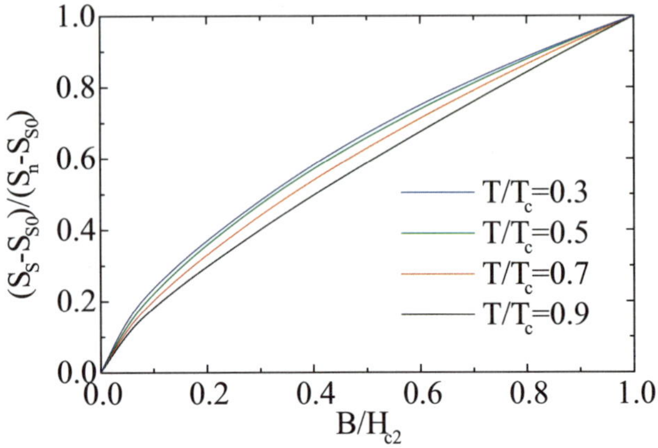

(a)

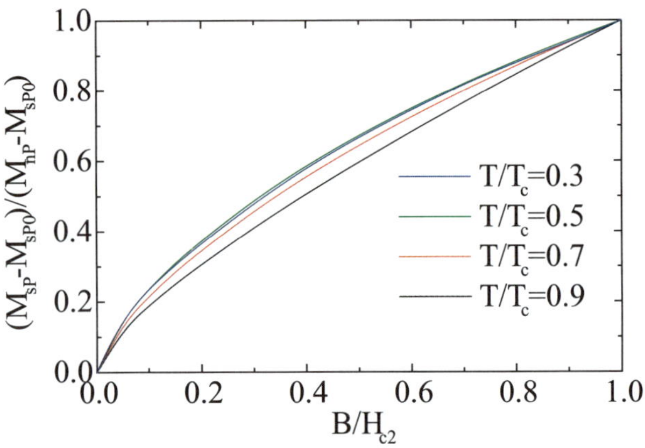

(b)

Figure 1. (a) The entropy $S_{\rm s}$ and (b) the magnetization $M_{\rm sP}$ by Pauli paramagnetism as a function of B/H_{c2}. The temperatures are $T/T_{\rm c}$ = 0.9, 0.7, 0.5, and 0.3 from bottom to top. They are normalized to vary from 1 at $B = H_{c2}$ to 0 at $B = 0$.

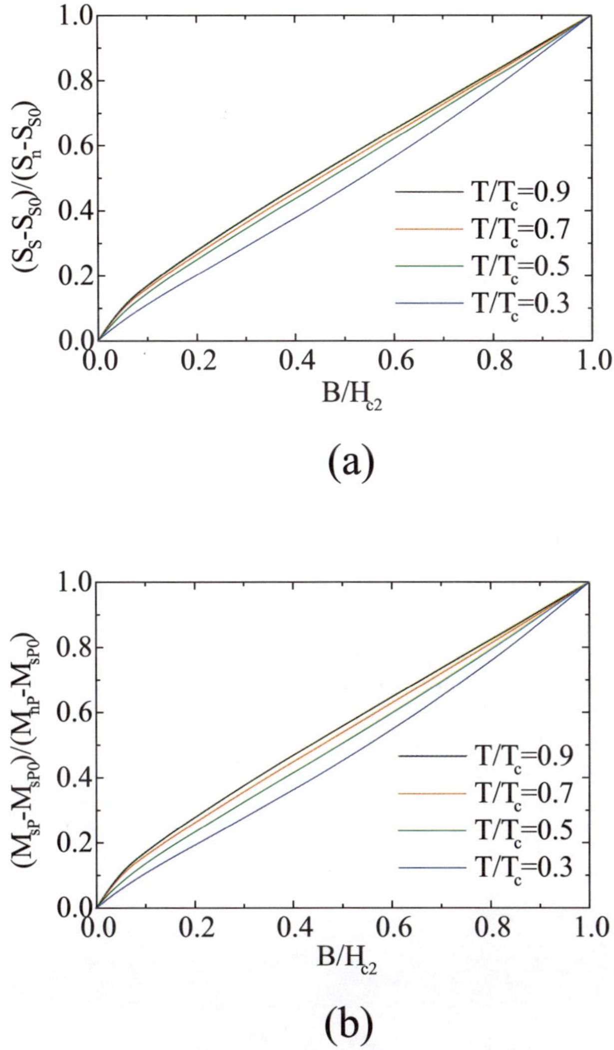

Figure 2. (a) The entropy $S_{\rm S}$ and (b) the magnetization $M_{\rm sP}$ by Pauli paramagnetism as a function of B/H_{c2} for the s-wave pairing with $\tau = 0.01\hbar/\Delta(0)$. The temperatures are $T/T_{\rm c} = 0.9$, 0.7, 0.5 and 0.3 from top to bottom.

behavior at low temperatures, in agreement with a previous calculation near H_{c2}.[8] This feature also appears in the field dependence of the zero-energy density of states as calculated recently by Miranović *et al.*[12] The convex-downward behavior at $T = 0.3T_c$ may become pronounced enough at lower temperatures to be experimentally observable. These results are obtained for the case with simple cylindrical Fermi surface. For real materials, due to more complicated Fermi surface, we must consider the effects of Fermi surface.

References

1. A. V. Sologubenko, J. Jun, S. M. Kazakov, J. Karpinski, and H. R. Ott, Phys. Rev. B **66**, 014504 (2001).
2. E. Boaknin, M. A. Tanatar, J. Paglione, D. Hawthorn, F. Ronning, R. W. Hill, M. Sutherland, L. Taillefer, J. Sonier, S. M. Hayden, and J. W. Brill, Phys. Rev. Lett. **90**, 117003 (2003).
3. K. Izawa, Y. Nakajima, J. Goryo, Y. Matsuda, S. Osaki, H. Sugawara, H. Sato, P. Thalmeier, and K. Maki, Phys. Rev. Lett. **90**, 117001 (2003).
4. H. Aoki, T. Sakakibara, H. Shishido, R. Settai, Y. Onuki, P. Miranović, and K. Machida, J. Phys. Condens. Matter **16**, L13 (2004).
5. K. Deguchi, Z. Q. Mao, H. Yaguchi, and Y. Maeno, Phys. Rev. Lett. **92**, 047002 (2004).
6. G. Eilenberger: Z. Phys. **214**, 195 (1968).
7. T. Kita, Phys. Rev. B **68**, 184503 (2003).
8. T. Kita, Phys. Rev. B **69**, 144507 (2004).
9. T. Kita and M. Arai, Phys. Rev. B **70**, 224522 (2004).
10. T. Kita, J. Phys. Soc. Jpn. **67**, 2067 (1998).
11. W. H. Press, S. A. Teukolsky, W. T. Vetterling and B. P. Flannery, *Numerical Recipes in C* (Cambridge University Press, Cambridge, 1988) Chap. 9.
12. P. Miranović, M. Ichioka, and K. Machida Phys. Rev. B **70**, 104510 (2004).
13. M. Nohara, M. Isshiki, F. Sakai, and H. Takagi, J. Phys. Soc. Jpn. **68**, 1078 (1999).

THREE-MAGNON-MEDIATED NUCLEAR SPIN RELAXATION IN QUANTUM FERRIMAGNETS OF TOPOLOGICAL ORIGIN *

HIROMITSU HORI

Division of Physics, Hokkaido University, Sapporo 060-0810, Japan
E-mail: hiromitsu@phys.sci.hokudai.ac.jp

SHOJI YAMAMOTO

Division of Physics, Hokkaido University, Sapporo 060-0810, Japan
E-mail: yamamoto@phys.sci.hokudai.ac.jp

Nuclear spin-lattice relaxation in the alternating-spin chains and the trimeric interwining double-chains is studied by means of a modified spin-wave theory. We consider the second-order process, where a nuclear spin flip induces virtual spin waves which are then scattered thermally via the four-magnon exchange interaction, where a nuclear spin directly interacts with spin waves via the hyperfine interaction. We point out a possibility of the three-magnon relaxation process.

1. Introduction

Design of molecule-based ferromagnets is a challenging topic in materials science and numerous quasi-one dimensional ferrimagnets have been synthesized in this context. The simplest example is a series of bimetallic chain compounds, which consists of two kinds of spins S and s alternating on a ring with antiferromagnetic exchange coupling between nearest neighbors, as is illustrated in Fig. 1 (a). These compounds were extensively synthesized by Verdaguer, Kahn and their coworkers [1,2], and were theoretically studied by numerous authors [3,4,5]. Besides such hetero spin ferrimagnets, isospin ferrimagnets of topological origin have recently been synthesized [6,7] and attracting further interest. These materials consist of homogeneous

*This work is supported by the 21st century coe program "topological science and technology".

spin centers, but exhibit noncompensating sub-lattice magnetizations originating in their topological crystalline structures. One of the prototypical compounds is Ca3Cu3(PO4)4 [6], the schematic structure of which is shown in Fig. 1 (b). The ground-state and the finite-temperature properties of this material have investigated both theoretically and experimentally but little is known about its dynamic properties.

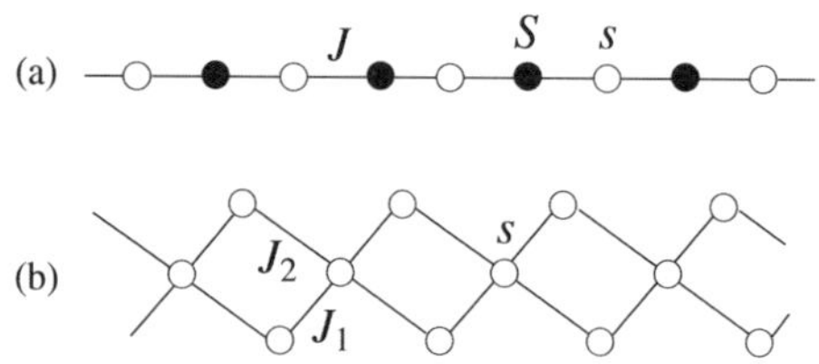

Figure 1. Schematic representation of the alternating-spin-(S, s) chain (a) and the trimeric interwining spin-s double-chain (b). Where J, J_1 and J_2 are the coefficients of exchange interactions.

Recently, the modified spin-wave (MSW) theory is particularly powerful to investigate both static [8,9,10] and dynamic [11,12] properties in ferrimagnetic chain. We also have predicted through the MSW calculations that the three-magnon relaxation process should predominate over the Raman one at high temperatures and weak fields on the nuclear spin-lattice relaxation rate $1/T_1$ of one-dimensional ferrimagnets [13]. We also have the experimental evidence of the major contribution to $1/T_1$ being made by the three-magnon scattering processes [14]. In such circumstances, we applied the modified spin-wave theory to the trimeric interwining double-chains, one of the topological ferrimagnets, and investigate thermodynamic properties and nuclear spin-lattice relaxations.

2. Formalism

We consider two kinds of the ferrimagnetic chains for comparison, whose Hamiltonian described by

$$\mathcal{H} = \sum_{n=1}^{N} \big[J\boldsymbol{S}_n \cdot (\boldsymbol{s}_{n-1} + \boldsymbol{s}_n) - (S_n^z + s_n^z) g\mu_{\rm B} H \big], \tag{1}$$

for the alternating ferrimagnetic chains, and

$$\mathcal{H} = \sum_{n=1}^{N} \big[J_1 \boldsymbol{s}_n \cdot (\boldsymbol{s}_{n,1} + \boldsymbol{s}_{n,2}) + J_2 \boldsymbol{s}_n \cdot (\boldsymbol{s}_{n-1,1} + \boldsymbol{s}_{n+1,2}) - (s_n^z + s_{n,1}^z + s_{n,2}^z) g\mu_{\rm B} H \big], \tag{2}$$

for the trimeric interwining double-chains. Introducing bosonic operators for the spin deviation in each sublattice and assuming that $O(S) = O(s)$, we expand the Hamiltonian with respect to $1/S$ as $\mathcal{H} = \mathcal{H}_2+\mathcal{H}_1+\mathcal{H}_0+O(S^{-1})$, where $\mathcal{H}_i$ contains the $O(S^i)$ terms. $\mathcal{H}_2$ is the classical ground-state energy, while $\mathcal{H}_1$ describes linear spin-wave excitations and is diagonalized in the momentum space. Thus, energy of a spin waves of ferromagnetic ($\sigma = -$) or antiferromagnetic ($\sigma = +$) aspect [10] is given by $\omega_k^\sigma = J[\omega_k + \sigma(S - s - g\mu_B H/J)]$ with $\omega_k \equiv \{(S+s)^2 - 4Ss\cos^2(k/2)\}^{1/2}$ for the alternating spin chains. On the other hand, energy of a spin waves of ferromagnetic ($\sigma = -$) or antiferromagnetic ($\sigma = +$) aspect is given by $\omega_k^\sigma = \omega_k + \sigma[s(J_1+J_2)/2 - g\mu_B H]$ with $\omega_k \equiv S\{(J_1+J_2)^2/4+8J_1J_2\sin^2(k/2)\}^{1/2}$ for the trimeric interwining double-chains. The trimeric interwining double-chains also have the ferromagnetic flat band aspect, whose energy is $\omega_k^0 = S(J_1 + J_2 + g\mu_B H)$ Minimizing the free energy under the condition of zero staggered magnetization [11], we obtain the optimum distribution functions.

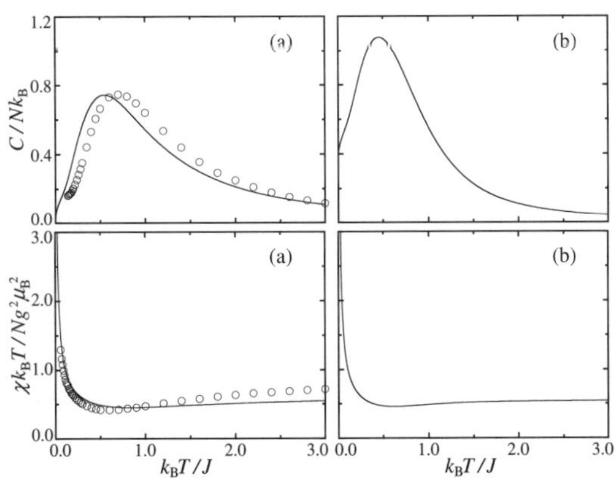

Figure 2. Modified spin-wave calculations of the specific heat (the upper two) and susceptibility temperature product (the lower two) for the alternating-spin chain (a) with QMC calculations (circles) and trimeric interwining double-chain (b).

3. Thermodynamics

We calculate the case of $(S, s) = (1, 1/2)$ for the alternating spin chains and the case of $J_1 = J_2(= J)$ and $s = 1/2$ for trimeric interwining double-chains. The specific heat and the susceptibility are compared to verify the validity of MSW method with quantum Monte Calro (QMC) scheme in the alternating-spin chains in Fig. 3 (a). MSW scheme is good tool to estimate static properties of one dimensional ferrimagnets, because this method well reproduces the specific heat and the susceptibility. In the trimeric interwining double-chain, the peak of specific heat slightly shifts toward low temperature due to a flat band. On the other hand, there is no effective difference for the susceptibility temperature product.

4. Nuclear Spin-lattice Relaxation Rate

The hyperfine interaction between nuclear spins and electoric spins in a sub-lattice is generally expressed as

$$\mathcal{H}_{\rm hf} = g\mu_{\rm B}\hbar\gamma_{\rm N}I^{+}\sum\nolimits_{n}\left(\tfrac{1}{2}B_{n}^{-}s_{n}^{-} + B_{n}^{z}s_{n}^{z}\right), \tag{3}$$

where B_n^σ is the dipolar coupling tensor between the nuclear and electronic spins. Since $\mathcal{H}_0$ and $\mathcal{H}_{\rm hf}$ are both much smaller than $\mathcal{H}_1$, they act as perturbative interactions to the linear spin-wave system. If we consider up to the second-order perturbation with respect to $\mathcal{V} \equiv \mathcal{H}_0 + \mathcal{H}_{\rm hf}$, the probability of a nuclear spin being scattered from the state of $I^z = m$ to that of $I^z = m+1$ is given by

$$W = \frac{2\pi}{\hbar}\sum_{f}\left|\left\langle f\left|V + \sum_{m(\neq i)}\frac{\mathcal{V}|m\rangle\langle m|\mathcal{V}}{E_i - E_m}\right|i\right\rangle\right|^2 \delta(E_i - E_f), \tag{4}$$

where i and f designate the initial and final states of the unperturbed system. Then we find that $T_1 = (I-m)(I+m+1)/2W$. Equation (4) contains various relaxation processes. We diagrammatically show them in Fig. 3. Due to the considerable difference between the nuclear and electronic energy scales, $\hbar\omega_{\rm N} \ll J$, the direct process, involving a single spin wave, is rarely of significance. Within the first-order mechanism, $1/T_1^{(3)}$ is much smaller than $1/T_1^{(2)}$ [15]. However, the first-order relaxation rate is generally enhanced through the second-order mechanism. We consider

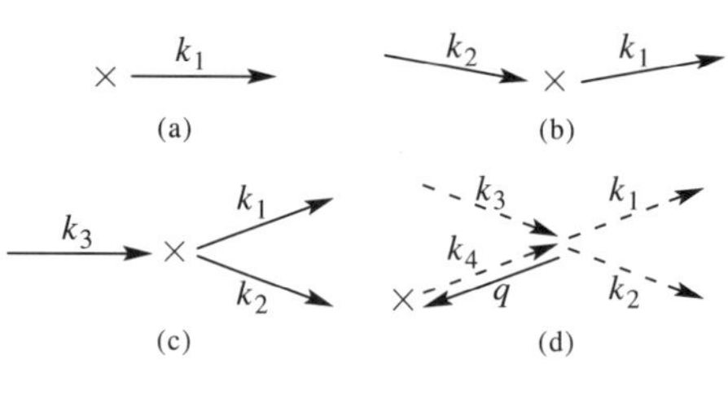

Figure 3. Illustration of the elementary nuclear spin-lattice relaxation processes. Solid arrows designate ferromagnetic, antiferromagnetic or flat band spin waves inducing a nuclear spin flip ($\times$). Broken arrows denote the four-magnon exchange interaction. (a) The direct process. (b) The first-order Raman process. (c) The first-order three-magnon process. (d) The second-order three-magnon process, where $q = -k_4 \equiv k_3 - k_2 - k_1$.

the leading second-order process, that is, the exchange-scattering-induced three-magnon relaxation, as well as the first-order process.

We assume that the Fourier components of the coupling constants have little momentum dependence [16] as $\sum_n e^{ikn}B_n^\tau \equiv B_k^\tau \simeq B^\tau$ $(\tau = -, z)$ and $B^-/B^z = 4$. Figure 4 shows $1/T_1$ as a function of temperature and an applied field. The exchange-scattering-enhanced three-magnon relaxation rate roughly grows into a major contribution to $1/T_1$ with increasing temperature and decreasing field. The field dependence of $1/T_1^{(2)}$ is initially logarithmic and then turns exponential with increasing field. Equation (4) claims that $1/T_1^{(3)}$ should exhibit much stronger power-law diverging behavior with decreasing field. Therefore, the three-magnon relaxation process predominates over the Raman one at weak fields. Temperature dependence

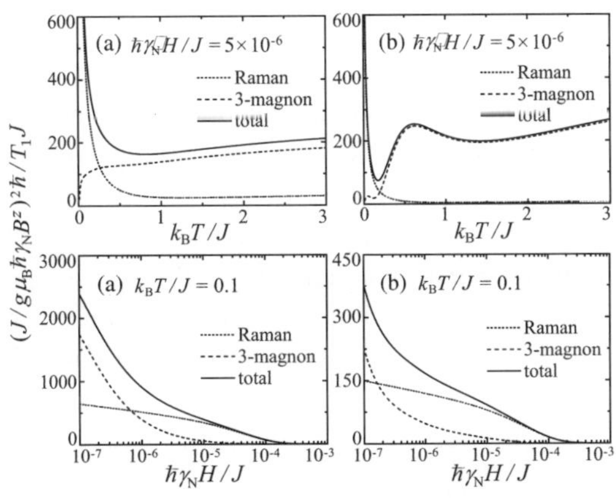

Figure 4. Modified spin-wave calculations of the nuclear spin-lattice relaxation rates for the alternating-spin chains (a) and the trimeric interwining double-chains (b). The upper two are $1/T_1$ as a function of temperature and the lower two are $1/T_1$ as a function of an applied field.

of the three-magnon relaxation process for the trimeric interwining double-chains is more interested due to peak at middle temperature. In Fig. 5, the three-magnon relaxation process are resolved to only ferromagnetic and antiferromagnetic spin-waves scattering process and the process including flat-band spin-waves scattering. The former behave like a three-magnon relaxation process for the alternating spin chains. The latter give the peak at the middle temperature due to flat-band spin waves.

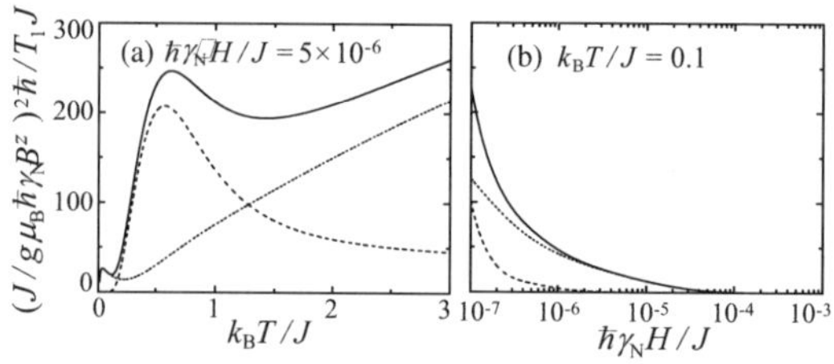

Figure 5. Temperature (a) and field (b) dependence of the three-magnon relaxation process in the trimeric interwining double-chains. Solid line is total three-magnon relaxation process. Broken line is the process including the flat band spin-waves scattering. Dotted line is the only ferromagnetic and antiferromagnetic spin-waves scattering process.

5. Concluding Remarks

Thermodynamic properties and nuclear spin-lattice relaxation of the trimeric interwining double-chains have been investigated in comparison with the alternating-spin chains. Thermodynamic properties mutually have no qualitatively difference. Nuclear spin-lattice relaxation in their systems is interested. We point out a possibility of the three-magnon relaxation process predominating over the Raman one. Temperature dependence of the three-magnon relaxation process in the trimeric interwining double-chains give the peak at middle temperature range differing from that of the alternating-spin chain. Such behavior will facilitate the experiment detection of the three-mugnon relaxation process. We hope our study will stimulate further experimental investigations into topologocal ferrimagnetism.

References

1. O. Kahn et al., *J. Am. Chem. Soc.* **110**, 782 (1988).
2. P. J. van Koningsbruggen et al., *Inorg. Chem.* **29**, 3325 (1990).
3. M. Drillon et al., *Phys. Rev. B* **40**, 10992 (1989).
4. S. Brehmer, H.-J. et al, *J. Phys.: Condens. Matter* **9**, 3921 (1997).
5. S. K. Pati et al., *Phys. Rev. B* **55**, 8894 (1997).
6. M. Drillon et al., *J. Magn. Magn. Matter.* **128**, 83 (1993).
7. A. Vicente et al., *Inorg. Chem.* **37**, 4466 (1998).
8. S. Yamamoto, *Phys. Rev. B* **59**, 1024 (1999).
9. T. Nakanishi and S. Yamamoto, *Phys. Rev. B* **65**, 214418 (2002).
10. S. Yamamoto, S. Brehmer and H-J. Mikeska, *Phys. Rev. B* **57**, 13610 (1998).
11. S. Yamamoto and T. Nakanishi, *Phys. Rev. Lett.* **89**, 157603 (2002).
12. H. Hori and S. Yamamoto, *Phys. Rev. B* **68**, 054409 (2003).
13. H. Hori and S. Yamamoto, *J. Phys. Soc. Jpn.* **73**, 1453 (2004).
14. H. Hori and S. Yamamoto, *J. Phys.: Condens. Matter* **16**, 9023 (2004).
15. T. Oguchi and F. Keffer *J. Phys. Chem. Solids* **25**, 405 (1964)
16. N. Fujiwara and M. Hagiwara *Solid State Commun.* **113**, 433 (2000)

TOPOLOGICAL ASPECTS OF WAVE FUNCTION STATISTICS AT THE ANDERSON TRANSITION

H. OBUSE* AND K. YAKUBO

Department of Applied Physics, Graduate School of Engineering, Hokkaido University, Sapporo 060-8628, Japan.

We investigate wave function statistics at the critical point of the Anderson transition in two-dimensional symplectic systems with different topologies. It is found that the distribution function of critical wave function amplitudes depends on the topology even in infinite systems, while the typical value of the correlation dimension does not change for alternating the system topology. Topological effects, however, survive even in very large systems.

1. Introduction

The disorder induced metal-insulator transition at absolute zero temperature, namely, the Anderson transition, shows similar behaviors to those of thermal critical phenomena near the critical point. In the critical region of the Anderson transition, physical quantities show power law behaviors with specific critical exponents. It is well known that values of these critical exponents depend only on the basic symmetry and the spatial dimension of the system (universality). It is also a fact that boundary conditions of systems play an important role in the Anderson transition. Indeed, the Thouless number quantifying a sensitivity of energy levels to changing boundary conditions gives crucial information on spatial extent of wave functions.[1,2] In a topological sense, a two-dimensional system with periodic boundary conditions has a torus topology, while a system with fixed boundary conditions possesses a square topology. Since phase transitions occur in infinite systems, topologies of systems do not seem to influence critical properties. However, considering the divergent correlation length at the critical point, this expectation cannot be accepted so naively. The

*Present address: Condensed Matter Theory Laboratory, RIKEN, Wako, Saitama 351-0198, Japan

relation between topologies of systems and their critical properties has been studied in previous works.[3–10] These works revealed that the critical level statistics,[3–5] the conductance distribution at the critical point,[6–9] and the scaling function of quasi-one dimensional localization lengthes[10] depend on the topology, while the universality class is maintained by changing the system topology.[10] Although boundary conditions do not affect the universality class, fractality of the critical wave function in a finite system may depend on the topology of the system. It is quite interesting to study how fractalities of critical wave functions in finite systems with different topologies become to coincide each other as increasing system size.

In this paper, we investigate wave function statistics at the critical point of the Anderson transition in systems with torus and square topologies. We found that the distribution function of critical wave function amplitudes depends on the topology even in infinite systems and the typical value of the correlation dimension does not change by alternating the system topology.

2. System and Methods

The system we treat here is described by the SU(2) model.[11] This model belongs to the symplectic class which shows the Anderson transition even in two dimensions. The Hamiltonian of the SU(2) model is compactly written in a quaternion form as

$$\boldsymbol{H} = \sum_i \epsilon_i \boldsymbol{c}_i^\dagger \boldsymbol{c}_i - V \sum_{i,j} \boldsymbol{R}_{ij} \boldsymbol{c}_i^\dagger \boldsymbol{c}_j, \tag{1}$$

where $\boldsymbol{c}_i^\dagger$ ($\boldsymbol{c}_i$) is the creation (annihilation) operator acting on a quaternion state vector and ϵ_i represents the on-site random potential distributed uniformly in the interval $[-W/2, W/2]$. (Quaternion-real quantities are denoted by bold symbols.) The strength of the hopping V is taken to be the unit of energy. The quaternion-real hopping matrix element $\boldsymbol{R}_{ij}$ between the sites i and j is given by

$$\begin{aligned} \boldsymbol{R}_{ij} &= \cos\alpha_{ij}\cos\beta_{ij}\boldsymbol{\tau}^0 + \sin\gamma_{ij}\sin\beta_{ij}\boldsymbol{\tau}^1 \\ &\quad - \cos\gamma_{ij}\sin\beta_{ij}\boldsymbol{\tau}^2 + \sin\alpha_{ij}\cos\beta_{ij}\boldsymbol{\tau}^3, \end{aligned} \tag{2}$$

for the nearest neighbor sites i and j, and $\boldsymbol{R}_{ij} = 0$ otherwise. $\boldsymbol{\tau}^\mu (\mu = 0, 1, 2, 3)$ is the primitive elements of the quaternion algebra. Random quantities α_{ij} and γ_{ij} are distributed uniformly in the range of $[0, 2\pi)$, and β_{ij} is distributed according to the probability density $P(\beta)d\beta = \sin(2\beta)d\beta$ for

$0 \leq \beta \leq \pi/2$. The critical disorder W_c of this model is known to be 5.952 at the energy $E = 1.0$.[11]

Critical wave functions $\psi_c(r)$ of the SU(2) model have been calculated by using the forced oscillator method[12] extended to eigenvalue problems of quaternion-real matrices. 10^4 critical wave functions for systems with torus and square topologies are calculated for system sizes $L = 12 - 120$.

3. Results

From the ensemble of critical wave functions, we calculate the distribution function of wave function amplitudes $F[\ln(t)]$, where $t = |\psi_c(r)|^2 L^2$, as shown in Fig. 1. Solid lines and dashed lines represent for systems with torus and square topologies, respectively, and colors represent system sizes. It is found that $F[\ln(t)]$ depends on the system topology at least for finite systems. To examine the influence of the topology in infinite systems, we introduce a quantity defined by

$$\alpha = \int_{-\infty}^{-2} F[\ln(t)] d[\ln(t)]. \tag{3}$$

The inset of Fig. 1 shows the system size dependence of the difference of α between different topologies, namely, $\Delta\alpha = \alpha^s - \alpha^t$, where α^t and α^s are

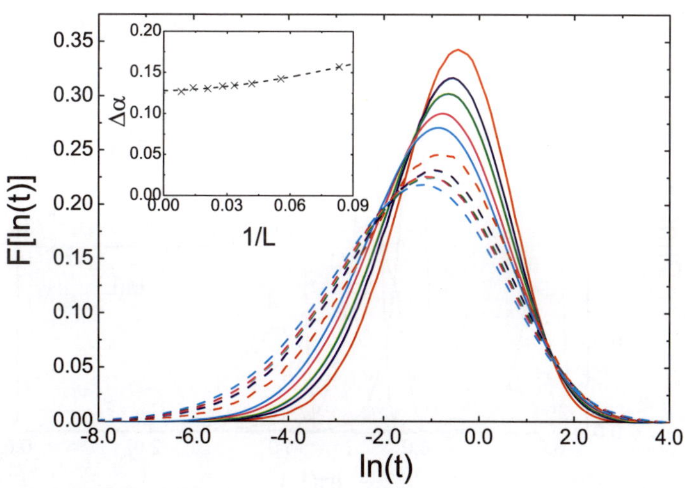

Figure 1. Distribution functions of wave function amplitudes $F[\ln(t)]$ for $L = 12$ (red), 24 (blue), 36 (green), 72 (pink), 120 (azure) with the torus topology (solid line) and the square topology (dashed line). The inset shows the system size dependence of $\Delta\alpha$. The dashed line represents a scaling correction at criticality, $\Delta\alpha = \Delta\alpha_\infty + aL^{-y}$.

calculated for $F[\ln(t)]$ with the torus and square topologies, respectively. The dashed line representing the scaling correction leads to $\Delta\alpha \neq 0$ for $L \to \infty$. The distribution function $F[\ln(t)]$, therefore, depends on the topology even in infinite systems.

The distribution function of the inverse participation ratio (IPR) $P[\ln(I_2)]$, where $I_2 = \int |\psi_c(r)|^4 d^d r$, is illustrated in Fig. 2, which exhibits that $P[\ln(I_2)]$ also depends on the topology. It is known that the typical value of the IPR behaves as $\langle I_2\rangle_{\text{typ}} = cL^{-(d+\tilde{D}_2)}$, where d and $\tilde{D}_2$ are the spatial dimension and the correlation dimension characterizing multifractality of critical wave functions, respectively. The inset of Fig. 2 suggests that values of the coefficient c for the square and torus topologies are different each other while the value of $\tilde{D}_2$ does not depend on the system topology.

In the context of the multifractal analysis, the correlation dimension is defined in a different way, namely,[13]

$$D_2 = \lim_{l \to 0} \frac{\log \sum_b \left[\sum_{i \in b(l)} |\psi_c(i)|^2\right]^2}{\log l}, \tag{4}$$

where the summation $\sum_{i \in b(l)}$ is taken over sites i within a small box b of size

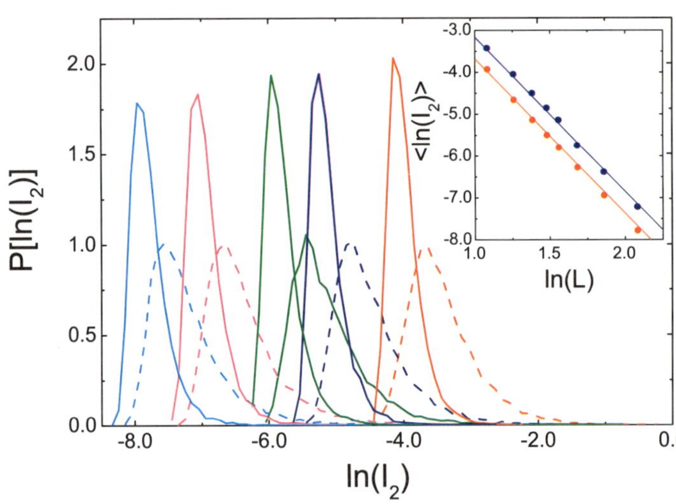

Figure 2. Distribution functions of IPR $P[\ln(I_2)]$. The meanings of line types and colors are the same as those in Fig. 1. The inset shows L dependences of $\ln\langle I_2\rangle_{\text{typ}}$ which is equal to the arithmetic mean of $\ln(I_2)$ for the system with the torus topology (red circles) and the square topology (blue circles). Lines represent power law fittings.

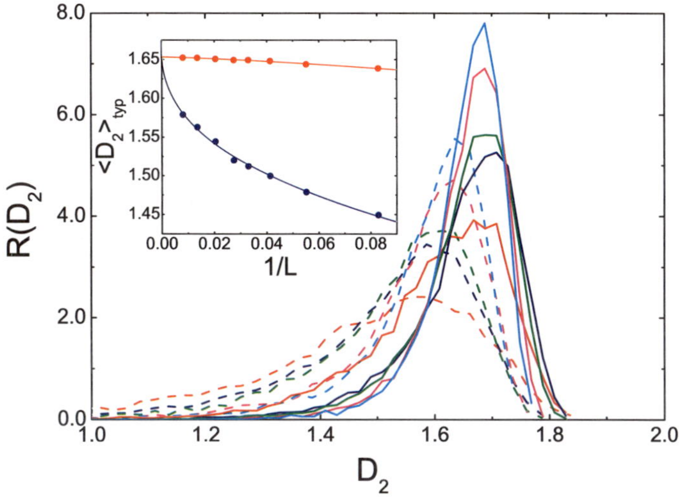

Figure 3. Distribution functions of the correlation dimension D_2. The meanings of line types and colors are the same as those in Fig. 1. The inset shows L dependences of the typical value of D_2 for the torus (red circles) and square (blue circles) topologies. Lines show power law fittings $\langle D_2 \rangle_{\rm typ} = D_2^\infty + bL^{-\gamma}$.

l, and $\sum_b$ is taken over such boxes. This exponent D_2 is generally different from $\tilde{D}_2$ defined via the IPR. The correlation dimension given by Eq. (4) is defined for individual wave functions, while $\tilde{D}_2$ stands for an ensemble of wave functions. Values of D_2 and $\tilde{D}_2$ coincide each other only when D_2 does not depend on disorder realization. We focus our attention to the system size dependence of D_2. Distribution functions of the correlation dimension D_2 are depicted in Fig. 3. The correlation dimension D_2 depends clearly on the sample, the system size, and the topology. However, as increasing system sizes, typical values of D_2 for both topologies come close each other. The inset of Fig. 3 shows the system size dependence of typical values of D_2. The L dependence of $\langle D_2 \rangle_{\rm typ}$ for the square topology is much stronger than that for the torus one. The solid lines represent power law fittings $\langle D_2 \rangle_{\rm typ} = D_2^\infty + cL^{-\gamma}$. We obtain that the values of D_2^∞ which gives the typical value of D_2 for $L \to \infty$ are 1.65 ± 0.01 for the torus topology and 1.69 ± 0.04 for the square topology, respectively. Therefore, the typical value of the correlation dimension is not affected by the topology for $L \to \infty$. However, the power law behavior of $\langle D_2(L) \rangle_{\rm typ}$ implies that topological effects survive even in very large systems.

Acknowledgments

We are grateful to T. Nakayama for helpful discussions. This work has been supported by the 21st Century COE program on "Topological Science and Technology" from the Ministry of Education, Culture, Sport, Science and Technology of Japan. Numerical calculations in this work have been mainly performed on the facilities of the Supercomputer Center, Institute for Solid State Physics, University of Tokyo.

References

1. D.J. Thouless, *Phys. Rep.* **C13**, 93 (1974).
2. D.C. Licciardello and D.J. Thouless, *J. Phys.* **C8**, 4157 (1975); Phys. Rev. Lett. **35**, 1475 (1975).
3. D. Braun, G. Montambaux, and M. Pascaud, *Phys. Rev. Lett.* **81**, 1062 (1998).
4. V.E. Kravtsov and V.I. Yudson, *Phys. Rev. Lett.* **82**, 157 (1999).
5. L. Schweitzer and H. Potempa, *Physica* **A266**, 486 (1999).
6. C.M. Soukoulis, X. Wang, Q. Li, and M.M. Sigalas, *Phys. Rev. Lett.* **82**, 668 (1999).
7. K.Slevin and T. Ohtsuki, *Phys. Rev. Lett.* **82**, 669 (1999).
8. D. Braun, E. Hofstetter, G. Montambaux, and A. MacKinnon, *Phys. Rev.* **B64**, 155107 (2001).
9. M. Rühländer, P. Markos, and C.M. Soukoulis, *Phys. Rev.* **B64**, 172202 (2001).
10. K. Slevin, T. Ohtsuki, and T. Kawarabayashi, *Phys. Rev. Lett.* **84**, 3915 (2000).
11. Y. Asada, K. Slevin, and T. Ohtsuki: *Phys. Rev. Lett.* **89**, 256601 (2002).
12. T. Nakayama and K. Yakubo, *Phys. Rep.* **349**, 239 (2001).
13. T. Nakayama and K. Yakubo, *Fractal Concepts in Condensed Matter Physics.* (Springer-Verlag, Berlin, 2003).

METAL-INSULATOR TRANSITION IN 1D CORRELATED DISORDER

HIROYUKI SHIMA AND TSUNEYOSHI NAKAYAMA

Department of Applied Physics, Graduate School of Engineering, Hokkaido University, Sapporo 060-8628, Japan

E-mail: shima@eng.hokudai.ac.jp

We investigate numerically critical properties of one-dimensional (1D) electron systems in the presence of long-range correlated diagonal disorder. Calculated results for the localization length ξ of eigenstates indicate the existence of the metal-insulator transition in 1D systems and elucidate non-trivial behavior of ξ as a function of the disorder strength. Finite-size scaling analysis is employed to obtain a precise value of transition points and the critical exponent ν, which reveals that every ν disobeys the Harris criterion $\nu > 2/d$.

1. Introduction

Non-interacting electron systems with uniformly distributed disorder usually exhibit the metal-insulator transition from localized to extended states; this is called the Anderson transition[1]. The localization of electron eigenstates induced by disorder originate from destructive quantum interference due to incoherent backscattering. It has long believed that the Anderson transition occurs for spatial dimension greater than two, that is, all eigenstates of a non-interacting electron in one-dimensional (1D) disordered systems are localized in the thermodynamic limit[2–4]. However, this is the case only for systems with spatially *uncorrelated* disorder. In fact, short-range correlated disorder was unexpectedly found to cause the appearance of extended states in 1D systems at spatial resonance energies[5,6]. Subsequently, short-range correlation in disordered potentials were put forward to explain unusual conducting properties of polymers[6] as well as those of semiconductor superlattices grown with correlated quantum well sequences[7].

The abovementioned findings have motivated the studies of the nature of 1D systems with *long-range* correlated disorder[8–12]. Particularly noteworthy is the system in which the sequence of on-site potentials $\{\varepsilon_i\}$ has a power-law spectral density of the form $S(k) \propto k^{-p}$; this type of disorder is also being studied in biophysics in connection with the large-distance transport in DNA sequence[10,13,14]. For exponents p greater than 2.0, there is a finite range of energy values with extended eigenstates. This indicates the presence of the Anderson transition in 1D disordered systems against the conclusion of the well-known scaling theory[1,2]. Nevertheless, there is only a few attempt to reveal quantitatively the critical properties of the Anderson transition in 1D systems, although it is crucial for a better understanding on the nature of the disordered system.

The purpose of the present work is to investigate critical properties of electron eigenstates in 1D systems with long-range correlated disorder. Numerical studies on the localization lengths ξ have elucidated the non-trivial behavior of ξ as a function of disorder width W and the exponent p. A series of critical points, W_c and p_c, and that of critical exponents, ν and μ, are determined by finite-size scaling analysis. Remarkably, the results of critical exponents disobey the Harris criterion, $\nu, \mu > 2/d$[15,16], which is believed to be satisfied in general disordered systems.

2. Model and method

The standard 1D Anderson model is described by a tridiagonal Hamiltonian

$$H = \sum_j \varepsilon_j |j\rangle\langle j| + t \sum_j \left(|j\rangle\langle j+1| + |j\rangle\langle j-1| \right), \tag{1}$$

where disorder is introduced on the site energies ε_i. The hopping energy t is taken as a unit of energy hereafter. A sequence of long-range correlated potential $\{\varepsilon_i\}$ is produced by the Fourier filtering method (See Ref. 17 for details). The resulting potentials ε_i are spatially correlated and produce the power-law spectral density $S(k) \propto k^{-p}$. In the following, the mean value $\langle \varepsilon_i \rangle$ is set to be zero and the periodic boundary condition is imposed.

Localization lengths of eigenstates in a potential field $\{\varepsilon_i\}$ are computed using the conventional transfer matrix method[4]. The localization length ξ at a given energy E is defined by the relation

$$\xi = L \left[\ln \frac{|\Pi_{i=1}^N \boldsymbol{Q}_i z(0)|}{|z(0)|} \right]^{-1}, \quad \boldsymbol{Q}_i \equiv \begin{pmatrix} E - \varepsilon_i & -1 \\ 1 & 0 \end{pmatrix}. \tag{2}$$

Since the calculated result of Eq. (2) depends on the choice of the potential field $\{\varepsilon_i\}$, we take its geometrical mean on more than 10^4 samples to obtain a typical value of ξ for a given L. Energy E is fixed at the band center $E = 0$ throughout this paper.

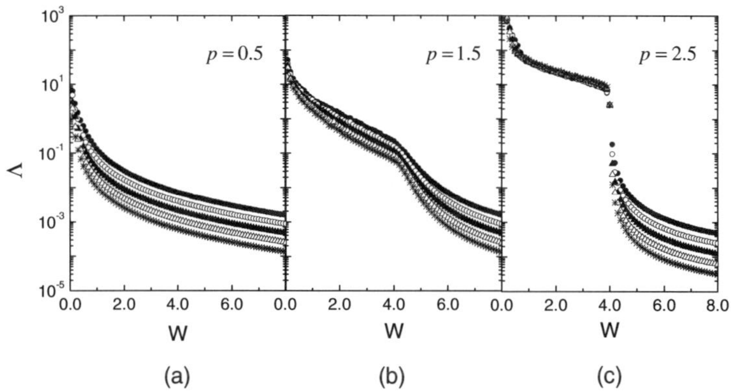

Figure 1. The normalized localization length $\Lambda \equiv \xi_L/L$ as a function of the distribution width W. The exponent p of the power-law spectral density $S(k) \propto k^{-p}$ is varied as displayed in the figure. The system size L is varied to $L = 2^{13}$ (solid circles), 2^{14} (open circles), 2^{15} (solid triangles), 2^{16} (open triangles), and 2^{17} (asterisks).

3. Localization lengths

Figure 1 plots the normalized localization length $\Lambda(W, L)$ as a function of the distribution width W defined by the relation $\varepsilon_i \in [-W/2, W/2]$. The exponent p of the power-law spectral density $S(k)$ is varied from $p = 0.5$ to 2.5. The system size L is increased from $L = 2^{13}$ to 2^{17} incrementally. When p is less than unity ($p = 0.5$), $\Lambda(W, L)$ is a monotonously decreasing function with W and L. This indicates that the eigenstates for $p = 0.5$ is spatially localized at any W. On the other hand, curves of Λ for larger p exhibit a kink at $W = 4.0$, which sharpens as p increases. For $p = 2.5$, the function Λ eventually shows a shoulder structure at around $W = 4.0$, engendering a plateau-like shape within the interval $0.5 < W < 4.0$. We have confirmed that the shoulder of $\Lambda(W)$ appears whenever $p > 2.0$, i.e., when p is large enough to yield extended eigenstates[8,18,19].

Note that the curves of $\Lambda(W, L)$ shown in Fig. 1 (c) merge together for $W < 4$, namely, the calculated results of Λ for $W < 4$ is invariant to the change in the system size L. This implies the eigenstate within a plateau-like region ($W < 4$ and $p > 2.0$) to be extended over the system, because, for extended states, the right hand side in Eq. (2) gives a quantity proportional to L. In contrast, the data of Λ for $W > 4$ monotonically decrease with increasing the system size L, indicating the eigenstates for $W > 4$ to be localized. We thus conclude that the critical point W_c separating localized and extended phases locates at around $W \sim 4$.

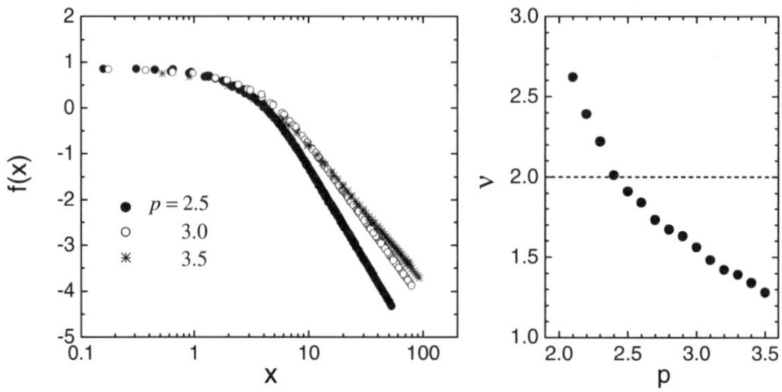

Figure 2. (a) Scaling plots of $\ln \Lambda = f(x)$. (b) The critical exponent ν for various values of p. The dotted line indicates the lower bound of ν given by the Harris criterion $\nu > 2/d$.

The behavior of Λ similar to that of Fig. 1 (c) has also been observed in 2D electron systems with long-range correlated disorder[20]. In the latter system, the curve of Λ for various L merge together for $W < W_c$, suggesting the presence of a line of critical points for $W < W_c$ (See Ref. 20 for details). This indicates that 2D system with long-range correlated disorder undergoes a disorder-driven Kosterlitz-Thouless (KT) metal-insulator transition. At this stage, we may not rule out the possibility that the same picture is applied to 1D systems we have considered, namely, the L-independent $\Lambda(W)$ within a plateau region indicates the occurrence of a KT-type transition. We believe, nevertheless, that the above similarity between 1D and 2D systems is fortuitous; preliminary studies based on the multifractal analysis have suggested that the eigenstates for $W < 4$ and $p > 2$ are extended[21], ensuring the presence of a single transition point $W = W_c$ in 1D systems.

4. Finite-size scaling analysis

The critical properties of the system are addressed by using the finite-size scaling analysis for the normalized localization length[4] $\Lambda = f(L/\xi)$. Figure 2 (a) shows scaling plots of $\ln \Lambda$ with the argument $x \equiv (W - W_c)L^{1/\nu}$. For each p, data of $\ln \Lambda$ for various values of W and L fit well onto a single curve. The resulting value of W_c is $W_c = 4.001 \pm 0.001$ for all ps. We have numerically confirmed that, at least for $2.0 < p < 4.5$, the critical point W_c is independent of p.

In contrast, numerical results of the critical exponent ν show a systematic dependence on p; the results are summarized in Fig. 2 (b). This

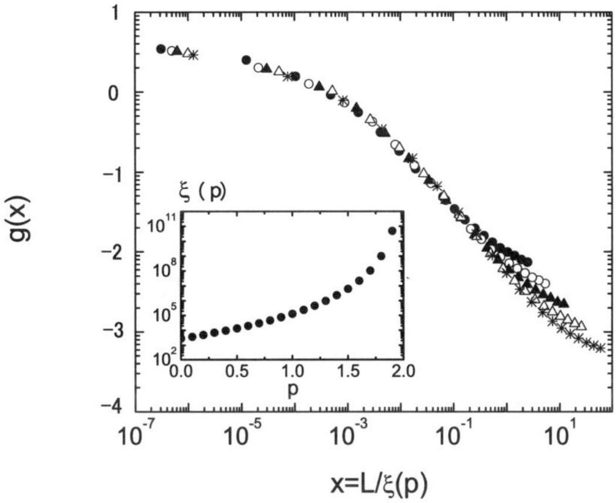

Figure 3. Scaling plots of $\ln \Lambda = g(x)$ with the argument $x = |p_c - p|L^{1/\mu}$. The disorder strength is fixed to $W = 3.0$, and the system size is varied from $L = 2^{13}$ (solid circles) to 2^{17} (asterisks) as same as the case of Fig. 1. Inset: Power-law behavior of the localization length $\xi \propto |p_c - p|^{-\mu}$ in the vicinity of the transition $p_c = 2.0$.

contradicts the principle of one-parameter scaling requiring that ν should be independent of a choice of parameters in the Hamiltonian of the system. The origin of the discrepancy remain unknown at present; a novel technique of scaling correction[22] would help to solve the problem. We have also obtained a scaling plot of $\ln \Lambda(p, L) = g(x)$ with the argument $x = (p_c - p)L^{1/\mu}$ as shown in Fig. 3, wherein the disorder strength is fixed to $W = 3.0$. The resulting transition point p_c, and the critical exponent μ defined by $\xi = \xi_0|p - p_c|^{-\mu}$, are estimated as $p_c = 2.0$ and $\mu = 6.7$, respectively. Similarly to the case of ν, The value of μ show a systematic dependence on W, which will be published elsewhere in detail[21].

Intriguingly, most results of the critical exponents ν and μ disobey the Harris criterion[15,16] $\nu, \mu > 2/d$. The inequality is widely believed to be satisfied in general disordered systems with any spatial dimension d, thereby apparently determining the lower bound both for ν and μ. It should be noted, however, that the Harris criterion was originally derived for a spatially *uncorrelated* disorder[15]. Therefore, the inequality may be violated in systems with long-range *correlated* disorder. In fact, for classical percolation model, the inequality must be modified in the presence of long-range correlation in the site or bond occupations[23]. To elucidate the lower bound of ν and/or μ for 1D quantum systems we have considered, we should generalize the argument in Ref. 16 for the correlated disorder producing the power-law spectral density $S(k) \propto k^{-p}$. Quite recently, numerical investigations on 3D disordered systems have suggested an extended Harris criterion that is valid for quantum systems with long-range correlated disorder[24].

Acknowledgment

This work is supported financially in part by a Grant-in-Aid for the 21st Century COE program "Topological Science and Technology". Numerical calculations were performed on the SR8000 of the Supercomputer Center, ISSP, University of Tokyo.

References

1. P. W. Anderson, Phys. Rev. **109**, 1469 (1958).
2. E. Abrahams *et al.*, Phys. Rev. Lett. **42**, 673 (1979).
3. P. A. Lee and T. V. Ramakrishnan, Rev. Mod. Phys. **57**, 287 (1987).
4. B. Kramer and A. MacKinnon, Rep. Prog. Phys. **56**, 1469 (1993).
5. D. H. Dunlap, H. -L. Wu, and P. Phillips, Phys. Rev. Lett. **65** (1990) 88.
6. H. -L. Wu and P. Phillips, Phys. Rev. Lett. **66** (1991) 1366; P. Phillips and H. -L. Wu, Science **252**, 1805 (1991).
7. V. Bellani *et al.*, Phys. Rev. Lett. **82** (1999) 2159.
8. F. A. B. F. de Moura and M. L. Lyra, Phys. Rev. Lett. **81**, 3735 (1998).
9. S. Russ *et al.*, Phys. Rev. B **64**, 134209 (2001).
10. P. Carpena *et al.*, Nature (London) **418**, 955 (2002)
11. F. A. B. F. de Moura *et al.*, Phys. Rev. B **68**, 012202 (2003).
12. F. M. Izrailev and N. M. Makarov, Phys. Rev. B **67**, 113402 (2003); Appl. Phys. Lett. **84**, 5150 (2004).
13. S. Roche *et al.*, Phys. Rev. Lett. **91**, 228101 (2003).
14. H. Yamada, Int. J. Mod. Phys. B, **18**, 1697 (2004), Phys, Lett. A, **332**, 65 (2004).
15. A. B. Harris, J. Phys. C **7**, 1671 (1974); Z. Phys. B **49**, 347 (1983).
16. J. T. Chayes *et al.*, Phys. Rev. Lett. **57**, 2999 (1986).
17. J. Feder, *Fractals* (Plenum Press, New York, 1988).
18. H. Shima, T. Nomura and T. Nakayama, Phys. Rev. B **70**, 075116 (2004).
19. H. Shima and T. Nakayama, Microelec. J. **36**, 422 (2005).
20. W-S. Liu, T. Chen, and S-J. Xiong, J. Phys.:Condens. Matter **11**, 6883 (1999).
21. H. Shima and T. Nakayama, *unpublished.*
22. K. Slevin and T. Ohtsuki, Phys. Rev. Lett. **82**, 382 (1999).
23. A. Weinrib and B. I. Halperin, Phys. Rev. B **27**, 413 (1983).
24. M. L. Ndawana, R. A. Römer and M. Schreiber, Europhys. Lett. **68**, 678 (2004).

SUPERCONDUCTIVITY IN URu_2Si_2 UNDER HIGH PRESSURE

K. TENYA, I. KAWASAKI AND H. AMITSUKA
Graduate School of Science, Hokkaido University, Sapporo 060-0810, Japan

M. YOKOYAMA
Faculty of Science, Ibaraki University, Mito 310-8512, Japan

N. TATEIWA
Advanced Science Research Center, Japan Atomic Energy Research Institute, Tokai 319-1195, Japan

T. C. KOBAYASHI
Department of Physics, Okayama University, Okayama 700-8530, Japan

Magnetization measurements of heavy fermion superconductor URu_2Si_2 have been performed under hydrostatic pressures. The superconducting transition temperature decreases linearly with pressure up to 0.8 GPa and disappears around 1.2 GPa. The hysteretic magnetization shows charasteristic behaviors at high pressures, suggesting that the pinning properties change due to the switch of the normal state from the hidden order phase to the antiferromagnetic phase.

1. Introduction

The heavy fermion system which is defined by a large electronic specific heat γT at low temperatures has the rich variety of different types of ground states; normal Fermi liquid, non-Fermi liquid, magnetic ordering, quadrupolar ordering and superconductivity. Among heavy fermion compounds, URu_2Si_2 has been attracting much interest because of the coexistence between unconventional superconducting phase and so-called hidden order (HO) phase with tiny antiferromagnetic moments (0.02 - 0.04 μ_B/U) at ambient pressure. The origin of the latter HO phase is unclear up to now.

Recent microscopic experiments under hydrostatic pressure have revealed that inhomogeneous antiferromagnetic (AF) phase with normal-size moments develops between $\sim$ 0.4 GPa and $\sim$ 1.2 GPa,[1] indicating that (i) the AF phase is separated from the coexisting HO phase by a first-order transition and (ii) the AF phase with a small volume fraction is the origin of the weak antiferromagnetism at ambient pressure.[2] On the other hand, there are a few reports on the pressure dependence of the superconductivity in URu_2Si_2, and it is still open whether the superconductivity coexists with the inhomogeneously-developed AF phase. When the pressure-induced AF phase is spatially separated from the superconducting phase, it should work as additional pinning centers in the superconducting mixed state. The hysteretic magnetization, which reflects the flux-pinning properties, is expected to change drastically as the volume fraction of the inhomogeneous AF phase increases with pressure.

In order to investigate both superconducting and pinning properties under hydrostatic pressure, static magnetization measurements have been performed under pressures up to 1.5 GPa.

2. Experimental

A single crystalline URu_2Si_2 was prepared by the Czochralski pulling method in the tetra-arc furnace and was annealed at 1000 °C for a week. The superconducting transition temperature T_c was 1.30 K at ambient pressure. Magnetization measurements were performed at temperatures ranging from 0.1 K to above T_c and in the field parallel to the c-axis, using a Faraday force capacitive magnetometer installed in the ^{3}He-^{4}He dilution refrigerator. Hydrostatic pressure was applied by means of a piston cylinder device made of copper-beryllium and tungsten-carbide.

3. Results and Discussions

Pressure variations of the isothermal magnetization curve are displayed in Fig. 1. Here the normal state contributions, which are insensitive to the pressure, are subtracted. As seen from the figure, the superconductivity exists at the pressure of 1.1 GPa where the AF phase is almost dominant in the normal state above the upper critical field H_{c2}.[1] The volume fraction of the superconducting phase, which can be estimated from the initial slope of the zero-field-cooled magnetization, decreases to much less than 40 % at 1.1 GPa.

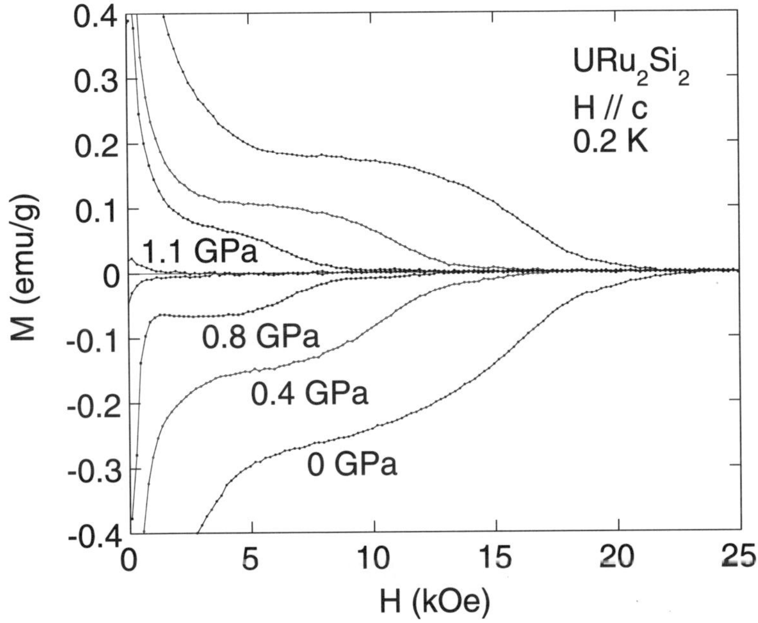

Figure 1. Pressure variations of the magnetization curve in URu_2Si_2 where the normal state contributions are subtracted.

The strong irreversibility appearing at low fields is due to the ordinary flux pinning effect. The hysteresis ΔM decreases as the field increases and becomes reversible just above H_{c2}. There is no peak effect in ΔM that is often observed in heavy fermion superconductors such as UPt_3.[3] At ambient pressure H_{c2} is 23.2 kOe. With increasing pressure both H_{c2} and ΔM decrease rapidly. It should be noted that the shapes of ΔM curves are unchanged at low pressures; ΔM has a plateau in the intermediate field range between $1/3H_{c2}$ and $2/3H_{c2}$. Furthermore, the plots of ΔM vs. H/H_{c2} at ambient pressure show similar behavior to those at 0.4 GPa. Since ΔM is proportional to the pinning force, the above results suggest that no additional pinning centers are developed by applying pressure at least up to 0.4 GPa. On the other hand, ΔM has no plateau at 1.1 GPa and rapidly decreases with field.

According to the μSR measurements of the single crystalline URu_2Si_2, the volume fraction of the AF phase abruptly develops up to ~ 80 % in the narrow pressure range 0.4 GPa $\leq P \leq$ 0.8 GPa; the normal state above H_{c2} is the HO (AF) phase at low (high) pressures.[4] The change of ΔM behavior

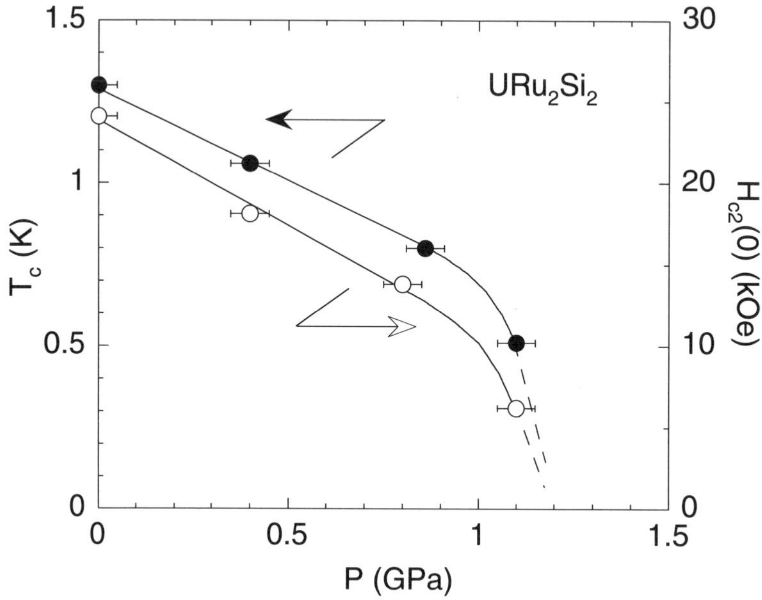

Figure 2. Pressure dependences of the superconducting critical temperature T_c and the extrapolated upper critical field $H_{c2}(0)$ for $H \parallel c$.

at 1.1 GPa seems to originate from the switch of the normal state to the AF state[1], reflecting the topology change of the pinning center.[5] It should be noted that the drastic change of ΔM behavior at high pressure as well as the reduction of the superconducting volume fraction with pressure does not suggest that the superconductivity coexists with the AF phase.

The superconducting transition temperature T_c as well as the upper critical field extrapolated to 0 K, $H_{c2}(0)$, decreases almost linearly with pressure up to 0.8 GPa, as shown in Fig. 2. The superconductivity seems to disappear around 1.2 GPa, in good agreement with the previous results obtained from the resistivity measurements.[6] The linear T_c decrease in the pressure range 0.4 GPa $\leq P \leq$ 0.8 GPa where the AF volume fraction abruptly develops indicates that the AF volume fraction in the normal state has little effects on the onset of the superconductivity.

Figure 3 shows pressure variations of the upper critical field H_{c2} for $H \parallel c$. The transition between superconducting state and normal state is of second order in the whole pressure range. At ambient pressure H_{c2} is strongly suppressed at low temperatures; the extrapolated $H_{c2}(0)$ is much smaller than the orbital critical field estimated from the slope at T_c.[7] At 0.4

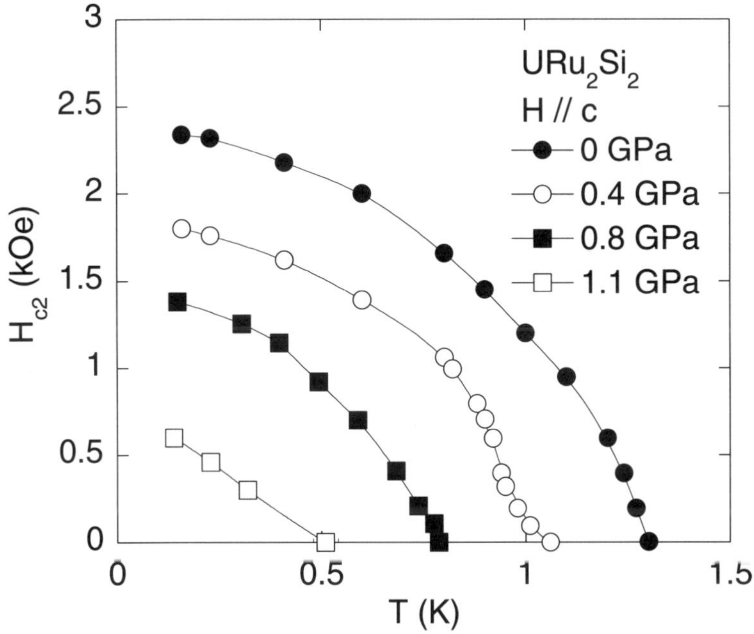

Figure 3. Pressure variations of $H_{c2}(T)$ for $H \parallel c$ in URu_2Si_2.

GPa $H_{c2}(T)$ is strongly suppressed at low temperatures as well; the ratio $H_{c2}(0)/T_c$ is almost same as that at ambient pressure. On the other hand, at 1.1 GPa $H_{c2}(0)/T_c$ is strongly reduced while $H_{c2}(T)$ is not so suppressed at low temperatures.

4. Summary

Magnetization measurements in the single crystalline URu_2Si_2 have been performed under hydrostatic pressures. The hysteretic magnetization ΔM which reflects the pinning properties behaves differently at high pressures where AF phase is dominant above H_{c2}, suggesting that the superconductivity does not coexist with the pressure-induced AF phase. The superconducting transition temperature T_c linearly decreases with pressure up to 0.8 GPa, suggesting that the inhomogeneous AF phase has little effects on the superconductivity.

This work has been partially supported by the 21COE program on "Topological Science and Technology" from the Ministry of Education, Culture, Sport, Science and Technology of Japan.

References

1. H. Amitsuka, K. Tenya, M. Yokoyama, A. Schenck, D. Andreica, F. N. Gygax, A. Amato, Y. Miyako, Ying Kai Huang and J. A. Mydosh, *Physica* **B326**, 418 (2003).
2. K. Matsuda, Y. Kohori, T. Kohara, K. Kuwahara and H. Amitsuka, *Phys. Rev. Lett.* **87**, 087203 (2001).
3. K. Tenya, M. Ikeda, T. Tayama, T. Sakakibara, E. Yamamoto, K. Maezawa, N. Kimura, R. Settai and Y. Onuki, *Phys. Rev. Lett.* **77**, 3193 (1996).
4. H. Amitsuka and M. Yokoyama, *Physica* **B329-333**, 452 (2003).
5. A. M. Campbell and J. E. Evetts, *Adv. Phys.* **21**, 199 (1972).
6. M. W. McElfresh, J. D. Thompson, J. O. Willis, M. B. Maple, T. Kohara and M. S. Torikachvili, *Phys. Rev.* **B35**, 43 (1987).
7. N. R. Werthamer, E. Helfand and P. C. Hohenberg, *Phys. Rev.* **147**, 295 (1966).

VI

Topology in Optics

OPTICAL VORTICULTURE

M. V. BERRY

H H Wills Physics Laboratory, Bristol University, Tyndall Avenue, Brisrol BS8 1TK, UK

Lines of topological singularity in the phase and polarization of light are being intensively studied now [1], motivated in part by a theoretical paper published thirty years ago [2]. However, the subject has a very long prehistory, that is not well known.

In puzzling over Grimaldi's observations of edge diffraction in the 1660s, Isaac Newton narrowly missed discovering phase singularities in light. The true discovery of phase singularities was made by William Whewell [3] in 1833, not in light but in the pattern of ocean tides. The first polarization singularity was observed (but not understood) by Arago in 1817, in the pattern of polarization of the blue sky. A different polarization singularity was predicted by Hamilton in the 1830s, in the optics of transparent biaxial crystals (this was also the first 'conical intersection' in physics).

After reviewing this history, the general structure of the singularities, as we understand them today, will be presented. Phase singularities have several aspects [4,5]: as vortices, around which the current (lines of the Poynting vector) circulates; as lines on which the phase of the light wave is undefined; as nodal lines, where the light intensity is zero; and as dislocations [2], where the wavefronts possess singularities closely analogous to the edge and screw dislocations of crystal physics. Polarization singularities are lines [5,6] of two types: C lines, where the polarization is purely circular, and L lines, where the polarization is purely linear.

Then, three modern applications of optical singularities will be described. The first [7] is the pattern of optical vortices behind a spiral phase plate, which is a device, commonly used to study phase singularities, that introduces a phase step into a light beam. The intricate dance of the vortices as the height of the step is varied (especially complicated near halfinteger multiples of 2π) is a surprising illustration of how vortices behave in practice. Experiment confirms the theory [8].

The second application is to knotted and linked vortex lines. A mathematical construction [9,12] leads to solutions of the wave equation whose vortices have the topology of any chosen knot on a torus. The knots are described by two integers m, n (if m and n have a common factor N, the ' knot ' consists of N linked loops). The construction can be implemented experimentally [11]. Vortex knots and links also exist in quantum waves [12].

The third application is a prediction of quantum effects near the phase singularities of classical light. This is motivated by a philosophical aspect [13,14] of singularities in physics. They have a dual role: as the most important predictions from any physical theory, and also as a signal that the theory is breaking down. In light, the phase singularities are threads of darkness, offering a window through which can be seen the faint fluctuations of the quantum vacuum [15]; the radius of this 'quantum core' can be calculated. Analogous cores exist in sound waves.

References

1. M. V. Berry *et al*, J. Optics A 6, (Editorial introduction to special issue) (2004).
2. J. F. Nye and M. V. Berry, Proc. Roy. Soc. Lond. A336, 165 (1974).
3. W. Whewell, Phil. Trans. Roy. Soc. Lond. 123, 147 (1833).
4. M. V. Berry, in SPIE 3487, 1 (1998).
5. J. F. Nye, Natural focusing and fine structure of light: Caustics and wave dislocations. Institute of Physics Publishing, Bristol (1999).
6. J. F. Nye and J. V. Hajnal, Proc. Roy. Soc. Lond. A409, 21 (1987).
7. M. V. Berry, J. Optics. A 6, 259 (2004).
8. J. Leach *et al*, New Journal of Physics 6, 71 (2004).
9. M. V. Berry and M. R. Dennis, Proc. Roy. Soc. Lond. 457, 2251 (2001).
10. M. V. Berry and M. R. Dennis, J. Phys. A 34 (2001).
11. J. Leach *et al*, Nature 432, 165 (2004).
12. M. V. Berry, Found. Phys. 31, 659 (2001).
13. M. V. Berry, in Proc. 9th Int. Cong. Logic, Method., and Phil. of Sci., edited by D. Prawitz, B. Skyrms, and D. Westerståhl (1994), pp. 597.
14. M. V. Berry, Physics Today, May, 10 (2002).
15. M. V. Berry and M. R. Dennis, J. Optics A 6, S178 (2004).

THE TOPOLOGY OF VORTEX LINES IN LIGHT BEAMS

M. J. PADGETT, K. O'HOLLERAN, J. LEACH AND J. COURTIAL

Department of Physics and Astronomy,
University of Glasgow,
Glasgow G12 8QQ, UK
E-mail: m.padgett@physics.gla.ac.uk

M. R. DENNIS

School of Mathematics
University of Southampton,
Southampton SO17 1BJ, UK

Optical vortices, or phase singularities, are places in a scalar optical wave field where the phase is not defined. Within any beam cross-section they are observed as points of darkness. However, in 3D, these singularities are lines of darkness embedded within the light. The topologies of these dark lines can be imaged by finding the phase singularities in successive cross-sections.

1. Introduction

When light waves traveling in different directions intersect, destructive interference can lead to complete cancellation giving places of zero intensity in the light field. Whenever there are three or more waves, these places of zero intensity form lines of darkness embedded within the light[1]. The resulting effect is the typified by the scattering of a spatially coherent ($\approx$ plane wave) laser beam, giving many plane waves propagating in various directions which interfere to produce laser speckle, and hence a complicated 3D structure of dark lines[2]. On these lines, the phase is ill-defined; around each line, the phase of the optical field advances or retards by 2π. The optical energy flow associated with the phase circulates around the dark lines, leading them to be called optical vortices [3]. They are also called phase singularities or wave dislocations[1].

The resulting beams are examined by imaging the intensity and/or phase distribution in successive cross-sections, within which the vortices are identified as points. The $\pm 2\pi$ phase change around the singular points

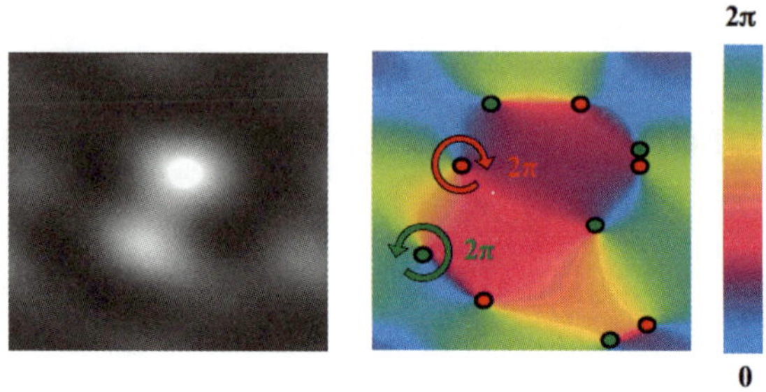

Figure 1. When multiple plane waves interfere (seven in this example), the interference pattern (left) contains many phase singularities (right).

is called the vortex strength (or topological charge) (Figure 1). When moving from one cross-section to the next, one observes that two neighbouring vortex points of opposing signs can spontaneously appear or disappear[4]. Consequently, such features have previously been referred to respectively as the birth and annihilation of optical vortices. This choice of language implies a temporal dynamical evolution; however, as the fields are monochromatic, the interference patterns and vortex positions are independent of time. Rather than describing vortex pairs in terms of birth and annihilation, it is better to visualize the vortex structure as dark lines embedded within the light field, which are free to bend back on themselves to form open or closed loops (Figure 2). The sign of a vortex is defined with respect to the observation plane; in the 3D picture, the optical phase does not undergo abrupt change[5], but the vortex sign reverses where the vortex line tangent is in the observation plane. The sense of phase circulation, and hence energy flow, around the vortex line is preserved both at these turning points and indeed along the whole length of the vortex line.

Within crystallography screw, edge and mixed dislocations are encountered. Examining the phase cross-section of light similar patterns are found. This has lead to the same language being adopted [1,6]. However, in the case of light, these terms simply describe the orientation of vortex lines with respect to propagation direction: parallel (screw), perpendicular (edge) and intermediate (mixed). In general, this dislocation classification cannot be applied without a well-defined propagation direction [7]. Recognition of the 3D nature of these vortex lines is particularly relevant as we progress to consider their topological properties.

In a field made by superposing three plane waves, the vortex lines are

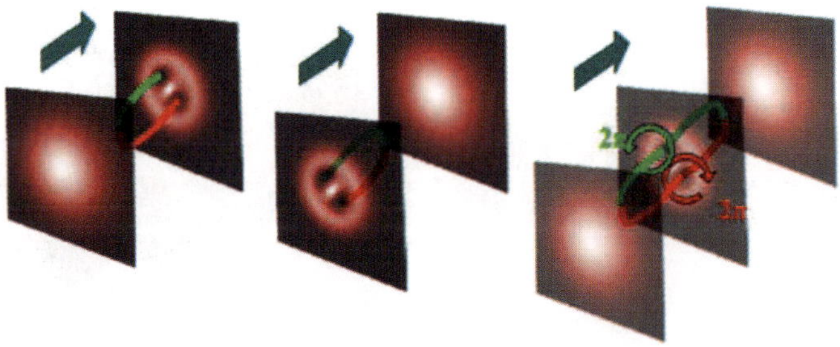

Figure 2. The birth (left) and annihilation (centre) of vortices can be understood in terms of vortex loops (right) embedded within the light.

straight and parallel. For more waves this is not the case, and in general the lines are curved and can form closed loops[6]. Recently[8] we experimentally verified predictions of Berry and Dennis[9], showing that vortex loops can be linked or even knotted. In those experiments we used higher charge vortices where the phase around the vortex line changed by a integer multiple of 2π. However, higher-strength beams are never found in nature and require specifically designed holograms for their generation[10] (Figure 3).

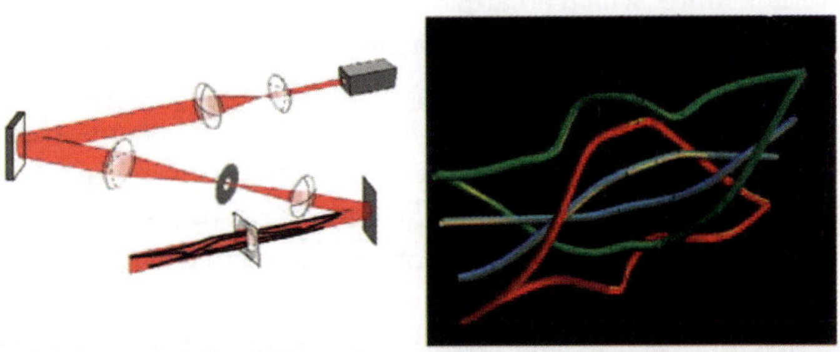

Figure 3. The link structure formed from two vortex loops within a light beam produced using a computer-generated hologram implemented on a spatial light modulator (SLM), and imaged in several planes along the propagation direction.

As the link/knot experiment demonstrates, light beams containing optical vortices are readily produced by using diffractive optical components (computer generated holograms) to modify the output beam from a conventional laser. All diffractive components can be considered as phase gratings where the phase distribution of the target beam is added to that of a diffraction grating, which itself is blazed to enhance the power in the first diffraction order (Figure 4). Typically, diffraction efficiencies of over 50%

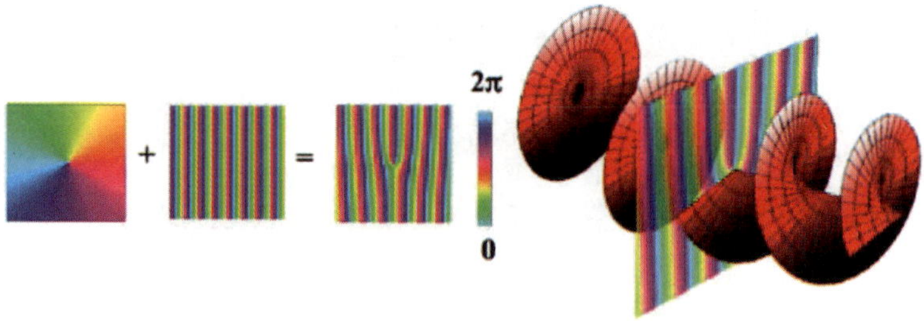

Figure 4. A hologram form producing a beam containing an optical vortex is formed by adding a phase singularity to a diffraction grating.

can be achieved from commercially available devices.

2. Numerical investigation into vortex topology

More recently we have reconsidered the general case of interference between multiple plane waves. Specifically we have used both numerical modeling and experimental investigation to understand the topological properties of vortex lines. Numerically, the volume over which we calculate the interference pattern is finite, which creates ambiguity at the volume edge regarding whether or not the vortex line forms a closed loop. We overcome this problem by restricting the possible plane wave directions to lie on a regular, rectangular grid in k-space, with spacing Δk. This ensures that the resulting interference pattern is also periodic with an axial repeat period of $4\pi k_0/(\Delta k)^2$ (i.e. the Talbot effect [11]) and a transverse repeat period of $2\pi/\Delta k$. The resulting interference pattern calculated within this 'Talbot cube' can be tiled to give the vortex topology over all space (Figure 5).

Using this approach we can investigate the vortex topology as the number of interfering plane waves is increased above three. As stated above, when three equal-amplitude plane waves interfere, regardless of their angles of intersection, the resulting interference pattern contains parallel vortices in the direction of the average k-vector. For this direction, k_r and k_z are the same for each of the beams meaning that the intensity distribution of transverse position of the vortex lines is propagation invariant. In general, for four plane waves, the values of k_r and k_z cannot all be the same, giving a beam cross-section which changes with propagation. Interestingly, when the four waves have equal amplitudes, the vortex pattern forms a characteristic 'criss-cross' topology of straight lines, with two characteristic vortex directions and intersecting vortex lines. Increasing the amplitude

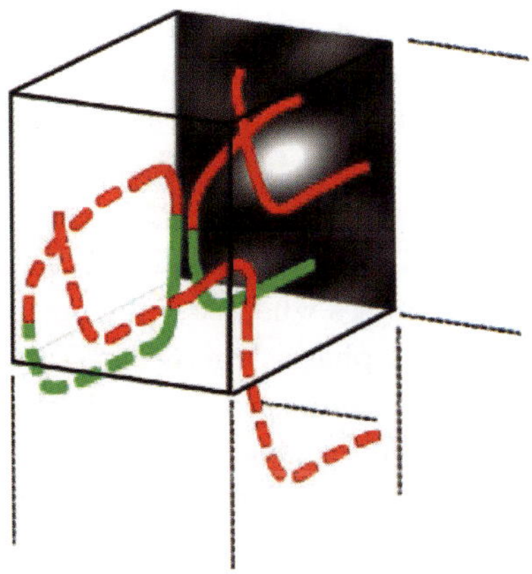

Figure 5. Tiling the interference pattern calculated over the 'Talbot cube' gives the topology over all space.

of any plane-wave component breaks this pattern into isolated vortices or individual vortex loops. Increasing the number of waves to five, in general, produces combinations of vortex curves and loops (Figure 6).

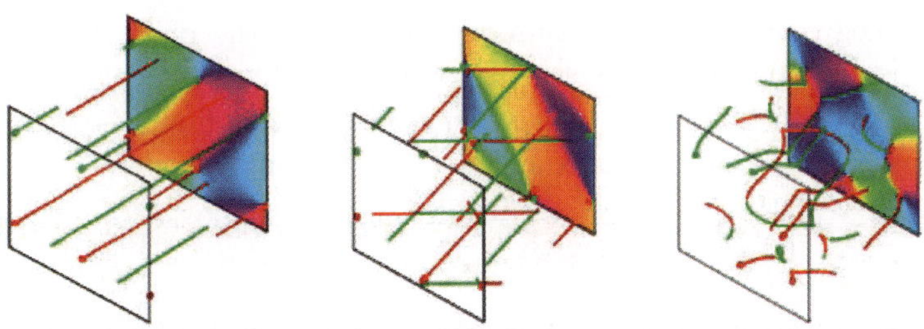

Figure 6. The numerically calculated topologies of vortex lines formed by the interference of three (left), four (centre) and five (right) plane waves.

As the number of wave increases, adjustment of the relative weights (amplitude and phase), as well as the directions of the plane wave components, results in a great variety of topologies. However, it seems that even if loops can be formed in this way, they are rarely (if ever) linked, knotted or even simply threaded by a vortex line. This leads us to speculate that in random light fields, linked or knotted loops are extremely rare topological features and that our experimentally produced links and knots are indeed only obtained over a narrow parameter range.

Numerically, the location of a vortex within a particular cross-section is found by examining the phase structure around each point, since the phase changes by $\pm 2\pi$ in a small circuit around a strength ± 1 vortex point. The knowledge of whether vortex points in neighbouring planes are on the same vortex line is ultimately limited by the spatial resolution to which the phase structure has been measured or calculated. However, since vortex lines are continuous, ambiguity only arises when two different vortex lines approach to combine or cross. Once the phase structure has been calculated over the whole Talbot cube, the vortex points are located in planes normal to each of the three axes. This resolves any ambiguity, and allows the configuration of the vortex lines to be found. A vortex line, traced though the Talbot cube, can leave and re-enter at the associated point on the opposite face (possibly many times) before arriving back at the starting position. Whether the associated vortex line forms an open or closed loop is deduced from the net number of cube crossings. If this is zero, the vortex line has returned to the same cube and hence formed a closed loop.

Vortex links are harder to identify since they do not depend on a single vortex line alone. However, the phase topology of the field can be used to establish whether a given vortex loop is threaded by other vortices. The total topological singularity strength threading a vortex loop affects the topology of the phase within the loop, since the phase change on a circuit just inside the loop is a nonzero multiple of 2π [9,7]. This topological feature of vortex threading opens the possibility for finding vortex loops using a numerical hunt of an extended, measured or calculated, interference pattern.

3. Experimental investigation into vortex topology

Spatial light modulators (SLMs) are readily available, providing video resolution control of the phase structure of a light beam. They are now being applied to holographic beam shaping in applications as diverse as optical tweezers[12], atom trapping and adaptive optics.

In all but a few special cases, when a phase-only SLM is programmed to produce a number of optical traps, the light beams due to the additional traps, located at intermediate and related positions, corrupt the desired pattern of optical traps in the Fourier plane. When used to create multiple-beam interference patterns, these additional beams result in reduced contrast and significantly degrade the quality of the Talbot image revival. Recently, we have reported an algorithm that gives control over both the phase and intensity of the diffracted beams even when using a phase-only SLM[8]. After designing the diffractive element (hologram) so that the first-order diffracted beam has the desired phase structure, the intensity at any point can be attenuated by adjusting the efficiency of the blazing in the corresponding region of the hologram.

Taking the plane of the SLM as the plane of interest, the initial stage in hologram design is to specify the desired weights, directions, and relative phases of the interfering beams, and calculate the phase distribution of the desired superposition, $\Phi(x, y)$. This phase distribution is then added to that of a blazed diffraction grating $\Phi(x, \Lambda)$, of period Λ. The first-order diffracted beam is then angularly separated from the other orders, which are subsequently blocked using a spatial filter. The intensity distribution of the desired superposition, $I(x, y)$, is also calculated and normalized so the maximum intensity is one. This intensity distribution is applied as a multiplicative mask to the phase distribution of the hologram, acting as a selective beam attenuator imposing the necessary intensity distribution on the first order diffracted energy. Thus the phase pattern applied to the SLM to create the desired hologram is

$$\Phi_{\text{holo}}(x, y) = \left[\left((\Phi(x, y) + \Phi(x, \Lambda))_{\text{mod}2\pi} - \pi\right)\right] \times \text{sinc}^2(\pi(1 - I(x, y))) + \pi. \tag{1}$$

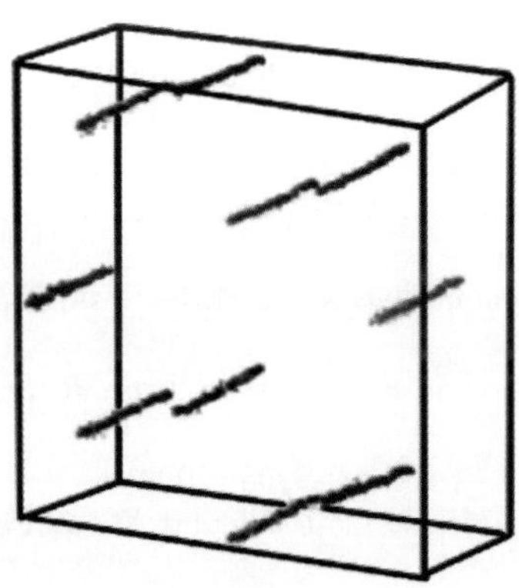

Figure 7. Example of an experimentally determined vortex structure.

Alternatively, the SLM can be programmed with a phase distribution corresponding to a randomised surface, producing an interference pattern akin to a speckle pattern. Hence, the apparatus allows comparison of calculated topologies to those obtained for both interfering plane waves and more generalised patterns. The 3D intensity of the resulting interference pattern is observed directly using a camera mounted on a motorised stage. Introducing an additional plane wave gives an interference pattern from which the phase of the original superposition of beams can be deduced, again over three dimensions (Figure 7).

4. Conclusions

We have preformed numerical and experimental studies of the topologies of vortex lines created within both simple and complex interference patterns and this work is continuing. Specifically we are trying to understand the generic condition by which the vortex lines form different topologies. Our ultimate goal is to discover if a particular statistics of a scattering surface result in the vortex loops of the resulting interference pattern to be linked or knotted.

Acknowledgements

We are grateful to Michael Berry for useful discussions. JC and MRD are supported by the Royal Society of London.

References

1. J. F. Nye and M. V. Berry, *Proc. R. Soc. Lond. A* **336**, 165–190 (1974).
2. M. V. Berry and M. R. Dennis, *Proc. Roy. Soc. Lond. A* **456**, 2059–2079 (2000).
3. M. J. Padgett and L. Allen, *Opt. Commun.* **121**, 36-40 (1995).
4. D. Rozas, C. T. Law and G. A. Swartzlander, *J. Opt. Soc. Am. B* **14** 3054–65 (1997).
5. M. V. Berry, in *Singular Optics*, ed M. S. Soskin, *SPIE Proc* **3487** pp 1–5 (1998).
6. M. V. Berry, J. F. Nye and F. J. Wright, *Proc. R. Soc. Lond. A* **291** 453–484 (1979).
7. M. R. Dennis, *J. Opt. A* **6** S202-S208 (2004).
8. J. Leach, M. R. Dennis, J. Courtial and M. J. Padgett, *Nature* **432** 165 (2004).
9. M. V. Berry and M. R. Dennis, *Proc. R. Soc. Lond. A* **457** 2251–2263 (2001).
10. J. Leach, M. R. Dennis, J. Courtial, and M. J. Padgett, *New J. Phys.* **7** 55 (2005).
11. M. V. Berry and S. Klein, *J. Mod. Opt.* **43** 2139–2164 (1996).
12. D. G. Grier, *Nature* **424** 810–816 (2003).

OPTICAL SPIN VORTEX: TOPOLOGICAL OBJECTS IN NONLINEAR POLARIZATION OPTICS

HIROSHI KURATSUJI AND SHOUHEI KAKIGI

Department of Physics, Ritsumeikan University-BKC, Kusatsu City, Shiga 525-8577, Japan

A theoretical study is given of a new type of optical vortex accompanying nonlinear birefringence that is induced by the Kerr effect. This is called the optical spin vortex. To treat this object we start with the two-component nonlinear Schrödinger equation. The vortex is inherent in the spin texture caused by an anisotropy of dielectric tensor, for which a role of spin is played by the Stokes vector (or pseudo-spin). By using the effective Lagrangian for the pseudo-spin field, we give an explicit form for the vortex solution. We also examine the evolution equation of the new vortex with respect to the propagation direction.

1. Introduction

The optical vortex has been explored since the advent of nonlinear optics. (For example, see Refs. 1 and 2). In this report we give an overview of a new class of vortex that is expected to occur in nonlinear polarization optics. Specifically, we are concerned with the vortex in nonlinear optical media exhibiting "birefringence". This is called "optical spin vortex" (OSV). Amazing is that there is a close resemblance between OSV and a vortex that occurs in superfluid He3A or ferromagnet. The apperance of OSV is inherent in the polarizaiton degree of light wave which is described by the "Stokes parameters" (just an analogue of conventional spin), so it is natural to expect the vortex in analogy with the usual spin vortex condensed matter physics.

The basic equation is given by the two-component nonlinear Schrödinger equation (abbreviated as TCNLS) , which is derived from the Maxwell equation by adopting an envelope approximation. This is reduced to the field equation written in terms of the Stokes parameters by constructing the effective Lagrangian. This form is shown to naturally incororate the solution of OSV. We address two features of the OSV. (i): The first is concerning the profile for a single vortex; the profile function will be numerically given.

(ii): The second is to examine the evolutional behavior of the vortex; that is described by the equation of motion of the vortex with respect to the propagation distance. It is shown that the topological invariant is incorporated in the equation of motion of vortex.

2. Field induced nonlinear birefringence and TCNLS

We start with the brief sketch for the TCNLS for the light wave traveling through the nonlinear substance. First thing to be done is to fix the anisotropy induced by the Kerr effect. Following the standard procedure[4], the dielectric tensor has the form:

$$\epsilon_{ij} = n_0\delta_{ij} + g(E_i^* E_j + E_j^* E_i) \tag{1}$$

Here $\boldsymbol{E}$ stands for the electric field written in complex form and g being some factor that plays a role the coupling constant. It should be noted that the matrix $\hat{\epsilon}$ is symmetric matrix, which is required by the symmetry principle and the application of the electric field. Now as far as the weak field is concerned, the field $\boldsymbol{E}$ can be approximated by the form

$$\boldsymbol{E}(x, y, z) = \boldsymbol{f}(x, y, z)\exp[ikn_0 z] \tag{2}$$

namely, $\boldsymbol{f}$ gives the deviation from the plane wave of the wave number $k = \frac{\omega}{c}$ that propagates along z direction in the isotropic medium of the refractive index $n_0(\equiv \sqrt{\epsilon_0})$. (x, y) means the coordinate that is perpendicular to z direction. In order to represent the polarization, we write the amplitude $\boldsymbol{f}$ in terms of the basis of polarization, namely, $\boldsymbol{f}(x, y, z)$ is written as $\boldsymbol{f} = {}^t(f_1, f_2) = f_1\boldsymbol{e}_1 + f_2\boldsymbol{e}_2$, where $\boldsymbol{e}_1$ and $\boldsymbol{e}_2$ denotes the basis of linear polarization. Here the z-axis is taken to be perpendicular to the (x, y) plane which is chosen as the optical axises (that correspond to the eigenvalues of the dielectric tensor). In this way $\hat{\epsilon}$ tensor is reduced to 2×2 matrix:

$$\hat{v} = \begin{pmatrix} v_0 + \alpha & \beta \\ \beta & v_0 - \alpha \end{pmatrix}. \tag{3}$$

with

$$v_0 = g(|f_1|^2 + |f_2|^2), \alpha = g(|f_1|^2 - |f_2|^2), \beta = g(f_1^* f_2 + f_2^* f_1)$$

Here we adopt the well known procedure of the "para-axial approximation". The basic equation is the field equation for the electric field $\boldsymbol{E}$, which is reduced from the Maxwell equation:

$$\frac{\partial^2 \boldsymbol{E}}{\partial z^2} + \nabla^2 \boldsymbol{E} + \left(\frac{\omega}{c}\right)^2 \hat{\epsilon}\boldsymbol{E} = 0 \tag{4}$$

Here we adopt the paraxial approximation; namely, by substituting (2) into (4) and noting the slowly varying nature of $\boldsymbol{f}$ i.e., $|\frac{\partial^2 \boldsymbol{f}}{\partial z^2}| \ll k|\frac{\partial \boldsymbol{f}}{\partial z}|$, we can derive the equation for the amplitude $\boldsymbol{f}$, namely, we retain only the first order derivative $\frac{\partial \boldsymbol{f}}{\partial z}$ as well as the Laplacian with respect to (x, y) [6]. Here instead of the linear basis, $(\boldsymbol{e}_1, \boldsymbol{e}_2)$, we use the circular basis, that is,$\boldsymbol{e}_\pm = \left(1/\sqrt{2}\ \right)(\boldsymbol{e}_1 \pm i\boldsymbol{e}_2)$, which is written as $(\boldsymbol{e}_+, \boldsymbol{e}_-) = U(\boldsymbol{e}_1, \boldsymbol{e}_2)$, where U is given by 2×2 unitary matrix. Thus for the wave function $\psi = U\boldsymbol{f} = {}^t(\psi_1^*, \psi_2^*)$, we have the Schrödinger equation for ψ:

$$i\lambda \frac{\partial \psi}{\partial z} = \hat{H}\psi \tag{5}$$

with the transformed "Hamiltonian"

$$\hat{H} = UhU^{-1} = -\frac{\lambda^2}{n_0}\nabla^2 + V. \tag{6}$$

where λ is the wavelength divided by 2π and the "field-dependent" potential V is written as

$$V = g\begin{pmatrix} |\psi_1|^2 + |\psi_2|^2 & \psi_2^*\psi_1 \\ \psi_1^*\psi_2 & |\psi_1|^2 + |\psi_2|^2 \end{pmatrix}. \tag{7}$$

3. Effective Lagrangian for the pseudo-spin field

We now introduce the "quantum" action leading to the Schrödinger type equation, which is given by

$$I = \int \psi^\dagger \left(i\lambda \frac{\partial}{\partial z} - \hat{H}\right)\psi d^2x dz. \tag{8}$$

By making use of variation equation $\delta I = 0$ recovers the Scrödinger equation. Having defined the Lagrangian for the two-component field ψ, we rewrite this in terms of the Stokes parameters: This is defined as $S_i = \psi^\dagger \sigma_i \psi, S_0 = \psi^\dagger 1 \psi$ with $i = x, y, z$ [5]. We see that the relation $S_0^2 = S_x^2 + S_y^2 + S_z^2$ holds, namely, S_0 gives the field strength; $S_0 \equiv |\boldsymbol{E}|^2$. Using the spinor representation,

$$\psi_1 = \sqrt{S_0}\cos\frac{\theta}{2},\ \psi_2 = \sqrt{S_0}\sin\frac{\theta}{2}\exp[i\phi], \tag{9}$$

we have the polar form for the Stokes vector

$$\boldsymbol{S} = (S_x, S_y, S_z) \equiv (S_0 \sin\theta\cos\phi, S_0\sin\theta\sin\phi, S_0\cos\theta) \tag{10}$$

which forms a pseudo-spin and is pictorially given by the point on the Poincaré sphere. In terms of the angle variables, L_C is written as

$$L_C = \int \frac{S_0 \lambda}{2}(1 - \cos\theta)\frac{\partial \phi}{\partial z} d^2x \tag{11}$$

and the Hamiltonian becomes $H = H_T + \tilde{V}$, the potential $\tilde{V}$ is given in terms of the pseudo-spin:

$$\tilde{V} = \int \left(v_0^2 + \sum_{i=1}^{2} S_i^2 \right) d^2x. \tag{12}$$

and he kinetic energy term H_T is given as a sum of three terms; $H_T = \frac{\lambda^2}{n_0} \int \nabla\psi^\dagger \nabla\psi d^2x \equiv H_1 + \tilde{H}$ where $H_1 = \int \frac{\lambda^2}{4S_0 n_0} \left(\nabla S_0\right)^2 d^2x$, which gives the energy that is needed for space modulation of the field strength, and the remaining terms are written as

$$\tilde{H} = \int \frac{S_0\lambda^2}{n_0} \left\{ (\nabla\theta)^2 + \sin^2\frac{\theta}{2}(\nabla\phi)^2 \right\} d^2x, \tag{13}$$

which is also given by the sum of two terms; $\tilde{H} = H_2 + H_3$, where

$$H_2 = \int \frac{S_0\lambda^2}{4n_0} \{(1-\cos\theta)\nabla\phi\}^2 d^2x$$

$$H_3 = \int \frac{S_0\lambda^2}{4n_0} \{(\nabla\theta)^2 + \sin^2\theta(\nabla\phi)^2\} d^2x. \tag{14}$$

Here if we define the "velocity field" $\boldsymbol{v} = (1-\cos\theta)\nabla\phi$, the first term is regarded as fluid kinetic energy inherent in spin structure, while the last term represents an intrinsic energy for the pseudo-spin which exactly coincides with a continuous Heisenberg spin chain [7].

4. Vortex solution and its numerical evaluation

We are now concerned with getting an explicit form of a vortex solution for the TCNLS. The solution we want here is a "static" solution, namely, we look for the solution that is independent of the variable z. The potential energy corresponding to the nonlinear birefringence $\tilde{V} = \int \sum_{i=1}^{2} v_i' S_i d^2x = S_0^2 \int g d^2x - S_0^2 \int g \cos^2\theta d^2x$, where the first term is constant and should be discarded. In what follows, we confine our argument to the case that S_0 becomes constant. Physically, this corresponds to the constant background field with a proper core which is controlled by the profile of the angle functions (θ, ϕ). A static solution for the one vortex is obtained by choosing the phase function $\phi = n\tan^{-1}\left(\frac{y}{x}\right)$, with $n = 1, 2, \cdots$. being the winding number, together with the profile function θ that is given as a function of the radial variable $r(=\sqrt{x^2+y^2})$. Note that such a vortex becomes non-singular, namely, the velocity field $\boldsymbol{v} = (1-\cos\theta)\nabla\phi$ does not bear the singularity due to the behavior of $\theta(r)$ near the origin (see below). The static Hamiltonian is thus written in terms of the field $\theta(r)$:

$$H = \frac{S_0 \lambda^2}{4n_0} \int \left[\left\{ \left(\frac{d\theta}{dr} \right)^2 + \frac{n^2}{r^2} \sin^2 \frac{\theta}{2} \right\} - g \cos^2 \theta \right] rdr \tag{15}$$

where $g' = \frac{4gn_0 S_0}{\lambda^2}$. The profile function $\theta(r)$ may be derived from the extremum of H', namely, the Euler-Lagrange equation leads to

$$\frac{d^2\theta}{d\xi^2} + \frac{1}{\xi}\frac{d\theta}{d\xi} - \frac{n^2}{4\xi^2} \sin\theta - \frac{1}{2}\sin 2\theta = 0 \tag{16}$$

where we adopt the scaling of the variable: $\xi = \sqrt{g'}r$. In order to examine the behavior of $\theta(\xi)$, we need a specific boundary condition at $\xi = 0$ and $\xi = \infty$. We impose $\theta(0) = 0$, whereas at $\xi = \infty$, there are two options: a) $\theta(\infty) = \pi$ and b) $\theta(\infty) = \pi/2$. We first consider the behavior near the origin $\xi = 0$, for which the differential equation behaves like the Bessel equation, so we see $\theta(\xi) \simeq I_{n/2}(\xi)$, which satisfies $\theta(0) \simeq 0$. We examine

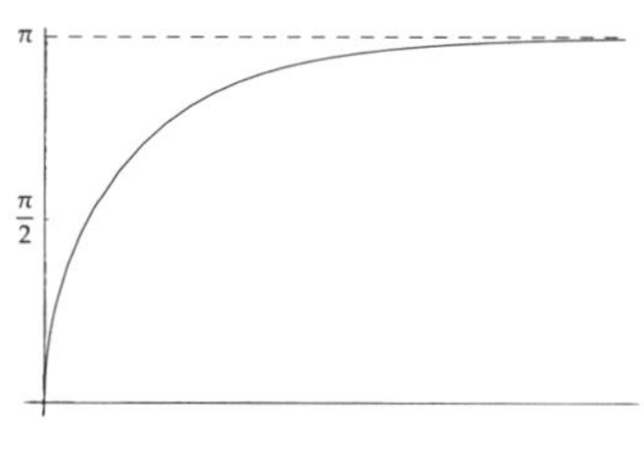

Figure 1. The profile of the non-singular vortex; (a) The case of $\theta(\infty) = \pi$.

the behavior at $\xi = \infty$. This is simply performed by checking the stationary feature of the solution for two cases mentioned above: (a) and (b). Now for the case (a), if putting $\theta(\xi) = \pi + \alpha$, with α the infinitesimal deviation, then we have the linearized equation $\alpha'' - \alpha \simeq 0$ near $\xi = \infty$, which results in $\alpha \simeq \exp[-\xi]$. This means that the solution with $\theta(\infty) = \pi$ converges to the stationary solution. On the other hand, for the case (b) we have $\alpha'' + \alpha \simeq 0$, which gives $\alpha \simeq \exp[\pm i\xi]$ meaning the oscillatory behavior. This simply implies that the solution with $\theta(\infty) = \frac{\pi}{2}$ does not converge to the stationary solution, so that the case (a) is not allowed as a relevant solution.

If introducing the vector $\boldsymbol{m}(\xi) \equiv \boldsymbol{S}/S_0$, we have $m_3(0) = 1$ for both cases a), b) and we have $m_3(\infty) = -1$ for case a) and $m_3(\infty) = 0$ for

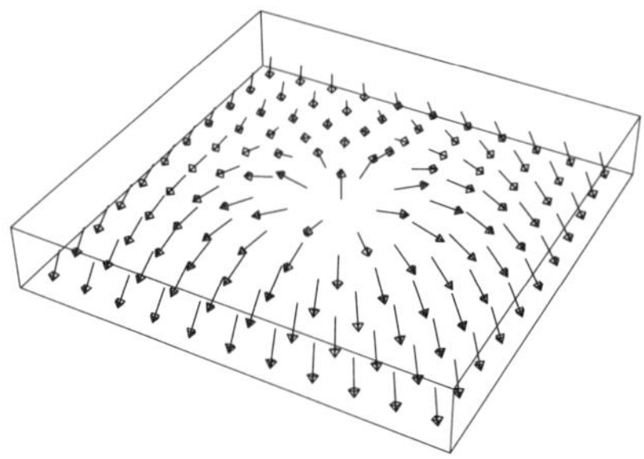

Figure 2. The profile of the function $\theta(\xi)$for two cases of boundary conditions: (a) $\theta(\infty) = \pi$.

case b). The stable solution just obtained in the above indicates that the pseudo-spin field which directs *upward* (left-handed circular polarization) at the origin changes to the state of *downward*(right handed polarization). This feature is schematically given by the distribution of the reduced spin $\boldsymbol{m}$ in Fig.2.

5. Evolution equation for vortex

Having constructed the explicit form for the vortex solution, we now consider the evolutional behavior for a single vortex with respect to the propagation direction z. Following the procedure used in the magnetic vortex [7], let us introduce the coordinate of the center of vortex, $\boldsymbol{R}(z) = (X(z), Y(z))$, by which the vortex solution is parameterized such that $\theta(\boldsymbol{x} - \boldsymbol{R}(z))$ and $\phi(\boldsymbol{x} - \boldsymbol{R}(z))$. By using this parametrization, the canonical term L_C (the first term in the action function) is written as

$$L_C = \frac{S_0\lambda}{2} \int \boldsymbol{v} \cdot \dot{\boldsymbol{R}} d^2x \tag{17}$$

where we have used the relation: $\frac{\partial\phi}{\partial z} = \frac{\partial\phi}{\partial \boldsymbol{R}}\dot{\boldsymbol{R}}$, $\frac{\partial\phi}{\partial \boldsymbol{R}} = -\nabla\phi$ with $\dot{\boldsymbol{R}} \equiv \frac{d\boldsymbol{R}}{dz}$. By using the Euler-Lagrange equation, we get the "balance of force":

$$\frac{S_0\lambda}{2}\sigma\left(\boldsymbol{k} \times \dot{\boldsymbol{R}}\right) = -\frac{\partial \tilde{H}}{\partial \boldsymbol{R}} \tag{18}$$

Here the $\boldsymbol{k}$ is the unit vector perpendicular to the xy-plane and σ is defined as

$$\sigma = \int_{R^2} \left(\frac{\partial v_y}{\partial x} - \frac{\partial v_x}{\partial y}\right) d^2x. \tag{19}$$

(In the derivation of this, we have used the relation $\frac{\partial v_x}{\partial X} = -\frac{\partial v_x}{\partial x}$). The integrand of σ is nothing but the vorticity which we put ω. Using the expression for the velocity field, ω can be written in terms of the angular functions:

$$\omega = (\nabla \times \boldsymbol{v})_z = \sin\theta(\nabla\theta \times \nabla\phi) \tag{20}$$

or in terms of the spin field $\boldsymbol{m}$

$$(\nabla \times \boldsymbol{v})_z = \boldsymbol{m} \cdot \left(\frac{\partial \boldsymbol{m}}{\partial x} \times \frac{\partial \boldsymbol{m}}{\partial y} \right). \tag{21}$$

The equation (19) is an optical counterpart of a topological invariant of hydrodynamical origin [8], which is rewritten as

$$\sigma = \int_S \sin\theta d\theta \wedge d\phi \tag{22}$$

where S stands for the area in the pseudo-spin space (θ, ϕ). The quantity σ has a topological meaning; it depends on the boundary condition for $\theta(r)$. Namely, for the case a) corresponding to the boundary condition $\theta(\infty) = \pi$, the vortex configuration gives the mapping from the compactified two-dimensional space $R^2 \cup \infty \simeq S^2$ to the pseudo-spin space S_2. Hence σ has a meaning of the degree of mapping for $S_2 \to S_2$ leading to the topological invariant σ: $\sigma = n$ (n=integer). The appearance of such two types of topological invariant is characteristics of a new type vortex presented here.

Acknowledgements

One of the authors (H.K) would like to thank professor S.Tanda and professor T.Matsuyama for having invited him to TOP2005. He also thanks professor M.V.Berry for his pointing out relation between optical spin vortex and polarization singularity.

References

1. R. Y. Chiao, E. Gamire, and C. H. Townes, Phys. Rev. Lett. **13**, 479 (1964).
2. G. A. Swartzlander, Jr. and C. T. Law, Phys. Rev. Lett. **69**, 2503 (1992).
3. M. Born and E. Wolf, *Principle of Optics* (Pergamon, Oxford, 1975).
4. L. Landau and E. Lifschitz, *Electrodynamics in Continuous Media*, chapter 11, Course of Theoretical Physics Vol.8 (Pergamon Oxford, 1968).
5. H. Kuratsuji and S. Kakigi, Phys. Rev. Lett. **80**, 1888 (1998) and references cited therein.
6. See S.A.Akhmanov, *Physical Optics*, Chapter 14 (Clarendon press, Oxford 1997).
7. H. Ono and H. Kuratsuji, Phys. Lett. **186A**, 255(1994),
8. H.Lamb, Hydrodynamics, Cambridge University Press.

COHERENT DYNAMICS OF COLLECTIVE MOTION IN THE $NbSe_3$ CHARGE DENSITY WAVE STATE*

Y. TODA,† K. SHIMATAKE, T. MINAMI, AND S. TANDA

Department of Applied Physics, Hokkaido University, Sapporo 060-8628, Japan.

We have observed a coherent collective excitation of the charge density wave (CDW) of the quasi-one-dimensional metal, $NbSe_3$, by means of ultrafast pump-probe measurement. The temperature dependence of the transient signal reveals that the collective amplitude mode (AM) disappears around 10K below a Peierls transition temperature (T_p), at which the inverse of single-particle (SP) relaxation time τ_{sp} becomes smaller than the AM frequency. In addition, the time-frequency analysis of the AM exhibits a time-developed frequency change. These results suggest that the instantaneously excited SPs deform the CDW gap and generate the coherent AM motion reflecting gap formation within their relaxation.

1. Introduction

Low-dimensional electronic materials sometimes show a phase transition associated with a formation of charge density wave (CDW). One of the collective excitations in CDW is the amplitude mode (AM), which have a finite energy at wave number $q \sim 0$. Thanks to the development of the femtosecond pulse laser, we are now able to study the coherently excited AM oscillation in real time, which offers the detailed study of ultrafast dynamics of CDW.[1] Since the CDWs are formed by the displacements of ions and electrons, the generation of coherent CDW motion should be associated with the single-particle (SP) excitation. Based on this, the generation mechanism of AM can be identified as a kind of displacive excitation of coherent phonons (DECP), in which the instantaneous SP excitation changes the equilibrium positions of electrons and ions, resulting in a coherent motion. As another plausible candidate for coherent motion, Kenji et al pointed out the breakdown of the CDW state. High-density excitation of SPs will

*This work is partly supported by Grant-in-Aid for the 21st Century COE program "Topological Science and Technology".

†Also at PRESTO Japan Science and Technology Agency, Saitama 332-0012, Japan.

induce the abrupt destruction of the CDW state as closing the gap.[3] However, exact mechanism of the coherent AM generation is still an unsettled issue. Furthermore, little attention has been paid to the time-development of the AM at sufficiently short time scales.

In this work, we show that the coherent AM motion in the quasi-one-dimensional $NbSe_3$ chains exhibits the correlations not only with the SPs excitation but also with their relaxation. Note that this is the first report of the time-resolved experiment performed on transition metal chalcogenides of the type MX_3. The SP relaxation time of $NbSe_3$ shows a strong divergence when approachig to a Peierls transition temperature T_p from below, and then abruptly drops down above T_p. On the other hand, AM disappears around 10K below T_p, at which the inverse of SP relaxation time becomes smaller than the AM frequency. In addition, the data analyzed in time-frequency space also suggests the deformation of the CDW gap, leading to a remarkable softening of AM frequency in the initial time regions. These results suggest that the instantaneously excited SPs deform the CDW gap within their relaxation and generate the coherent AM motion reflecting gap formation.

2. Experiments and Sample

Time-resolved data for both SPs and collective excitations serve as snapshots of reflectivity changes based on a standard pump-probe setup.[4] For the excitation source, we used a mode-lock Ti:sapphire laser with an FWHM of 130 fs pulse centered at an energy of 1.56 eV with a repetition rate of 76 MHz. The pump and probe pulses were orthogonally polarized and were focused by an objective lens onto a single crystal region with a large domain size. The overlapping spot size of the pulses was estimated to be about 10 μm in diameter. The pump pulse was chopped and the reflectivity change of the probe was detected at the chopping frequency by a lock-in amplifier. The material used in the study is a quasi-one-dimensional compound $NbSe_3$, which consists of three pairs of metallic chains along the chain direction. By lowering temperature, $NbSe_3$ undergoes two Peierls transition at T_{p1} =145 K and T_{p2} =59 K.[2] It is shown elsewhere that the transient reflectivity changes show a characteristic behavior around T_{p2}, but no significant changes occur around T_{p1}.[5] We concentrate our discussions on the results in the temperature range around T_{p1} and below.

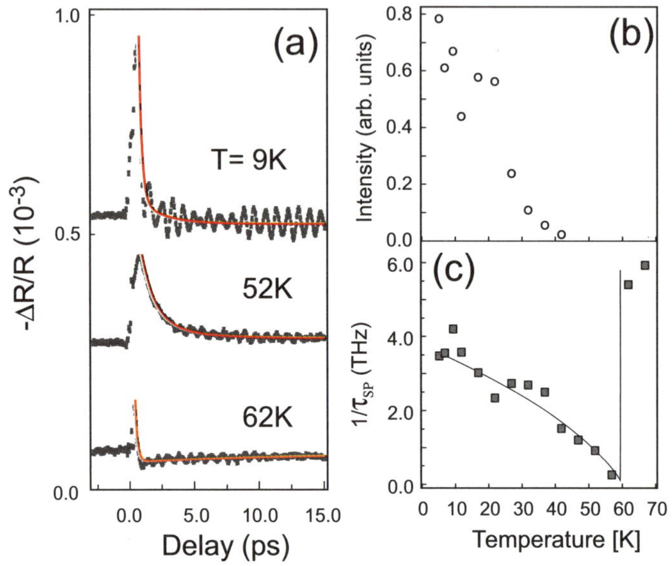

Figure 1. (a) Transient reflectivity changes at various temperatures across the T_{p2}. Temperature dependences of the amplitude of AM (b) and the inverse of τ_{sp} (c).

3. Results and Discussions

Figure 1 (a) shows the transient reflectivity changes at various temperatures below and just above T_{p2}. Since the excitation energy of 1.56 eV can excite the SPs into continuum states far above a CDW gap, the outline of the data reflects a transient response of the SP transitions. Below T_{p2}, the SP transition consists of an abrupt change within 0.2 ps and subsequent decay signal. We employed a sum of two exponential functions for the fitting to the data as decay components (the solid line in the figure). The data show that the SP relaxation time (τ_{sp}) and its amplitude are strongly dependent on the temperature around T_{p2}. When approaching to T_{p2} from below, ?s shows a pronounced increase, and the lifetime becomes as long as 10ps. The main contribution to the recombination across the gap is given by phonon emission and absorption. As the gap closes near T_{p2} from below, lower-energy phonons become available for reabsorption, resulting in a divergence of τ_{sp}. Subtracting the exponential parts from the data, we can obtain a strong modulation signal. The corresponding Fourier transform (FT) spectrum clearly reveals several oscillation modes. The dominant oscillation around 1.0 THz can be attributed to a CDW AM mode from the temperature dependence of its frequency and amplitude.

The temperature dependence of τ_{sp} is shown in Fig. 1 (b), where the inverse values of τ_{sp} are plotted for the sake of comparison with the temperature dependence of AM frequency in Fig. 1 (c). In contrast to the SP transition, AM disappears well below T_{p2} when warming up from the lowest temperature. The data clearly indicate that the disappearance of AM occurs when the inverse of SP relaxation time becomes smaller than the AM frequency. Therefore, the necessary condition for generating the coherent AM motion in the present sample is that both excitation and relaxation time of SPs are faster than the oscillation time of AM; the instantaneous SP excitation acts as a driving force for generating the coherent motion, and the relaxation preserves the coherence of the motion.

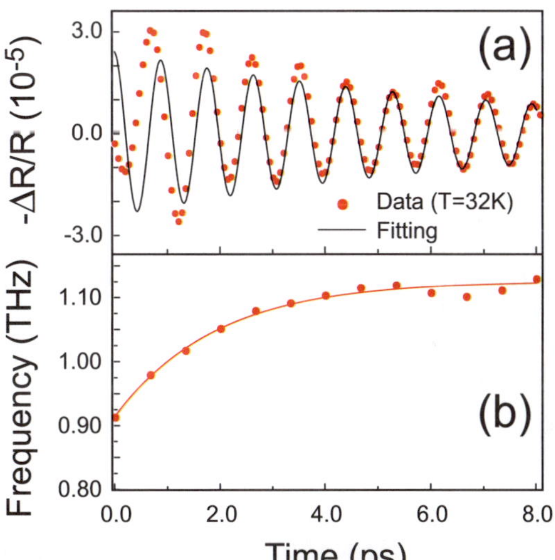

Figure 2. (a) Transient reflectivity changes of the AM mode at T = 32K. (b) Variation of the AM frequency as a function of the time delay.

To discuss the coherent dynamics of AM motion, we first tried to fit the data with a single exponentially damped sinusoidal function, in which the parameters are optimized for the AM oscillation during time interval 4-10 ps. The result indicates that the fitting function cannot reproduce the experimental data in the early time. There is a large deviation in the phase of the oscillation in the initial time regions, suggesting that the mode frequency exhibits a pronounced softening just after the SP excitation. In order to gain deep insight into the time-evolution of AM frequency in the initial time, next a time-frequency analysis were carried out. Figure 2 shows peak frequency changes of AM spectra evaluated by a wavelet transformation.[6] The AM peak shifts first toward lower-frequencies below 1THz, and then toward a ordinary frequency. Note that the spectra show almost the same peak frequencies throughout the temperature range where AM is observed. Therefore, the softening of AM frequency in the temperature dependence does not reflect the temperature dependence of the CDW gap, but the time-developed deformation of the gap. These results reveal the ultrafast dynamics of coherent AM motion in the $NbSe_3$ compound as follows: the instantaneously excited SPs deform the CDW gap and generate the anharmonic AM motion reflecting gap formation within their relaxation due to the presence of the SPs in the band edge. As a result, deformation of the gap makes the motion incoherent when the SP relaxation time exceeds the time period of AM.

References

1. J. Demsar, K. Biljakovic, and D. Mihailovic, *Phys. Rev. Lett.* **83**, 800 (1999).
2. N. P. Ong and P. Monceau, *Phys. Rev.* **B16**, 3443 (1977).
3. K. Kisoda, M. Hase, H. Harima, S. Nakashima, M. Tani, H. K. Sakai, Negishi, M. Inoue, *Phys. Rev.* **B58**, R7484 (1998).
4. Y. Toda, K. Tateishi, and S. Tanda, *Phys. Rev.* **B70**, 033106 (2004).
5. K. Shimatake, Y. Toda, T. Minami, and S. Tanda, unpublished.
6. M. Hase, M. Kitajima, A. Constantinescu, and H. Petek, *Nature* **426**, 51 (2003).

COHERENT COLLECTIVE EXCITATION OF CHARGE-DENSITY WAVE IN THE COMMENSURATE PHASE OF THE TaS_3 COMPOUND

T. MINAMI[1] , K. SHIMATAKE[1] , Y. TODA[1,2] AND S.TANDA[1]

[1] *Department of Applied Physics,*
Hokkaido University,
Sapporo 060-8628, Japan.
E-mail: south-fs@eng.hokudai.ac.jp

[2] *PREST Japan Science and Technology Agency,*
Saitama 332-0012, Japan.

Coherent collective excitations of charge density wave (CDW) in the quasi-one-dimensional metals TaS_3 were observed as transient reflectivity changes with ultrafast time resolved spectroscopy. We have also successfully grown large TaS_3 crystal that is several times larger than the typical one. Possibility of phase soliton excitation in commensurate CDW phase was pointed out. We identified the coherent oscillation as phase soliton based on its' temperature dependence and anisotropy.

1. Introduction

Charge density wave (CDW) exhibits two types of collective excitations: one is a sliding of phase of static CDW, "phason", and the other is a modulation of the amplitude of CDW, "amplitudon". Since the CDWs are formed by the displacements of ions and electrons, instantaneously excited electrons reveals the dynamics of a nonequilibrium state, resulting in coherent excitations of collective motions. The observation of such coherent motions in real time has now become possible by the ultrafast optical spectroscopy with femto-second pulse lasers.[1]

The quasi-one-dimensional (quasi-1D) metals TaS_3 is a member of transition metal chalcogenides of the type MX_3 and undergoes phase transitions associated with formations of CDW. In orthorhombic TaS_3 (o-TaS_3), the Peierls transition occurs only once at T_p=220 K and removes the whole Fermi surface, resulting in characteristic semiconductor-like properties below T_p. On the other hand, CDW becomes commensurate below T_c=100 K, where the CDW is pinned to the underlying crystal lattice. Takoshima, et.al have observed nonlinear conductivity below Tc and pointed out the presence of phase soliton excitation associated with pinned CDW condensates.[2] Recently, it has been shown that MX_3 compounds including TaS_3 forms a variety of topological structures.[3] The possibility of phase soliton excitation may give rise to unusual properties of CDW depending on their crystal topology. For example, when the 1D chains are closed, phase solitons may rotate around the ring, the topology of which imposes additional constraints on their allowed frequencies and phases.

Motivated by these issues, we have made a study of coherent oscillation in a commensurate CDW phase of TaS_3. This is the first report of ultrafast optical response in TaS_3 compound. We show that one of the coherent oscillation modes driven by instantaneous single-particle (SP) excitation is a soft-mode and exists only below T_c. By increasing the temperature from 5.5 K, the mode shows a pronounced softening as well as dephasing. The results indicate that the soft-mode can be attributed to a collective excitation connected with the commensurability of CDW phase. In addition, the mode exhibits remarkable anisotropy to the polarization of the incident pulses with respect to the chain direction (b-axis).

2. Experimental Details

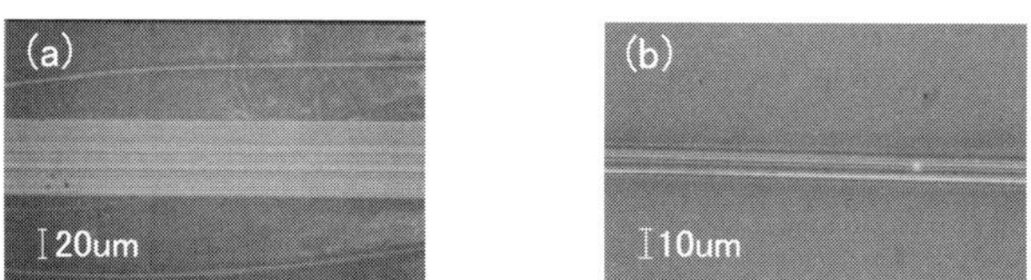

Figure 1. (a) SEM image of a whisker crystal made from Ta rod. (b) Optical microscope image of a TaS_3 crystal made from Ta powder.

The sample used in this work was TaS_3 (o-TaS_3) crystals grown by a chemical vapor transport technique. As a material source, we substituted Ta rod for Ta powder for ingredient. As a result, we have successfully obtained large TaS_3 crystals with chain width of 40-60 μm and length of 2-5 cm. Figure 1 (a) shows the SEM image of a whisker crystal made from Ta rod. For a comparison, the optical microscope image of a TaS_3 crystal made from Ta powder is also shown. Both crystals were reacted in electric furnace at 600 ℃ for 1 week.

The ultrafast optical response including dynamics of SP and collective excitations was measured using a standard pump-probe reflection configuration with 1.56 eV excitation. The temporal resolution of the experiment was 130 fs. The samples were mounted in a liquid-helium-flow cryostat. In order to minimize the background due to the scattering from the surface, the pump and probe pulses were orthogonally polarized and were focused by an objective lens onto a large single crystal region. The overlapping spot size of the incident pulses was estimated to be about 10 μm in diameter.

3. Experimental Results and Discussion

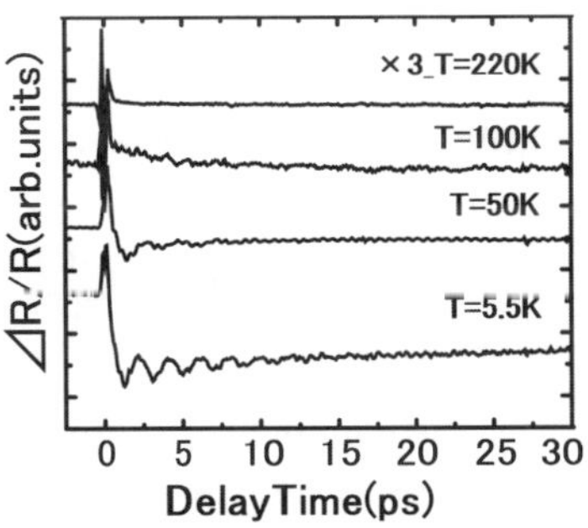

Figure 2. Transient reflectivity changes at various temperatures.

Figure 2 shows the transient reflectivity changes at various temperatures below and around T_p. The measured transient reflectivity change consists of two distinct components: one is exponential components predominantly reflecting the SP dynamics, and another is coherent oscillations due to the collective motions including lattice vibrations. The exponential components, again, can be divided into three parts: an initial rise-up within the temporal resolution, subsequent fast decay within 1 ps, and a longer-lived component. Especially, the last component revealed three kinds of remarkable changes depending on temperature, namely, when lowering the temperature from 240 K, the component appears around T_p, then becomes dominant around T_c, and then shows a change of sign below T_c. At the lowest temperature, the lifetime of the slow decay reaches longer than 100 ps. These results suggest that the slow component of SP reflects the CDW phase transitions.

Below T_c, the transient signals clearly reveal a oscillation component

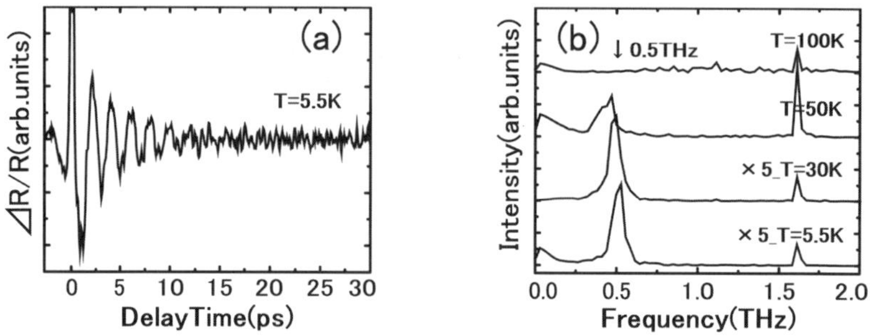

Figure 3. (a) Oscillatory component in a time-resolved reflectivity changes at 5.5K. (b) FFT spectra of the oscillatory component at various temperatures.

superimposed on the exponential slow decay. Subtracting the exponential contributions, we can obtain a strong oscillation signal as shown in Fig. 3 (a). Figure 3 (b) shows several Fourier transform (FT) spectra obtained below T_c, in which the peak existing around 0.5 THz shows down-shift on warming the sample from 5.5 K up to T_c. In addition, this soft-mode disappears around T_c, suggesting the mode attributable to the collective excitation characterized by the commensurate phase. As another example of characteristic behavior of the mode, the polarization dependence of the transient signal is shown in Fig. 4, where the signal is obtained at 5.5 K.

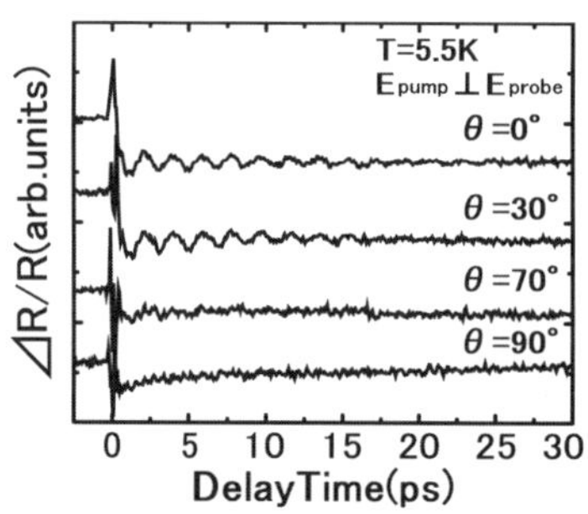

Figure 4. A set of transient reflectivity changes with different polarization configuration for the incident pulses at 5.5K, where θ represents degree of the angle between polarization configulation of pump pulse and chain axis (b-axis) of the sample.

The data successfully show a strong anisotropy with respect to the chain direction. Note that we used orthogonally polarized pump and probe pulses. Therefore, the observed anisotropy does not reflect the symmetry of the mode but the excitation probability of the mode.

From these results, the coherent oscillation with a frequency around 0.5 THz can be identified as a phase soliton. Although both AM and PM in 3D-CDW have 3D degree of freedom, the phase soliton can propagate only along the chain axis. Therefore, the phase soliton has a strong anisotropy and exists only in the commensurate phase. Further investigations are needed to clarify the exact origin of the observed soft mode. However, we stress that the behaviors of the soft-mode observed in Ta_3 are quite different from those obtained in other quasi-1D materials.

Acknowledgments

This work is partly supported by Grant-in-Aid for the 21st Century COE program "Topological Science and Technology".

References

1. J. Demsar, K. Biljakovic, and D. Mihailovic, *Phys. Rev. Lett.* **83**, 800 (1999).
2. T. Takoshima, M. Ido, K. Tsutsumi, T. Sambongi, S. Honma, K. Yamaya, Y. Abe, *Solid State Commun.* **35**, 911 (1980).
3. S. Tanda, T. Tsuneta, Y. Okajima, K. Inagaki, K. Yamaya, N. Hatakenaka,. *Nature* **417**, 397 (2002).

REAL TIME IMAGING OF SURFACE ACOUSTIC WAVES ON TOPOLOGICAL STRUCTURES

H. YAMAZAKI, O. B. WRIGHT AND O. MATSUDA

Department of Applied Physics, Hokkaido University, Sapporo 060-8628, Japan
E-mail: hirotoy@eng.hokudai.ac.jp

Surface acoustic waves are imaged in real time on an isotropic plane surface and also on isotropic spherical surfaces using an ultrafast optical technique. The effect of the difference in surface topology on the wave propagation is discussed.

1. Introduction

The propagation of acoustic waves in both isotropic and anisotropic solids is heavily dependent on spatial topology. Surface acoustic wave (SAW) propagation on a flat surface of an isotropic solid from a point source gives rise to circular wave fronts. The situation for flat crystalline anisotropic substrates is more complicated. The crystal anisotropy produces caustic effects and phonon focusing, in which the acoustic flux becomes very high in certain directions (see Ref. 1). Such focusing not involving any spatially dependent change in acoustic properties is somewhat surprising, but can be explained by the angular variation of the SAW velocity dependent on the fourth-order elastic constant tensor. SAW propagation on isotropic spheres gives rise to topological dispersion effects that result in a curvature dependence of the SAW velocity.[2,3] When the radius of curvature (a) is constant, as on a spherical surface, the SAW velocity v can be written as $v = v_R(1 + [\epsilon/ka])$, where v_R is the SAW velocity on a flat surface of the same material, k is the acoustic wave number and ϵ is a constant (taking the value ~ 2). This SAW dispersion due to curvature therefore increases as the acoustic frequency decreases. The propagation of SAW on anisotropic crystalline spheres is even more complicated and has not been completely understood theoretically to date.[4] In this paper we apply a real-time SAW imaging technique to the simpler case of isotropic flat and spherical surfaces and compare the results.

2. Imaging of surface acoustic waves on flat surfaces

We first present an example of real time SAW imaging for a flat isotropic surface. We make use of an ultrafast optical technique that allows SAW up to 1 GHz to be detected. This technique can sense SAW in two different ways: 1) through variations in surface displacement [5] or 2) through the photoelastic effect (changes in refractive index of the sample induced by acoustic strain).[6] In this paper we make use of the latter technique because of its simplicity and ease of application to curved surfaces.

The sample is a thin polycrystalline aluminium film of thickness 60 nm sputtered onto an isotropic crown glass substrate. Sub-picosecond excitation optical pulses in the infrared, so-called pump pulses, of central wavelength 830 nm and repetition rate 80 MHz (period 12.5 ns) from a mode-locked Ti:sapphire pulsed laser are focused to a circular spot of diameter $D \sim 5$ μm on the aluminium side of the sample through a microscope objective lens at normal incidence. The pump pulse absorption in the metal film launches SAW and bulk acoustic waves through the thermoelastic effect. The wavelength of the SAW is governed by the lateral spot size of the pump beam and by thermal diffusion. We find that the dominant SAW wavelength is of the order of $2D$ (~ 10 μm here), corresponding to a frequency ~ 300 MHz. Sub-picosecond probe optical pulses of central wavelength 415 nm generated in an optical second harmonic generation crystal are focused to a ~ 3 μm diameter spot on the substrate side of the sample at normal incidence using a microscope objective lens mounted on a scanning stage. After each pump optical pulse hits the sample a probe optical pulse arrives at a fixed time delay ($0 \leq \tau \leq 12.5$ ns). These probe pulses monitor the strain-induced reflectivity change of the sample, dependent on the photoelastic effect. The apparatus is described in detail in Ref. 6. For a homogeneous sample, in general, in reduced suffix notation,

$$\Delta\epsilon_I = P_{IJ}\eta_J \tag{1}$$

where, $\Delta\epsilon_I$ is the strain-induced change in the electric permittivity tensor, P_{IJ} is the photoelastic constant tensor and η_J is the strain tensor. For a portion of a circular surface acoustic wave front travelling in the x direction on a semi-infinite isotropic solid in vacuum, for example, the relevant strain tensor components are η_1, η_2, η_3 and η_5 (η_2 arising because of the circular wave front shape). In this case all components of P_{IJ}, that is P_{11}, P_{12} and P_{44} are involved in the optical detection process when probing from the vacuum side. However, in our detection geometry the probe light pulses are incident from the transparent substrate side. In this case contributions from both the thin film and the substrate influence the optical detection in general. We scan the probe beam spot laterally for a fixed delay time between the pump and probe pulses to produce an image. The delay time can then be changed and the process repeated, allowing an animation of the SAW propagation to be built up. Chopping the pump beam at 1 MHz and using lock-in detection allow relative reflectivity changes as small as 10^{-7} to be resolved.

Figure 1 shows an image of the reflectivity change for SAW propagation over a region of 100 μm $\times$ 100 μm on a crown glass substrate coated with a 60 nm Al film. The circular shape of the wave front arises because of the isotropic nature of the sample. The magnitude of the relative reflectivity change at this wave front is $\sim 10^{-6}$. The measured SAW velocity is approximately 3 km s^{-1}, in agreement with literature values for the substrate alone. For the propagation distances involved the wave front is not significantly broadened by the acoustic dispersion caused by the finite film thickness. The spot at the center of the image is caused by refractive index changes associated with temperature variations. Bright regions correspond to an increase and dark regions to a decrease in reflectivity. The slight deviations in SAW amplitude around the wave front and the white regions near the centre of the image are connected with the induced birefringence in the sample.[6]

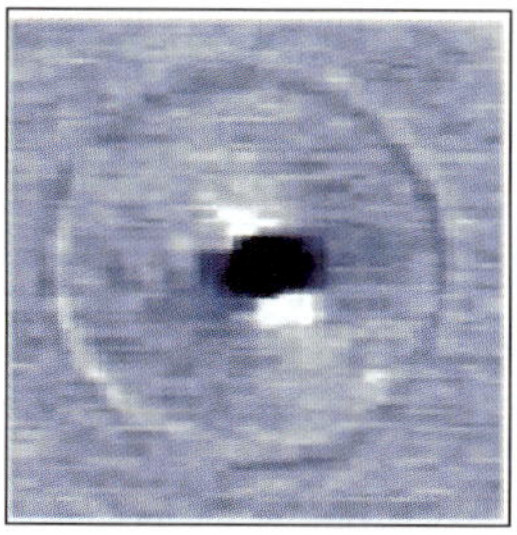

Figure 1. Reflectivity SAW image for a crown glass substrate coated with a 60 nm Al film. The SAW are generated at the centre (dark region).

It is interesting to make an approximate comparison of the present photoelastic detection method with the above-mentioned surface-displacement detection method, the latter involving the interferometric monitoring of optical phase changes.[5] The relative amplitude reflectance change for the surface-displacement detection is given by

$$\left(\frac{\delta r}{r}\right)_{\text{disp.}} \sim \frac{4\pi\delta z}{\lambda_{\text{probe}}} \sim \frac{4\pi\eta\lambda_{\text{SAW}}}{\lambda_{\text{probe}}} \tag{2}$$

where δz is the outward surface displacement, λ_{probe} is the optical probe wavelength, λ_{SAW} is the SAW wavelength and η is the magnitude of a typical strain tensor element. (This assumes that the temporal separation of the two probe pulses used in Ref. 5 is reasonably large, that is $\sim$ 500 ps or more.) The relative amplitude reflectance change for the present measurement method is given by

$$\left(\frac{\delta r}{r}\right)_{\text{photo.}} \sim p\eta \tag{3}$$

where, p is a typical photoelastic constant tensor element. The ratio of these is approximately

$$\left(\frac{\delta r}{r}\right)_{\text{photo.}} \Big/ \left(\frac{\delta r}{r}\right)_{\text{disp.}} \sim p\frac{\lambda_{\text{probe}}}{4\pi\lambda_{\text{SAW}}} \simeq 0.01 \tag{4}$$

where the last estimate applies to our SAW and optical wavelengths (assuming that $p \sim 1$). The relative amplitude reflectance changes for the reflectivity detection method are therefore considerably smaller than those for the surface-displacement detection method. However, the simpler experimental setup for the reflectivity detection method overides this disadvantage when imaging curved surfaces for which it is more difficult to control the precise distance between the sample and the microscope objective lens when spatially scanning.

3. Imaging of surface acoustic waves on spherical surfaces

We have also imaged SAW on isotropic spheres using a two-axis angle scanning system for the reflectivity change measurements.[7] Angular scanning of the sample stage (containing an attached micro-lens and flexible fibre) with respect to the optical probe spot allows SAW on spheres to be imaged in a similar way to the method for flat surfaces. The dominant SAW wavelength excited is similar to that for the flat surfaces above.

Figures 2(a) and (b) show respectively SAW images for 2 mm and 0.5 mm diameter BK7 glass balls half-covered with a thin film of Al of maximum thickness $\sim$ 70 nm, obtained in an experimental configuration in which the imaging is carried out near the acoustic source. The imaged regions correspond to $10° \times 10°$ and $40° \times 40°$ in Figs. 2(a) and (b), respectively. As was the case for the flat surface of the same isotropic material, the SAW wave fronts are circular. Significant birefringence effects (that is, deviations from circular symmetry in the wave front amplitude) are not present in these images.

Because of the broad frequency spectrum of the SAW, curvature dispersion should appear at sufficiently long propagation distances (of the order of the sphere radius in our case). In the present imaged area, however, this dispersion due to surface curvature and that due to the finite Al film thickness are not significant. These results indicate that SAW returning after traversing one round trip on the sphere are not strongly influencing the images. (Because of the periodic optical pulse generation, we expect consecutive SAW wavefronts to be separated by $\sim 8.6°$ for the 0.5 mm ball and $\sim 2.2°$ for the 2 mm ball, as observed in experiment, assuming a SAW velocity of 3 km s^{-1} mainly determined by the glass.) This implies that the acoustic attenuation length must be less than $\sim \pi d \sim 6$ mm, where $d = 2$ mm is the diameter of the larger sphere. (In contrast to the case for the larger sphere, for the smaller sphere with $d = 0.5$ mm a series of images for different delay times was not obtained. This prevented us from verifying that no returning SAW are present for this case.) SAW on bulk silica have a similar attenuation to that of shear waves in bulk silica;[2] this allows one to estimate the attenuation length ~ 50 mm at 300 MHz.[8,9] The deposited polycrystalline aluminium film and the effect of the spatial gradient of this film are presumably responsible for the extra loss.

We have also imaged SAW at the pole region, diametrically opposite the source, on a 1 mm diameter sphere [see Fig. 2(c)] with a spatial resolution $\sim$ 10 μm. The SAW departing from the excitation point converge to this pole, following the curvature of the ball, and then diverge. The superposition of these counterpropagating wave fronts produces a standing wave pattern that we have observed in real time. That the SAW reach the pole suggests that the acoustic attenuation length must be greater than $\sim \pi d/4 \sim 1.5$ mm (where $d = 1$ mm), as a rough estimate.

4. Conclusion

In conclusion we have shown that an ultrafast optical technique can be used to image surface acoustic wave propagation in real time on flat or spherical isotropic surfaces by monitoring variations in optical reflectivity mediated by the photoelastic effect. A detailed quantitative theory of the optical detection process still needs to be worked out, taking into account

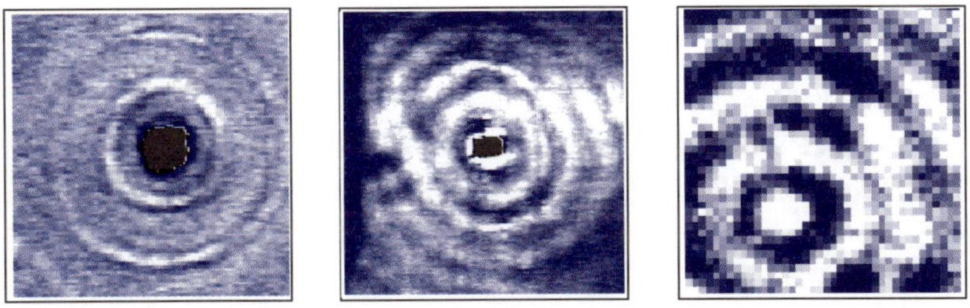

Figure 2. (a) Reflectivity SAW image near the pump source for a glass ball of a diameter 2 mm half coated with Al. (b) Reflectivity SAW image near the pump source for a glass ball of a diameter 0.5 mm half coated with Al. (c) Reflectivity SAW image near the opposite pole for a glass ball of a diameter 1 mm completely coated with Al.

all the tensor components of the SAW strain field. In future we hope to apply this method to the study of the complicated SAW wave fronts that are expected on crystalline spheres.

Acknowledgments

This work has been partially supported by the 21st century COE (Center of Excellence) program on "Topological Science and Technology" from the Ministry of Education, Culture, Sport, Science and Technology of Japan.

References

1. James P. Wolfe. Imaging Phonons, Cambridge University Press, 1998.
2. I. A. Viktorov, Rayleigh and Lamb Waves, Plenum, New York, 1967.
3. F. Jin, Z. Wang and K. Kishimoto, *Int. Jour. Eng. Sci.* **43**, 250 (2005).
4. S. Ishikawa, H. Cho, Y. Tsukahara, N. Nakaso and K. Yamanaka, *Ultrasonics* **41**, 1 (2003).
5. Y. Sugawara, O. B. Wright, O. Matsuda, M. Takigahira, Y. Tanaka, S. Tamura and V. E. Gusev, *Phys. Rev. Lett.* **88**, 185504 (2002).
6. H. Yamazaki, O. Matsuda and O. B. Wright, *phys. stat. sol.* (c) **1**, 2991 (2004).
7. H. Yamazaki, O. Matsuda, O. B. Wright, and G. Amulele, *phys. stat. sol.* (c) **1**, 2979 (2004).
8. B. A. Auld, Acoustic Fields and Waves in Solids, Krieger, Florida, 1990.
9. W. P. Mason, in Physical Acoustics, ed. by W. P Mason and R. Thurston, Vol. 1A, Academic, New York, 1965, p. 488.

OPTICAL VORTEX GENERATION FOR CHARACTERIZATION OF TOPOLOGICAL MATERIALS

Y. TOKIZANE, R. MORITA, K. OKA, A. TANIGUCHI, K. INAGAKI AND S. TANDA

Department of Applied Physics, Hokkaido University
Kita-13, Nishi-8, Kita-ku, Sapporo 060-8628, Japan
E-mail: morita@eng.hokudai.ac.jp

Generation of optical vortices with a simple reflective plate inducing multilevel spiral-phase distribution was demonstrated. Furthermore, a characterization technique of topological materials using optical vortices was proposed.

1. Introduction

Topological deformation of fields is the core of the description of a variety of natural phenomena. Its most extreme but universal manifestation occurs when the field is so strongly folded that it bends back upon itself to form a topological dislocation or defect. Screw wavefront dislocation looking like a helical surface is a common defect type, which possesses phase singularity. The multiplicity of the folding around the defect (winding number of the helicoid) determines its topological charge. For example, Laguerre-Gaussian (LG) modes LG_p^ℓ with non-zero index ℓ have phase dependence in the form $\exp[i(kz + \ell\phi)]$ providing a helical shape for the wavefronts. Here, ℓ is the azimuthal index that represents topological charge, that is, the number of 2π cycles in phase about circumference, p is the radial index that represents $(p+1)$ nodes along the radial direction, z is the longitudinal coordinate, ϕ is the azimuthal angle, and k is the wavenumber. A peculiarity of beams with helical wavefronts lies in the presence of transverse energy circulation, which is termed an optical vortex. The optical vortex carries a well-defined orbital angular momentum (OAM) $\ell\hbar$ per photon[1]. It is distinct from a spin angular momentum (SAM) which is carried by a circularly-polarized light beam. The SAM is given by $\sigma\hbar$, where σ=$\pm$1 corresponds to left- and right-handed circular polarization, respectively. Such a beam with the OAM recently attracted significant research interest because of its increasing applications, such as high-efficiency laser trapping[2–4], microstructure rotation in laser tweezers and spanners[5–7], and quantum information using multidimensional entangled states[8–10].

In the matter of materials, the growth technique of topological materials, such as ring-structure, Möbius-strip, and figure-of-eight single crystals of an inorganic conductor, has been developed recently[11]. However, their electronic or optical properties have not been well clarified so far in a macroscale where topological effects appear. In the present paper, we demonstrate generation of optical vortices with a simple reflective plate and propose a characterization technique of topological materials using them.

2. Fabrication of multilevel spiral phase mirrors

Several LG beam generation methods have been proposed; conversion from the Hermite-Gaussian (HG) mode using a cylindrical lens, a spiral phase plate, and a computer-generated phase hologram. We employed a mirror that has multilevel spiral phase distribution because of the capability for ultrashort laser pulses. The LG beam was generated by a spiral phase mirror (SPM), which had a thickness depending on the azimuthal angle. The multilevel SPMs (N=4 and 6) were fabricated on silica glass in a multistage Al-vapor deposition, as shown in Fig. 1. The relationship between Al thickness and a quartz resonator frequency was found by measuring Al thickness with a multiple-beam interferometer after many trials of deposition. The thickness of each level on the mirror was controlled with an error of ± 3 nm, by monitoring the quartz-resonator frequency.

3. Generation of optical vortices by multilevel spiral phase mirrors

Figure 2 shows the experimental setup for generation of an LG beam. A He-Ne laser (wavelength λ=632.8 nm, HG_{00} (TEM_{00} or Gaussian) mode) of output power 5 mW was used and its beam was linearly polarized by a polarizer P. The beam passed through the first beam splitter BS1, creating reflection and transmission arms. The reflected part of the beam was guided onto the SPM that modified the wavefront, and after reflection by the SPM, it passed through BS1. The transmitted part of the beam was reflected by mirrors M1 and M2, and was recombined with the reflected part at BS2, creating an interference pattern. Only for an interference with a spherical reference beam, lens L (focal length f=90 mm) was inserted.

The intensity distribution pattern recorded when the transmitted arm

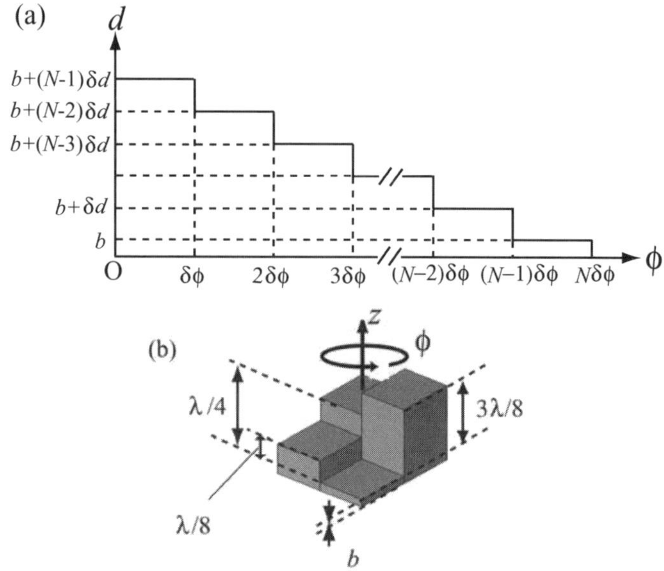

Figure 1. (a) Dependence of Al-thickness d on azimuthal angle ϕ for an N-level step phase mirror to generate an LG_0^{+1} beam. The parameter b is the thickness of the thinnest part (~60 nm) of the mirror, and $\delta\phi=2\pi/N$ and $\delta d=\lambda/2N$, where λ is the wavelength of light. (b) Schematic drawing of a four-level step spiral phase mirror designed for an LG_0^{+1} beam generation.

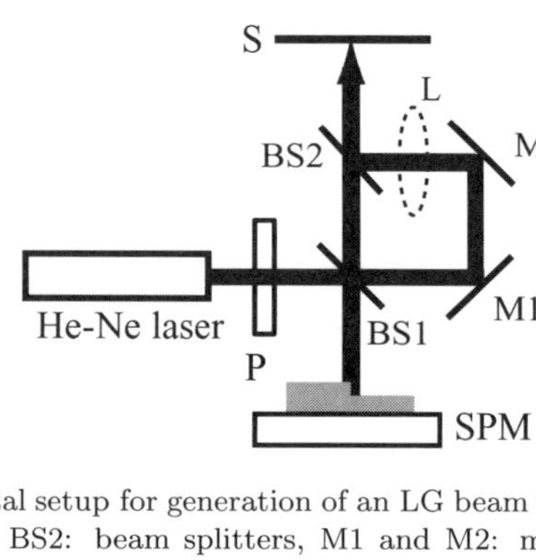

Figure 2. Experimental setup for generation of an LG beam with a spiral phase mirror. P: polarizer, BS1 and BS2: beam splitters, M1 and M2: mirrors, SPM: spiral phase mirror, and S: screen of a section paper. Only for an interference with a spherical reference beam, lens L (f=90 mm) was inserted.

was blocked are displayed in Fig. 3(a), indicating phase singularity in the center of the beam. It also implies the discreteness of the multilevel SPM

(N=4). The interference patterns with an inclined plane reference wave and a spherical reference wave are depicted in Figs. 3(b) and (c), respectively. The former shows the characteristic fork-like structure in the case of ℓ=1 mode and the latter manifests the characteristic vortex shape of ℓ=1. These results show that the simple method using a multilevel SPM enables us to generate an LG beam with phase singularity directly from a Gaussian beam. Especially for application to ultrafast laser pulses, our multilevel SPM of metal can avoid the group-delay dispersion effect that broadens the temporal profile of pulses unlike a transmission-type spherical phase plate.

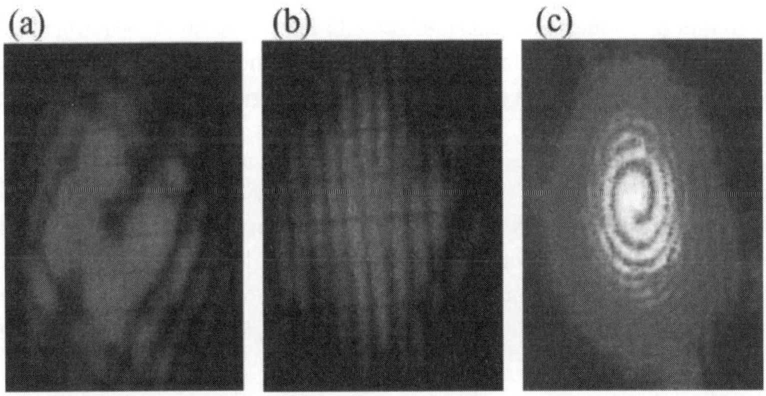

Figure 3. (a) Intensity disribution of a generated LG_0^{+1} beam (N=4), (b) interference pattern of a generated LG_0^{+1} beam (N=4) with an inclined plane reference beam, and (c) interference pattern of a generated LG_0^{+1} beam (N=6) with a spherical reference beam. All patterns were observed on the screen S from not normal direction.

4. Proposal of topological material characterization using optical vortices

For characterization of topological materials, for example, ring-structure crystals, using an LG_0^{-1} beam with left-handed circular polarization or an LG_0^{+1} beam with right-handed circular polarization whose annulus size (doughnut beam size) is adjusted to the ring diameter, electric field $\boldsymbol{E}$ that satisfies $\oint_C \mathrm{d}\boldsymbol{r} \cdot \boldsymbol{E} \neq 0$ can be applied, as shown in Fig. 4 (C: contour along the ring). Hence, femtosecond optical pulses of an LG_0^{-1} mode (ℓ=−1) with left-handed circular polarization (σ=+1) or an LG_0^{+1} mode (ℓ=+1) with

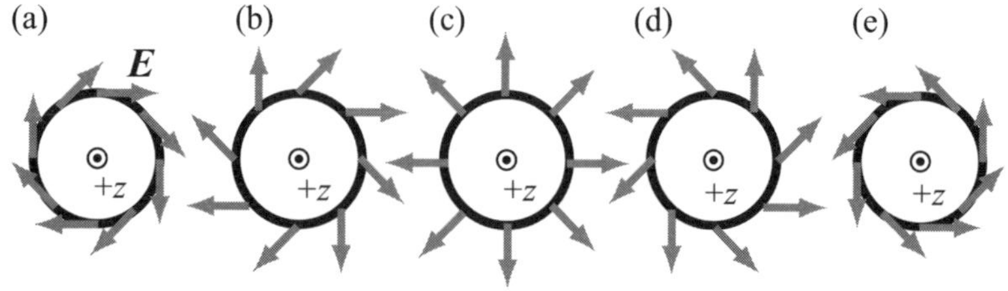

Figure 4. The spatial profile of electric field $\boldsymbol{E}$ applied by an LG_0^{-1} beam with left-handed circular polarization on a ring-structure crystal, at (a) t=0, (b) t=T/8, (c) t=T/4, (d) t=3T/8, and (e) t=T/2, where T=$2\pi/\omega$ is the optical cycle of light with an angular frequency ω. The beam propagation direction is along $+z$-direction.

right-handed circular polarization (σ=-1) can impulsively excite coherent phonon in ring-crystals along the ring contour and will enable us to clarify their charge-density wave dynamics with topological effects.

Acknowledgments

This work has been partly supported by Grant-in-Aid for the 21st Century COE program on "Topological Science and Technology" from the Ministry of Education, Culture, Sport, Science and Technology of Japan.

References

1. L. Allen, M. W. Beijersbergen, R. J. C. Spreeuw, J. P. Woerdman, *Phys. Rev.* **A45**, 8185 (1992).
2. A. Ashkin, *Biophys. J.* **61**, 569 (1992).
3. A. D. Mehta, M. Rief, J. A. Spudich, D. A. Smith, R. M. Simmons, *Science* **283**, 1689 (1999).
4. K. T. Gahagan, G. A. Swartzlander, Jr., *Opt. Lett.* **21**, 827 (1996).
5. N. B. Simpson, K. Dholakia, L. Allen, M. J. Padgett, *Opt. Lett.* **22**, 52 (1997).
6. L. Paterson, M. P. MacDonald, J. Arlt, W. Sibbet, P. E. Bryant, K. Dholaki, *Science* **292**, 912 (2001).
7. V. Garcés-Chávez, D. M. Mcgloin, M. J. Padgett, W. Dultz, H. Schmitzer, K. Dholakia, *Phys. Rev. Lett.* **91**, 093602 (2003).
8. A. Mair, A. Vaziri, G. Weihs, A. Zeilinger, *Nature* **412**, 313 (2001).
9. A. Vaziri, G. Weihs, A. Zeilinger, *et al.*, *Phys. Rev. Lett.* **89**, 240401 (2002).
10. G. Molina-Terriza, J. P. Torres, L. Torner, *Phys. Rev. Lett.* **88**, 013601 (2001).
11. S. Tanda, T. Tsuneta, Y. Okajima, K. Inagaki, K. Yamaya, N. Hatakenaka, *Nature* **417**, 397 (2002).

REAL TIME IMAGING TECHNIQUES FOR SURFACE WAVES ON TOPOLOGICAL STRUCTURES

T. TACHIZAKI, T. MUROYA, O. MATSUDA, AND O. B. WRIGHT

Department of Applied Physics, Graduate School of Engineering, Hokkaido University, Sapporo 060-8628, Japan
E-mail: assp@kino-ap.eng.hokudai.ac.jp

A novel optical scanning system for real time imaging of surface acoustic waves is developed that is well adapted to the investigation of topological structures. We demonstrate its operation by application to an opaque anisotropic crystal and to an embedded microscopic disc. The results indicate that the system allows the monitoring of ultrafast phenomena on opaque surfaces with micron-order lateral spatial resolution.

1. Introduction

Recent progress in microfabrication has lead to the production of single crystal topological structures in the shape of rings or distorted strips.[1] The effect of the geometry of the crystal on the wave propagation in such samples is a subject of much interest.

Here we present an optical technique for imaging surface acoustic wave propagation in real time on microstructures with high lateral spatial resolution using a single microscope objective for excitation and detection of the surface acoustic waves. Related techniques we have developed for real time surface acoustic wave imaging are limited in various ways.[2,3] One method allows high spatial resolution by the use of normal optical incidence from opposite sides of a sample,[2] but requires the use of either very thin samples on the order of the surface acoustic wavelength ($\sim$10 μm) or the use of a transparent substrate. In order to investigate thick opaque samples or samples accessible from only one direction we have developed a method that requires optical incidence from only one side of the sample,[3] but this requires non-normal optical incidence due to the physical size of the optics and thus results in limited spatial resolution. To overcome these problems we have developed a new optical scanning system. We apply this system to the investigation of two different opaque samples in this paper.

2. Surface Acoustic Wave Visualisation

The surface acoustic waves are excited thermoelastically with ultrashort light pulses at a wavelength of 415 nm, a repetition rate of 76 MHz (period 13.2 ns), and a fluence of ~1 mW/cm^2. The detection was performed with a pair of synchronous ultrashort light pulses at a wavelength of 830 nm, a repetition rate of 76 MHz, a fluence of ~0.3 mW/cm^2 and a temporal seperation of 1 ns. The excitation and detection conditions permit nondestructive investigation of the surface acoustic waves in the samples investigated. A common-path optical interferometer was used for monitoring the out-of-plane motion of the surface through optical phase changes.[5] We introduce a rotating mirror and two convex lenses to mechanically scan the probe beam angle of incidence when entering a microscope objective (×50) placed in front of the sample in order to build up two-dimensional images point by point. This system will be described in detail elsewhere.[4] The use of a chopped optical excitation beam at 1 MHz combined with lock-in detection allows the resolution of the surface motion to $\lesssim$1 pm accuracy (in the out-of-plane direction).

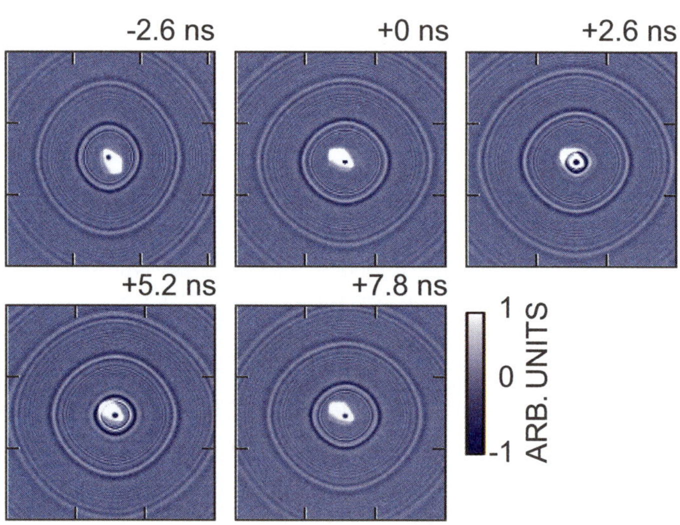

Figure 1. A series of surface acoustic wave propagation images for a 300 μm × 300 μm region of silicon (100) coated with a 60 nm polycrystalline gold film. Surface acoustic waves are generated at the centre of the image. At the top right of each image are shown the elapsed times after generation. Because of the periodic nature of the excitation, -2.6 ns is equivalent to +10.4 ns.

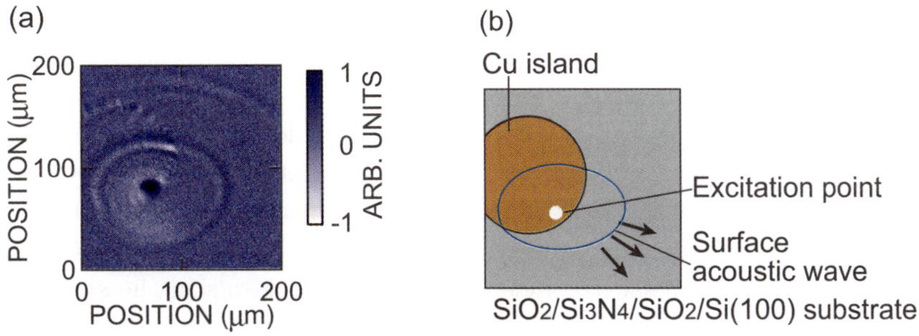

Figure 2. (a) A snapshot of the surface acoustic wave propagation for a 200 μm $\times$ 200 μm region of the sample including a disc of copper embedded in silica. The geometry of the sample is illustrated in (b). The copper disc, not completely imaged in (a), is on the left hand side of the image.

Here we demonstrate animations of surface acoustic wave propagation at acoustic frequencies up to 1 GHz. A series of snapshots of the surface acoustic wave propagation on the (100) surface of a single crystal of silicon coated with a 60 nm polycrystalline gold film is shown in Fig. 1 for a 300 μm $\times$ 300 μm region. The excitation point corresponds to the centre of the images. The gold film permits more efficient generation and detection of the surface acoustic waves. The surface acoustic waves form wave fronts with four-fold symmetry characteristic of the elastic anisotropy of the silicon substrate. The image has significantly higher lateral spatial resolution than that of the above-mentioned method for opaque samples,[3] allowing a more detailed map to be obtained of acoustic dispersion effects (small ripples) caused by the presence of the thin gold film

An image for the second sample is shown in Fig. 2(a), corresponding to a 50 μm radius polycrystalline copper disc embedded in a CVD (chemical vapour deposited) silica layer both of thickness 400 nm deposited on a 100 nm/550 nm SiN/SiO_2 bilayer on a (100) silicon substrate (Fig. 2(b)). Surface acoustic waves are excited near the edge of the Cu disc. The concentric wave fronts result from the point-excited surface acoustic waves in the regions covered by the isotropic Cu and silica that propagate with different velocities. The surface acoustic waves are refracted at the edge of the copper disc owing to the different acoustic impedances of Cu and SiO_2.

3. Conclusion

In conclusion, we have applied an optical scanning system to the imaging of surface acoustic wave propagation on single crystal silicon and in a sample consisting of a polycrystalline copper disc embedded in silica formed on silicon. These results indicate that this system is effective for imaging surface wave propagation in real time and real space on opaque surfaces. This optical scanning technique will in future be applied to topological structures such as films with thickness gradients, crystals in the shape of rings, and other topological crystals and structures such as looped micro-ribbons or Möbius crystals.[1]

4. Acknowledgement

This work has been partially supported by the 21st century COE (Centre of Excellence) program on "Topological Science and Technology" from the Ministry of Education, Culture, Sport, Science and Technology of Japan.

References

1. T. Tsuneta and S. Tanda, *J. Crys. Growth* **264**, 223 (2004).
2. Y. Sugawara, O. B. Wright, O. Matsuda, M. Takigahira, Y. Tanaka, S. Tamura, and V. E. Gusev, *Phys. Rev. Lett.* **88**, 185504 (2002).
3. Y. Sugawara, O. B. Wright, and O. Matsuda, *Rev. Sci. Instrum.* **74**, 519 (2003).
4. T. Tachizaki, T. Muroya, O. Matsuda, Y. Sugawara, D. H. Hurley, and O. B. Wright (to be published).
5. D. H. Hurley and O. B. Wright, *Opt. Lett.* **24**, 1305 (1999).

NONLINEAR OSCILLATIONS OF THE STOKES PARAMETERS IN BIREFRINGENT MEDIA

RYOHEI SETO
HIROSHI KURATSUJI
Department of Physics, Ritsumeikan University - BKC, Kusatsu City, Shiga 525-8577, Japan

ROBERT BOTET
Laboratoire de Physique des Solides, CNRS - UMR8502 / Universite Paris-Sud, Centre d'Orsay, F-91405 Orsay cedex (France)

We study the evolution of polarization of a transverse electromagnetic wave in nonlinear birefringent media, by solving the equation of motion for the Stokes parameters. Nonlinear oscillations occurring under the simultaneous effect of the linear and non-linear birefringence can be solved exactly. In particular, the forced oscillations of the circular polarization when the linear birefringence is periodically modulated along the wave propagation, are similar to the nonlinear Rabi resonance in NMR, and the influence of light intensity on the shape of the polarization-resonance line is studied for the case of incident wave with the linear polarization

1. Introduction

Since light was recognized as a transverse wave, polarization has been one of the major subjects in optics.[1–3] In a recent paper,[4] we proposed a new approach of polarization for a monochromatic light transmitted through anisotropic media. The theory begins from the equation of motion for a point – representing the Stokes parameters of the wave – on the Poincaré sphere. The basic idea is then to map the Maxwell equations onto a two state Schrödinger-like equation, and to consider the corresponding two level system. The dynamics of the Stokes parameters is naturally described by the motion of a "pseudo-spin" in a "pseudo-magnetic field", in analogy with the dynamics of the usual spin.[5] This gives a unified description for the linear birefringence, as classified through a few forms (e.g. Kerr and Faraday effects). Among several direct consequences of this approach of the polarization dynamics in inhomogeneous substances, an optical correspondence of the nuclear magnetic resonance (NMR) is expected to occur in media where anisotropy varies periodically.

The purpose of this note is to discuss an extension of the polarization dynamics to the media exhibiting *nonlinear* birefringence. This is a self-induced Kerr effect, since it comes from anisotropy induced by the electric-field component of light in the isotropic media. We will analyze the equation of motion of the Stokes parameters, and show that they are integrable. We will consider the nonlinear forced oscillations of the Stokes parameters.

2. Stokes parameters and their equation of motion

The starting point,[4] is the "envelope approximation" to represent the state of polarization of a plane wave of frequency ω_o at the distance z along its wave vector, in a medium of average refractive index n.

Complex amplitude for the electromagnetic wave is composed of two orthogonal components (ψ_+, ψ_-), which correspond generally to an elliptic polarization state. It is convenient to use the representation circular polarization (CP) basis, where (ψ_+, ψ_-) are the complex amplitudes on the CP basis. The envelope approximation equation for (ψ_+, ψ_-), with the evolution variable $\tau = \omega_o z/cn$, is:[6]

$$i\frac{d}{d\tau}\begin{pmatrix}\psi_+\\ \psi_-\end{pmatrix} = \hat{H}\begin{pmatrix}\psi_+\\ \psi_-\end{pmatrix}. \tag{1}$$

In this expression, $\hat{H}$ is a Hermitian and trace-less tensor. Thus, using the Pauli representation $\boldsymbol{\sigma} \equiv (\sigma_1, \sigma_2, \sigma_3)$, $\hat{H}$ can be expressed as

$$\hat{H} = -\frac{1}{2}\boldsymbol{\sigma}\cdot\boldsymbol{G}. \tag{2}$$

By this way, Eq.(1) is similar to a Schrödinger equation for a spin in the pseudo-magnetic field $\boldsymbol{G}$.

The Stokes parameters are defined as the expectation values of Pauli operators σ_i $(i = 1, 2, 3)$[10]

$$S_i \equiv \langle\sigma_i\rangle = (\psi_+^*, \psi_-^*)\sigma_i\begin{pmatrix}\psi_+\\ \psi_-\end{pmatrix}. \tag{3}$$

In analogy with the Heisenberg's equation of motion, the time evolution of a spin is given by

$$\frac{d\boldsymbol{S}}{d\tau} = \boldsymbol{S}\times\boldsymbol{G}. \tag{4}$$

Generally, two constants of motion for $\boldsymbol{S}$ can be obtained through the relations $\boldsymbol{S}\cdot d\boldsymbol{S}/d\tau = 0$ and $\boldsymbol{G}\cdot d\boldsymbol{S}/d\tau = 0$.

3. Optical spin resonance

Realization of birefringence One can write the previous Hamiltonians in terms of a single pseudo-magnetic field, as in (2). External and self anisotropy is represented by pseudo-magnetic field $\boldsymbol{G}$. Using (3), one obtains the respective three contributions to the pseudo-magnetic field :

$$\begin{aligned} \boldsymbol{G}_1^{\mathrm{l}} &= (\alpha, \beta, 0) \\ \boldsymbol{G}_2^{\mathrm{l}} &= (0, 0, \gamma) \\ \boldsymbol{G}_1^{\mathrm{nl}} &= \frac{g}{2}(S_1, S_2, -S_3) \end{aligned}$$

α and β show LP birefringence, and γ shows CP birefringence. g is non-linear parameter. We have treated non-linearity under the assumption that the medium was isotropic. When anisotropy is weak, the pseudo-magnetic field is approximately given by addition of the anisotropy term[7] : $\boldsymbol{G} \cong (\boldsymbol{G}_1^{\mathrm{l}} + \boldsymbol{G}_2^{\mathrm{l}}) + \boldsymbol{G}_1^{\mathrm{nl}}$.

Non-linear forced oscillation The main topics of the present paper concerns nonlinear extension of the phenomenon of "optical spin resonance", which was studied in Ref.[4] for the linear birefringent media. The linear pseudo-magnetic field under consideration is given by the combination of the constant field plus spatial sinusoidal oscillation of the external electric field :

$$\boldsymbol{G}^{l} = (\gamma_0 \cos kz, \gamma_0 \sin kz, \gamma) \tag{5}$$

with k a spatial frequency. We will see below that the corresponding equation of motion of spin is just the same as for the real spin submitted to the nuclear magnetic resonance (NMR).

Introducing the reduced *spatial* frequency : $\omega = kc/\omega_o$, one can write the following matrix equation for the evolution of the Stokes parameter :

$$\frac{d}{d\tau}\begin{pmatrix} S_1 \\ S_2 \\ S_3 \end{pmatrix} = \begin{pmatrix} -gS_2S_3 - \gamma_0 S_3 \sin\omega\tau + \gamma S_2 \\ gS_1S_3 - \gamma S_1 + \gamma_0 S_3 \cos\omega\tau \\ \gamma_0 S_1 \sin\omega\tau - \gamma_0 S_2 \cos\omega\tau \end{pmatrix}. \tag{6}$$

To solve this system of equations, it is convenient to make a transformation for the pseudo-spin as follows : $S_1' = S_1 \cos\omega\tau + S_2 \sin\omega\tau$, $S_2' = -S_1 \sin\omega\tau + S_2 \cos\omega\tau$, and $S_3' = S_3$, which is just a rotation about z axis. This leads to the formal *non-linear* evolution equation for the vector $\boldsymbol{S}'$:

$$\frac{d\boldsymbol{S}'}{d\tau} = \boldsymbol{S}' \times \boldsymbol{\Gamma}_{\mathrm{nl}} \tag{7}$$

with the vector $\boldsymbol{\Gamma}_{\text{nl}}$ given by :

$$\boldsymbol{\Gamma}_{\text{nl}} = (\gamma_0, 0, \omega + \gamma - gS_3')$$

From the above equation of motion, one can write down two constants of motion for the vector $\boldsymbol{S'}$:

$$S_1^{'2} + S_2^{'2} + S_3^{'2} = S_0^2, \tag{8}$$

$$\gamma_0 S_1' - \frac{g}{2} S_3^{'2} + (\omega + \gamma) S_3' = V_0, \tag{9}$$

indicating that the system is still integrable.

A resonance condition is realized whenever $V_0 = 0$ (e.g. initial polarization linear and perpendicular to the oscillating electric field). After Eqs.(7), (9), the ordinary differential equation for the S_3-Stokes parameter holds :

$$\left(\frac{dS_3}{d\tau}\right)^2 = -\frac{g^2}{4} S_3^4 + g(\omega + \gamma) S_3^3 - \Omega^2 S_3^2 + \gamma_o^2 S_o^2, \tag{10}$$

with $\Omega^2 = \gamma_o^2 + (\omega + \gamma)^2$.

Suppose first the linear case : $g = 0$. The solution is :

$$\frac{S_3}{S_0} = \frac{\gamma_o}{\Omega} \sin \Omega\tau \tag{11}$$

which is similar to the Rabi oscillations for NMR. Resonance takes place for $\omega = -\gamma$, and the width of the resonance is γ_o. Now, let us consider the non-

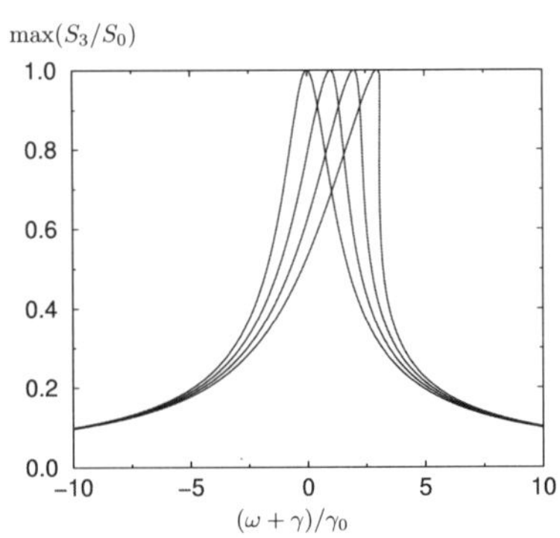

Figure 1. Resonance profiles of the CP amplitude *vs* the reduced frequency $(\omega + \gamma)/\gamma_0$, for four values of the non-linear parameter : $\Gamma = 0, 1, 2$ and 3. The resonance peak is shifted to the right when Γ increases.

linear terms, appearing when $g \neq 0$. The right hand term in (10) vanishes for two real values of S_3. So, the general solution is still an oscillating function – which can be written in terms of Jacobi elliptic functions – and the general resonance behavior is not destroyed by the non-linearities of the system. In this sense, the polarization resonance is *stable* with respect to the non-linearity of the optical medium.

To be more precise, the largest root of the right-hand term of (10) gives the amplitude of the S_3-Stokes parameter, since S_3 vanishes periodically. This value (or more precisely : $\max\{S_3/S_0\}$) can then be used as a measure of the average circular polarization. On Fig. 1 are shown resonance profiles of $\max\{S_3/S_0\}$ versus the reduced frequency $(\omega+\gamma)/\gamma_0$, for various values of the non-linearity parameter $\Gamma = gS_0/2\gamma_0$. Non-linearities shift progressively the resonance peak (the precise resonance frequency is : $\omega = -\gamma + gS_0/2$), and make its shape more asymmetric.

4. Summary

By use of a mapping of the Maxwell equations onto a two-level Schrödinger equation, we studied the dynamics of the Stokes parameters for the electromagnetic wave propagating in nonlinear birefringent media in presence of a static external magnetic field and an external electric field, which can be either static or spatially modulated.

Both cases (with the first relevant non-linear terms) can be solved exactly with use of the Jacobi elliptic functions. For the static electric field, non-linearity gives rise to a bifurcation in the circular polarization for a definite value of the light intensity. For the periodically modulated electric field, the circular polarization undergoes a resonance similar to the nuclear magnetic resonance. The effects of the non-linearities were seen to deform softly the resonance peak without changing its magnitude, indicating that this new kind of resonance should be observed in such transparent media.

References

1. M. Born and E. Wolf, *Principle of Optics* (Pergamon, Oxford, 1975).
2. L. Landau and E. Lifschitz, *Electrodynamics in Continuous Media*, chapter 11, (Pergamon Oxford, 1968).
3. A. Sommerfeld, *Lecture on Theoretical Physics*, Vol.4, (Academic Press. New York, 1964).
4. H. Kuratsuji and S. Kakigi, *Phys. Rev. Lett.* **80**, 1888 (1998).
5. H.Kuratsuji and T.Suzuki, *J. Math. Phys.* **21**, 471 (1980).
6. e.g.: B.E.A. Saleh and M.C. Teich, *Fundamentals of Photonics* (Wiley, New York 1991).

7. B. Daino, G. Gregori, and S. Wabnitz, *Opt. Lett.* **11**, 42 (1986).
8. See S.A.Akhmanov, *Physical Optics*, Chapter 14 (Clarendon press, Oxford 1997).
9. P.D. Maker, R.W. Terhune, and C.M. Savage, *Phys. Rev. Lett.* **12**, 507 (1967)
10. Explicitly : $S_1 = \psi_+^* \psi_- + \psi_-^* \psi_+$, $S_2 = i(\psi_-^* \psi_+ - \psi_+^* \psi_-)$, $S_3 = |\psi_+|^2 - |\psi_-|^2$, while the conventional definition of Stokes parameters is $S_0 \equiv |E_x|^2 + |E_y|^2, S_1 \equiv |E_x|^2 - |E_y|^2, S_2 \equiv E_x^* E_y + E_y^* E_x, S_3 \equiv i\left(E_x^* E_y - E_y^* E_x\right)$. Both sets of formulas are equivalent except for the sign of S_3.

PHONON VORTEX LOCALIZED IN A QUANTUM WIRE

NORIHIKO NISHIGUCHI*
Department of Applied Physics, Hokkaido University, Sapporo 060-8628, Japan
E-mail: nn@eng.hokudai.ac.jp

Acoustic phonon modes localized in an embedded wire structure are derived in terms of a scalar and vector potentials. The phonon modes comprise longitudinal and transverse waves, having angular momentum along the wire axis. Particle motion of the phonon modes shows peristaltic or rotating barber's pole motion, depending on the angular momentum. The dispersion relation and the density of states are also derived.

1. Introduction

In recent years, photon vortices having angular momentum has attracted attention and the properties have been strenuously studied[1,2] since they are supposed to be useful for information technology, planetary observation and so on. Phonons are also expected to have angular momentum.Supposing that photon and phonon vortices, and electrons coexist in a solid, their coupling is expected to bear new types of optical and phonon devices. In this paper, we derive phonon vortices in a structure where phonons and electrons can be confined. As a realizable system, we consider a wire structure embedded in another material, which is similar to optical fibers, where photons are confined in the core.

Phonons are extended in a bulk material, whose normal modes can be described in terms of longitudinal(LA) and transverse acoustic(TA) plane waves. The phonon modes are affected by system boundaries such as interface or surface. In particular, phonons confined in nano-structures are remarkably modified in characters including spatial distribution of displacement and stress fields and the spectra[3–7]. In this work, we show that

*Work partially supported by the 21COE program on "Topological Science and Technology" from the Ministry of Education, Culture, Sport, Science and Technology of JAPAN.

phonons localized in a quantum wire embedded in another material become vortex modes having angular momentum along the wire axis.

2. Model

We consider a cylindrical wire of radius R surrounded by another material, which has the same structure as an optical fiber. We put the wire axis on the z-axis. The core and cladding materials are assumed to be elastically isotropic. Then the system obviously is invariant under rotational operation along the wire axis, as a result, angular momentum along the wire axis becomes a good quantum number specifying the phonon modes. We note here that, because of the cylindrical interface, the LA and TA plane waves are no longer good pictures for phonons, being mixed into the phonon modes peculiar to the wire structure.

Thanks to the elastic isotropy of the materials, the phonon modes can be described in terms of a scalar and two vector potentials[1-5,8];

$$\mathbf{u}_\alpha = \chi_{\alpha,0}\nabla\phi_{\alpha,0} + \chi_{\alpha,1}\nabla\times\mathbf{\Psi}_{\alpha,1} + \chi_{\alpha,2}\nabla\times\nabla\times\mathbf{\Psi}_{\alpha,2}, \tag{1}$$

where the vector potentials $\mathbf{\Psi}_1$ and $\mathbf{\Psi}_2$ are given by $\mathbf{\Psi}_{\alpha,j} = \phi_{\alpha,j}\mathbf{e}_z$ for $j = 1$ and 2. The subscript α denotes the core (A) and the cladding (B), and $\mathbf{e}_z$ is the unit vector in the z direction. $\chi_{\alpha,j}$ are the coefficients of the potential functions in the core and the cladding. The potential functions $\phi_{\alpha,j}$'s obey the following scalar wave equation;

$$\rho_\alpha\frac{\partial^2\phi_{\alpha,j}}{\partial t^2} = \left[\delta_{j,0}\left(C_{\alpha,11} - C_{\alpha,44}\right) + C_{\alpha,44}\right]\nabla^2\phi_{\alpha,j}, \tag{2}$$

where ρ_α and $C_{\alpha,ij}$ denote the mass density and stiffness constants of the quantum wire and surrounding material in cylindrical coordinates, respectively. The potential functions are given by $\phi_{\alpha,j}(\mathbf{r},t) = f_{\alpha,j}(r)e^{i(n\varphi+qz-\omega t)}$, where $f_{\alpha,j}(r)$ is subject to the following equation;

$$\left(\frac{\partial^2}{\partial r^2} + \frac{1}{r}\frac{\partial}{\partial r} - \frac{n^2}{r^2} - q^2 + \frac{\omega^2}{v_{\alpha,j}^2}\right) f_{\alpha,j}(r) = 0. \tag{3}$$

$f_{\alpha,j}(r)$ becomes a Bessel or modified Bessel function, depending on both frequency ω and the longitudinal wave number q, or the lateral wave number k defined below. $v_{\alpha,j}$ denotes the sound velocities of the LA and TA waves given by $v_{\alpha,0} = v_{\alpha,LA} = \sqrt{\frac{C_{\alpha,11}}{\rho_\alpha}}$ and $v_{\alpha,1} = v_{\alpha,2} = v_{\alpha,TA} = \sqrt{\frac{C_{\alpha,44}}{\rho_\alpha}}$. Because of the coupling among the LA and TA waves at the wire surface, they have the common longitudinal wave number q along the wire axis as well as the same angular frequency ω, so that the waves have a different lateral wave vector $k_{\alpha,j}$ given by $k_{\alpha,j} = \sqrt{\left(\frac{\omega}{v_{\alpha,j}}\right)^2 - q^2}$. The coupling at the wire surface also requires the continuity of the displacement vector $\mathbf{u}$ and the stress vector $\boldsymbol{\sigma} = (\sigma_{rr}, \sigma_{r\varphi}, \sigma_{rz})$. The boundary condition determines χ's, $k_{\alpha,j}$ as well as ω for a given longitudinal wave number q.

3. Vortex Modes

We apply the model to a GaAs quantum wire buried in AlAs, and approximate the materials as isotropic media, redefining the stiffness constants C_{12}'s as $C_{12} = C_{11} - 2C_{44}$. Here, the stiffness constants and mass density for GaAs are $C_{11} = 11.88 \times 10^{11}$dyne cm^{-2}, $C_{44} = 5.94 \times 10^{11}$dyne cm^{-2}, and $\rho = 5.36$g cm^{-3}, and for AlAs $C_{11} = 12.02 \times 10^{11}$dyne cm^{-2}, $C_{44} = 5.89 \times 10^{11}$dyne cm^{-2}, and $\rho = 3.76$g cm^{-3}. The relations among the sound velocities of LA and TA waves are $v_{B,LA} > v_{A,LA} > v_{B,TA} > v_{A,TA}$ according to the material parameters. Considering the frequency region $q\,v_{B,TA} < \omega < q\,v_{A,TA}$, only the lateral wave vectors of TA waves in the wire are real, and those of TA waves for the surrounding material and LA waves for the wire and the surrounding material become imaginary. The potential functions are given by

$$\phi_{A,0}(\mathbf{r},t) = I_n(\kappa_{A,0}\ r)e^{i(n\phi+qz-\omega t)} \tag{4}$$

$$\phi_{A,1}(\mathbf{r},t) = J_n(k_{A,1}\ r)e^{i(n\phi+qz-\omega t)} \tag{5}$$

$$\phi_{A,2}(\mathbf{r},t) = \frac{1}{k_{A,2}} J_n(k_{A,2}\ r)e^{i(n\phi+qz-\omega t)} \tag{6}$$

and for the surrounding material

$$\phi_{B,0}(\mathbf{r},t) = K_n(\kappa_{B,0}\ r)e^{i(n\phi+qz-\omega t)} \tag{7}$$

$$\phi_{B,1}(\mathbf{r},t) = K_n(\kappa_{B,1}\ r)e^{i(n\phi+qz-\omega t)} \tag{8}$$

$$\phi_{B,2}(\mathbf{r},t) = \frac{1}{\kappa_{B,2}} K_n(\kappa_{B,2}\ r)e^{i(n\phi+qz-\omega t)}, \tag{9}$$

where $J_n(x)$, $I_n(x)$ and $K_n(x)$ are the first kind of Bessel, the first and second kind of modified Bessel function of order n. Here $\kappa_{\alpha,j}$ is given by the absolute value of the imaginary lateral wave vector $k_{\alpha,j}$, i.e. $\kappa_{\alpha,j} = |k_{\alpha,j}|$. $\chi_{\alpha,j}$'s, $\kappa_{\alpha,j}$'s and ω for a given q are determined by the boundary condition.

The acoustic phonon modes are characterized by the rotational symmetry order n. There are two azimuthally symmetric modes for $n = 0$, one of which is the torsional mode due to $\chi_1 \nabla \times \Psi_1$ that has only the azimuthal component u_φ. The other mode is the dilatational mode given by the sum of terms $\chi_0 \nabla \phi_0 + \chi_2 \nabla \times \nabla \times \Psi_2$, having the radial u_r and axial components u_z of the displacement vector. The displacement is shown in Fig. 1. For a finite integer n, all the waves are coupled into a phonon mode termed the flexural mode or *vortex modes*. Figure 2 shows the displacement of vortex mode with $n = 1$.

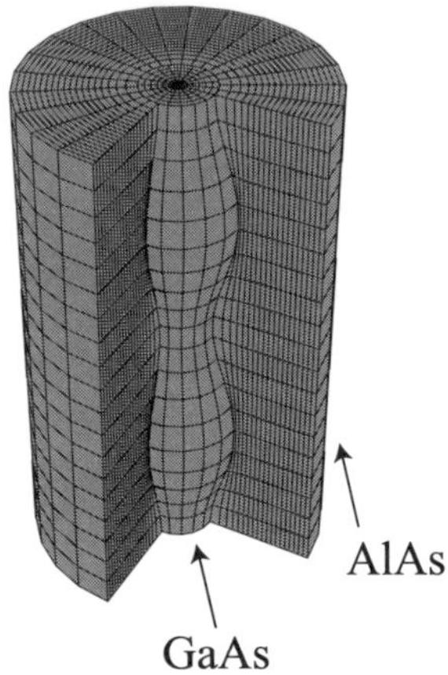

Figure 1. Dilatational mode with $n = 0$

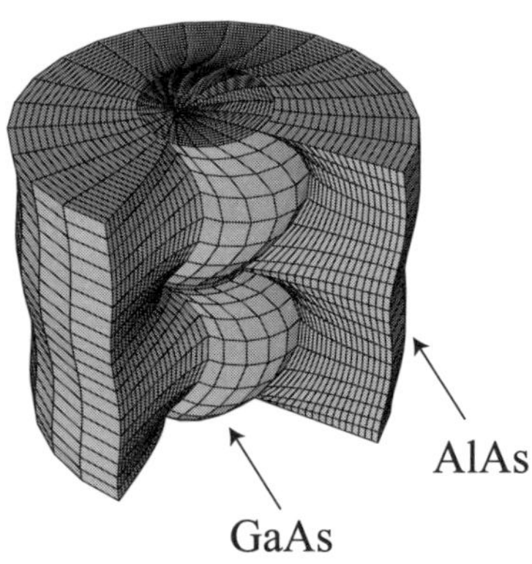

Figure 2. Vortex mode with $n = 1$

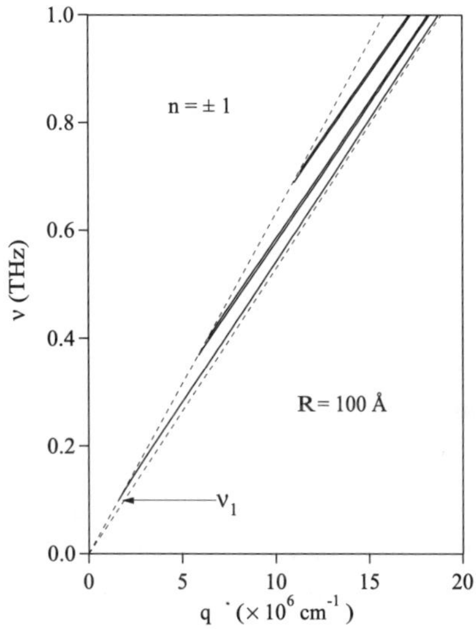

Figure 3. Dispersion relation of vortex mode with $n = \pm 1$

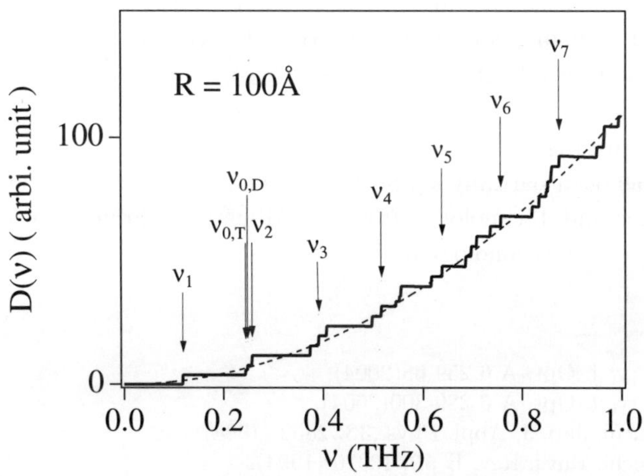

Figure 4. Density of states of vortex and dilatational modes

4. Dispersion Relations

The dispersion curves of the vortex mode with $n = \pm 1$ up to 1THz for $R = 100\text{Å}$ are plotted in Fig. 3. There are five dispersion curves with finite cutoff frequencies. Although the lowest curve is obviously separated from the others, the second and third curves, and the fourth and fifth curves, are close to each other. We note here that the dispersion curves are limited to the localized modes, and their dispersion curves link to those for extended modes below the cutoff frequencies. The mode with the negative integer $n'(= -n)$ has the same dispersion curves as the mode with the positive integer n because the characteristic equation is an even function of n. Hence the dispersion curves of the modes with the same absolute value of n are degenerate.

5. Density of States

The phonon subband structures shift to a higher frequency region when n increases. The lowest cutoff frequency ν_1 is the minimum among the lowest cutoff frequencies $\{\nu_n\}$. ν_n's are almost equally distributed with an increase of n. Such phonon subband structures with finite cutoff frequencies lead to staircase-like density of the confined phonon states. The lowest cutoff frequency among $\{\nu_n\}$ is ν_1 for $n = \pm 1$ and becomes 0.12 THz for $R = 100$Å, below which there is no confined phonon state. This figure also shows the increase of the cutoff frequencies with the rotational symmetry order $|n|$. The density of states tends to the ν^2 variation of the bulk 3D Debye model at high frequencies. The thin dashed line is drawn to show the ν^2 variation for reference.

Acknowledgments

This work has been partially supported by the 21COE program on "Topological Science and Technology" from the Ministry of Education, Culture, Sport, Science and Technology of Japan.

References

1. M. V. Berry, J. Opt. A 6 259-68(2004).
2. M. V. Berry, J. Opt. A 6 289-300(2004).
3. N. Nishiguchi, Jap. J. Appl. Phys.,**33**, 2852 (1994).
4. N. Nishiguchi, Phys. Rev. B **50**, 10970 (1994).
5. N. Nishiguchi, Phys. Rev. B **52**, 5279 (1995).
6. N. Nishiguchi, Y. Ando and M. N. Wybourne, J. Phys.:Condes. Matt. **9**, 5751(1997).
7. N. Nishiguchi,"Electron transport and phonons in quantum wires" in *Handbook of Semiconductor Nanostructures and Devices*, edited by A. A. Balandin et al. (the American Scientific Publishers in 2005).

VII

Topology in Quantum Device

QUANTUM DEVICE APPLICATIONS OF MESOSCOPIC SUPERCONDUCTIVITY*

P. J. HAKONEN †

Low Temperature Laboratory
Helsinki University of Technology
Espoo, 02015 HUT, Finland
E-mail: pertti.hakonen@hut.fi

A brief account is given on the possibilities of mesoscopic superconductivity in low-noise amplifier and detector applications. In particular, three devices will be described: 1) Bloch oscillating transistor (BOT), 2) Inductively-read superconducting Cooper pair transistor (L-SET), and 3) Quantum capacitive phase detector (C-SET). The BOT is a low-noise current amplifier while the L-SET and C-SET act as ultra-sensitive charge and phase detectors, respectively. The basic operating principles and the main characteristics of these devices will be reviewed and discussed.

Studies of phase coherence and solid-state quantum computation[1] have increased the interest in superconducting electronics and amplifiers working at milliKelvin temperatures. During the past few years, several novel devices and amplifiers utilizing mesoscopic Josephson junctions have been developed.[2] In a mesoscopic Josephson junction with a capacitance on the order of 1 fF or below, the quantum mechanical nature of charge has to be taken into account at sub-Kelvin temperatures. According to the quantum prescription, the charge Q is replaced by a differential operator

$$Q = \frac{2e}{i}\frac{\partial}{\partial\phi}, \tag{1}$$

where ϕ, the phase difference across the Josephson junction, is the canonical

*Work supported by the Academy of Finland and the Large Scale Installation Program ULTI III of the European Union (contract HPRI-1999-CT-00050)

†This work has been done in collaboration with J. Delahaye, J. Hassel, R. Lindell, M. Paalanen, L. Roschier, H. Seppä, and M. Sillanpää.

conjugate variable of Q. Therefore, the quantum behavior of a superconducting junction is described by the Schrödinger equation

$$4\frac{\partial^2\psi}{\partial\phi^2} + \left(\frac{E}{E_c} - \frac{E_J}{E_c}\cos\phi\right)\psi = 0, \tag{2}$$

where the Josephson coupling energy, $-E_J\cos(\phi)$, plays the role of a periodic potential for the state of the junction.[3] The solutions for this Mathieu equation form energy bands whose width depends strongly on the ratio of Josephson coupling energy to the Coulomb energy $E_C = e^2/2C$.

In double junction systems, in so called superconducting Cooper pair transistors (SCPT), the energy bands can be parametrized using the phase across the transistor φ and the gate-voltage-induced island charge $q_g = C_gV_g$, where C_g and V_g denote the gate capacitance and voltage, respectively. The lowest of the resulting bands $E_n(\varphi, q_g)$ has approximately sinusoidal energy

$$E_0(\varphi, q_g) \sim -E_0(q_g)\cos(\varphi) \tag{3}$$

and thus, the SCPT is effectively a gate-tunable single junction.[4]

We have used the band structure of a single mesoscopic Josephson junction as well as that of the SCPT to construct low-noise amplifiers and detectors. In the former case, our devices are based on the quantum dynamics of a Josephson junction, i.e., the interplay of interlevel transitions and the Coulomb blockade of Cooper pairs. In the latter case, we employ the equivalence between the second derivative of the energy bands and the inverse capacitance or inductance, depending whether the derivative is taken with respect to charge or phase, respectively.

1. Bloch oscillating transistor (BOT)

The control of quantum dynamics of a single Josephson junction allows us to construct transistor-like devices, Bloch oscillating transistors, with considerable current gain and high-input impedance.[5] In these transistors (see Fig. 1), the correlated supercurrent of Cooper pairs is controlled by a small base current made up of single electrons. Our devices reach current and power gains on the order of 30 and 5, respectively. The noise temperature is measured to be around 0.4 K,[6] but noise temperatures of less than 0.1 K can be realistically achieved. These devices provide new quantum-electronic building blocks that may be useful in low-noise applications with an intermediate impedance level of ~ 1 MΩ.

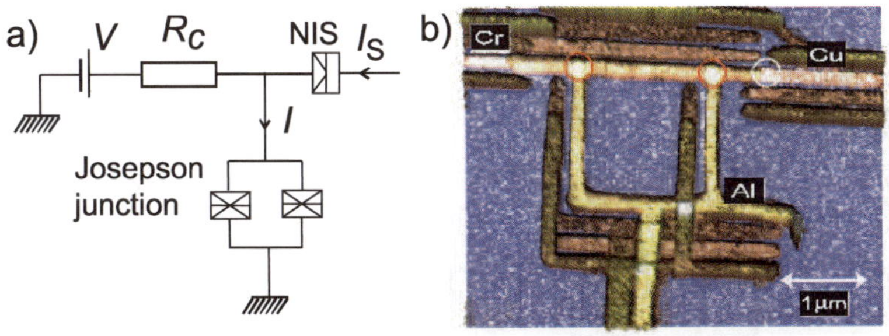

Figure 1. (a) Schematic illustration of the BOT: Josephson junction - a combination of two junctions in a tunable SQUID geometry, NIS - normal metal/insulator/superconductor junction, R_c - thin film chrome resistor on the order of 100 kΩ. A tunneling current of single electrons in the base junction (NIS) is amplified in to a sequence of coherent Cooper pair tunneling events in the Josephson junction. (b) Atomic force microscope image of a BOT. Using electron beam lithography and four-angle evaporation, the same pattern is deposited several times, thus forming shadows. The conducting parts are indicated by a brighter color. The active elements, the Josephson junctions, are circled by red in the picture.

2. Inductively-read superconducting Cooper pair transistor (L-SET)

The operating principle of the inductively-read SCPT, the L-SET, is illustrated in Fig. 2.[7] It is a device in which the resonant frequency of an LC oscillator is modulated by a parallel Josephson inductance of a SCPT

$$L_J^{-1}(\varphi, q_g) = (2\pi/\Phi_0)^2 \partial^2 E(\varphi, q_g)/\partial\varphi^2, \tag{4}$$

where $\Phi_0 = h/(2e)$ is the flux quantum. With a total shunting capacitance C, the circuit thus forms a harmonic oscillator for small amplitude oscillations with the plasma frequency $f_p = 1/(2\pi)(L_{TOT}C)^{-1/2}$ where $L_{TOT} = L_J||L$. The change in the resonant frequency as a function of q_g is traced using reflection measurements around 800 MHz as illustrated in Fig. 2b. The coupling capacitor C_c is chosen in such a way that the resonant circuit is rather well matched to 50 Ohms, which results in a rather small reflection amplitude at the resonance.

The $2e$-periodic gate charge modulation, characteristic to a separated superconducting island, is clearly visible as a vertical triangular modulation in Fig. 2 (b). The charge sensitivity of the device, $2.0 \cdot 10^{-3} e/\sqrt{\text{Hz}}$, was limited by the preamplifier noise.[7] The measurement band width amounted to 40 MHz which reflects the quality factor $Q \sim 20$ of the LC-circuit. At larger oscillation amplitude, the critical current of the SCPT is exceeded and the operation becomes anharmonic.[7] In this operation mode, the charge

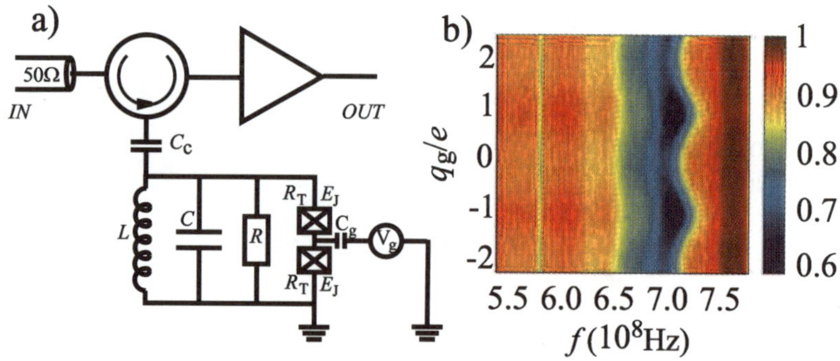

Figure 2. (a) Schematics of the experimental setup used in the L-SET measurements. (b) Variation of the magnitude of the reflected signal from the L-SET circuit using small, harmonic oscillation amplitudes in the tank circuit. The scale for the reflection magnitude is given on the right.

sensitivity could be enhanced to $1.4 \cdot 10^{-4}e/\sqrt{\text{Hz}}$ owing to the increased power level. By optimization of parameters, especially by reducing E_J/E_C ratio, it is quite possible to reach the quantum limit with a device like this.[8] The main motivation for pursuing this rather elaborate frequency read-out scheme is that the back-action noise is expected to be small, defined basically by the noise from the preamplifier. Thus, it may be possible to do quantum non-demolition measurements, e.g. on charge qubits [1].

3. Quantum capacitive phase detector (C-SET)

Capacitively-read superconducting Cooper pair transistor, the C-SET, is the dual circuit of the L-SET (see Fig. 3). Instead of the second derivative with respect to the phase that is used to obtain a charge-dependent inductance in the L-SET, the C-SET operation employs the second derivative with respect to the gate charge which yields a phase-dependent capacitance.[9] Unlike any previous considerations of single-electron or single Cooper-pair devices, the implementation of the C-SET device is a generally fast and sensitive phase detector. The sensitivity is estimated to be $30 \cdot 10^{-6}\text{rad}/\text{Hz}^{-1/2}$ for typical HEMT preamplifiers with a noise temperature of 3 K. In recent experiments, we have achieved a sensitivity of $1 \cdot 10^{-3}\text{rad}/\text{Hz}^{-1/2}$.[10]

In conclusion, mesoscopic Josephson junctions allow construction of different kinds of detectors that may find their way to various applications. Both L-SET and C-SET are potential devices for the read-out of supercon-

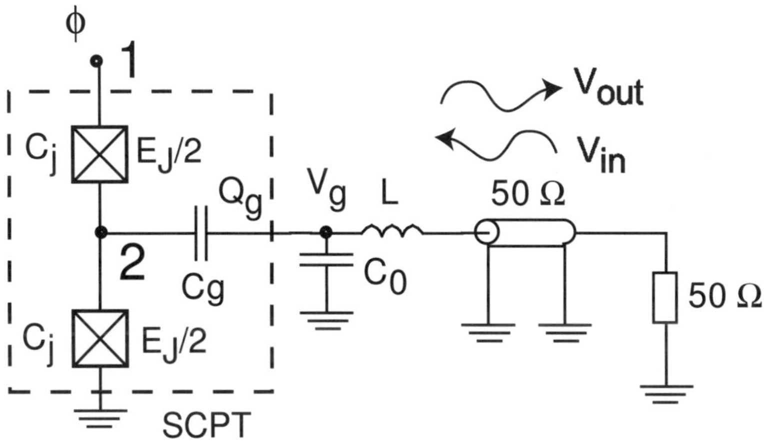

Figure 3. Principle of the C-SET. The phase of the reflected signal V_{out} is sensitive to the biasing phase over the SCPT.

ducting qubits. In general, their best characteristics is the low back action noise in combination with a fast response time. In addition, their back action noise can be engineered in such a way that a long relaxation time compared with the measurement time can be obtained, thereby facilitating QND measurements on qubits.

Acknowledgments

It is a pleasure to thank J. Delahaye, J. Hassel, R. Lindell, M. Paalanen, L. Roschier, H. Seppä, and M. Sillanpää for a very fruitful collaboration. This work was supported by the Academy of Finland and by the Large Scale Installation Program ULTI-3 of the European Union (contract HPRI-1999-CT-00050).

References

1. See, *e.g.* Yu. Makhlin, A. Schnirman, G. Schön, *Rev. Mod. Phys.* **73**, 357 (2001).
2. For a brief review, see, P. Hakonen, M. Kiviranta, and H. Seppä, J. Low Temp. Phys. **135**, 823 (2004).
3. See, e.g., D. Averin, K.K. Likharev, and A.B. Zorin, Sov. Phys. JETP **61**, 407 (1985).
4. A. Maassen van den Brink, L. J. Geerligs, and G. Schön, Phys. Rev. Lett. **67**, 3030 (1991).

5. J. Delahaye, J. Hassel, R. Lindell, M. Sillanpää, M. Paalanen, H. Seppä, and P. Hakonen, Science **299**, 1045 (2003); J. Hassel, J. Delahaye, H. Seppä, and P. Hakonen, J. Appl. Phys. **95**, 8059 (2004).
6. R. Lindell and P. Hakonen, Appl. Phys. Lett. **86**, 173507 (2005).
7. M. Sillanpää, L. Roschier, and P. Hakonen, Phys. Rev. Lett. **93**, 066805 (2004).
8. M. Sillanpää, L. Roschier, and P. Hakonen, cond-mat/0504122.
9. L. Roschier, M. Sillanpää, and P. Hakonen, Phys. Rev. B **71**, 024530 (2005).
10. M. Sillanpää, L. Roschier, and P. Hakonen, cond-mat/0504517.

THEORY OF CURRENT-DRIVEN DOMAIN WALL DYNAMICS

GEN TATARA

PRESTO, JST, 4-1-8 Honcho Kawaguchi, Saitama 332-0012, Japan
and
Graduate School of Science, Tokyo Metropolitan University
1-1 Minamiosawa, Hachioji, Tokyo 192-0397, Japan
E-mail: tatara@phys.metro-u.ac.jp

HIROSHI KOHNO

Graduate School of Engineering Science, Osaka University, Toyonaka, Osaka 560-8531, Japan

JUNYA SHIBATA

Frontier Research System (FRS), The Institute of Physical and Chemical Research (RIKEN), 2-1 Hirosawa, Wako, Saitama 351-0198, Japan

EIJI SAITOH

Department of Physics, Keio University, Yokohama, 223-8522, Japan

Equation of motion of a planar domain wall in a nanoscale ferromagnetic wire under electric current is derived based on microscopic description. The wall is shown to be driven by two mechanisms, spin transfer and momentum transfer, both arising from the exchange interaction, as has been argued by Berger. Domain wall under spin torque (spin transfer) is shown to have intrinsic pinning arising from hard-axis anisotropy energy, and that extrinsic pinning does not affect the threshold current. A resonating oscillation of domain wall occurs under AC current, and this was recently used for a spectroscopy of a single "domain wall particle". Nucleation of domain walls by spin torque is discussed. Fundamental mechanism of spin transfer effect is understood in terms of spin Josephson effect.

1. Introduction

The integration of magnetic and electric properties is highly important in the present technology context, since most of the magnetic devices are nowadays based on the magnetoresistive effect arising from modification of the electron transport by the magnetic configuration. In particular, the manipulation of magnetization by use of an electric current is promising for the next generation magnetic memories, where the information is written by an electric current. Manipulation of magnetization and magnetic domain wall by use of electric current of special interest recently [1,2,3,4,5,6] is also interesting as a basic physics in that it involves fascinating angular momentum dynamics.

Current-driven motion of a domain wall was studied in a series of pioneering works by Berger [7,8,9]. These theoretical works are based on his deep physical insight, but seems to lack transparency as a self-contained theory. Also, their phenomenological character makes the limit of applicability unclear. Here we reformulate the problem from microscopic point of view, and explore some new features such as intrinsic pinning[10,11,12] and resonance of domain wall[6].

2. Formulation

We start from a microscopic Hamiltonian with an exchange interaction between conduction electrons and spins of a domain wall. With a key observation that the wall position X and polarization ϕ_0 (the angle between spins at the wall center and the easy plane) are the proper collective coordinates to describe its dynamics[13,14], it follows straightforwardly that the electric current affects the wall motion in two different ways, in agreement with Berger's observation. The first is as a force on X, or momentum transfer, due to the reflection of conduction electrons [Fig. 1(a)]. This effect is proportional to the charge current and wall resistance, and hence negligible for thick walls[15]. The other is as a spin torque (a "force" on ϕ_0), arising when an electron passes through the wall with spin rotation [Fig. 1(b)]. It is also called as spin transfer [1] between electrons and wall magnetization. This effect is the dominant one for thick walls where the spin of the electron follows the magnetization adiabatically.

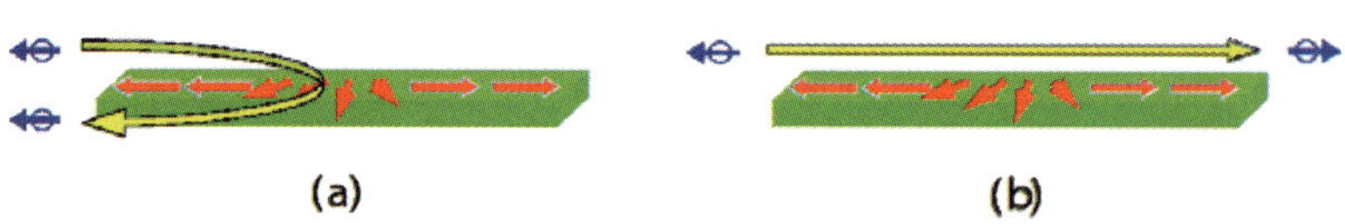

Figure 1. Momentum transfer arises from reflection of the conduction electron (a), while spin transfer arises from the spin flip of the transmitted electron (b).

The equations of motion are derived as[10,11,12]

$$\frac{\hbar NS}{\lambda}\left(\dot{\phi}_0+\alpha\frac{\dot{X}}{\lambda}\right)=F_{\rm pin}+F_{\rm el} \tag{1}$$

$$\frac{\hbar NS}{\lambda}(\dot{X}-\alpha\lambda\dot{\phi}_0)=\frac{NS^2K_\perp}{2}\sin 2\phi_0+T_{{\rm el},z}, \tag{2}$$

where $F_{\rm pin}=-(\partial V_{\rm pin}/\partial X)$ is the pinning force,

$$F_{\rm el}\equiv-\frac{\Delta}{S}\int d^3x\nabla_x \boldsymbol{S}_0(x-X)\cdot\boldsymbol{n}(x), \tag{3}$$

is the force (momentum transfer),

$$\boldsymbol{T}_{\rm el}\equiv-\frac{\Delta}{S}\int d^3x\,\boldsymbol{S}_0(x-X)\times\boldsymbol{n}(x), \tag{4}$$

is the torque (spin transfer). Here $\boldsymbol{S}_0$ denotes the classical domain wall solution (*e.g.*, $S_z=S\tanh\frac{x-X}{\lambda}$, S being magntude of local spin), Δ is the spin splitting, and $\boldsymbol{n}(x)$ is the spin density of electron. λ is the thickness of the wall, $K_\perp$ is the transverse anisotropy energy, and $N\equiv 2\lambda A/a^3$ is the number of spin in the wall (A is the area of the wire). α represents a standard damping torque (Gilbert damping). Note that the spin-transfer effect acts as a source to the wall velocity via

$$v_{\rm el}\equiv(\lambda/\hbar NS)\,T_{{\rm el},z}. \tag{5}$$

After some calculation, the force is obtained as

$$F_{\rm el}=enR_{\rm w}IA, \tag{6}$$

where $R_{\rm w}$ is the resistance due to a wall, I is the current, and n is the density of electron. In the adiabatic limit, $\lambda\gg k_{\rm F}^{-1}$, we immediately see that $F_{\rm el}=0$, whereas $v_{\rm el}=\frac{1}{2S}\frac{a^3}{e}j_{\rm s}$ remains finite ($e<0$ is the electron charge), where $j_{\rm s}$ is spin current, $j_{\rm s}\equiv\frac{e\hbar}{mV}\sum_{\boldsymbol{k}}k_x(f_{\boldsymbol{k}+}-f_{\boldsymbol{k}-})$. In the opposite case of thin wall (non-adiabatic case), $F_{\rm el}$ is finite, while spin transfer does not occur, $v_{\rm el}=0$. Hence the wall dynamics is very different for thick and thin walls. Also driving mechanism can depend whether the current is DC or AC as we see below.

3. Examples of wall dynamics

Here we show two typical examples of wall dynamics.

3.1. *Intrinsic pinning under DC spin current*

In the case of steady spin current, there exists a threshold value for wall motion[10],

$$j_{\rm s}^{{\rm cr}(1)} = \frac{eS^2}{a^3\hbar}K_\perp\lambda. \tag{7}$$

Below this threshold, $j_{\rm s} < j_{\rm s}^{{\rm cr}(1)}$, the spin torque from spin current can be absorbed by the transverse magnetization, ϕ, and so the wall does not move. Domain wall starts to move when $\phi \geq \pi/2$ (i.e., $j_{\rm s} > j_{\rm s}^{{\rm cr}(1)}$). The time-averaged velocity of wall as function of spin current is plotted in Fig. 2). Very important feature of spin torque-induced motion is that the dynamics

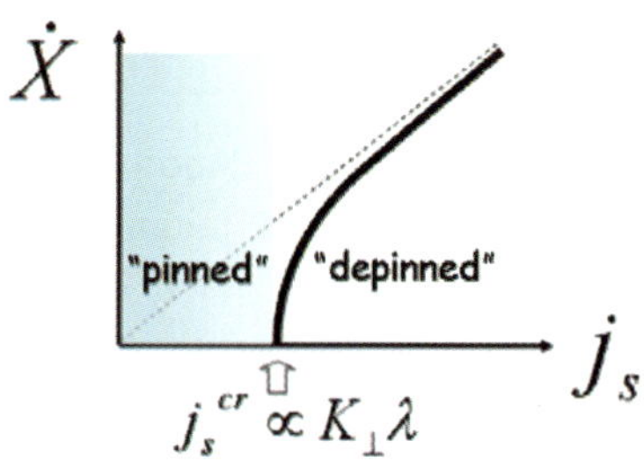

Figure 2. Time-averaged wall velocity as a function of spin current, $j_{\rm s}$.

is quite insensitive to extrinsic pinning such as impurities as long as the impurities are spinless[10]. This has been observed in experiments[16] and would be a great advantage in device application.

3.2. *Resonant motion of "domain wall particle"*

AC-current can drive domain walls quite effectively if the frequency is tuned to be close to the resonance with the pinning frequency. This resonance was realized in recent experiment by Saitoh *et al*[6]. They applied a small AC current in a wire with a domain wall in a weak pinning potential. The amplitude is well below the threshold, but the wall can shift slightly (for about a distance of μm). Equation of motion (2) indicates that ϕ remains small under small current, and as is easily seen, the wallthen satisfies a simple equation of motion of a "particle";

$$m\ddot{X} + \frac{m}{\tau}\dot{X} + m\Omega^2 X = f(t), \tag{8}$$

where $m \equiv \hbar^2 N/(K_\perp \lambda^2)$ is the wall mass, $\tau \simeq \hbar/(\alpha K_\perp)$ is a damping time, Ω is the pinning frequency, and $f(t) \simeq enAIR_{DW} - i\hbar^2 \omega I_s/(eK_\perp \lambda)$ is a force due to current (I and I_s are charge- and spin-current, respectively, R_{DW} being resistance due to domain wall)[6]. By measuring the energy dissipation (complex resistance), a resonance peak was observed when the frequency of the current, ω, is tuned to be Ω. From the resonance spectra, mass, friction constant, and resistivity of a single domain wall were identified to be $m = 6.6 \times 10^{-23}$kg, $\tau = 1.4 \times 10^{-8}$sec, $R_{DW} = 3 \times 10^{-4}\Omega$. What is more, driving mechanism of domain wall turned out to be the momentum transfer force, which is in contrast to the motion under DC current. This is quite surprisng considering a thickness of wall of $\sim$ 100nm, but is due to strong enhancement of momentum transfer force by resonance (spin torque is, in contrast, suppressed in MHz range)[6].

Domain wall, which is a soliton in ferromagnet, has been theoretically known to behave like a particle. This experiment is the first one to measure physical properties of this soliton "particle".

4. Domain nucleation by spin torque

The role of spin torque is represented by the following effective Hamiltonian of the local spin[17,18,19],

$$H_{\rm ST} = \int d^3x \frac{\hbar}{2e} \boldsymbol{j}_{\rm s} \cdot \nabla\phi(1 - \cos\theta), \tag{9}$$

where $\boldsymbol{j}_{\rm s}$ is the spin current density. As is seen in this term, spin current favors spatial variation of magnetization, $\nabla \boldsymbol{S}$, and thus the energy of domain wall can be lowered under spin current. In fact, the enegy of domain wall is estimated (in the case of large $K_\perp$) to be[19]

$$E_{\rm dw} \simeq NKS^2 \left[1 - \frac{1}{2}\left(\frac{\hbar a^3}{2eS^2\sqrt{KK_\perp}\lambda} j_{\rm s}\right)^2\right]. \tag{10}$$

Hence, for $j_{\rm s} > j_{\rm s}^{\rm cr(2)} \equiv \frac{2eS^2\sqrt{KK_\perp}\lambda}{\hbar a^3}$, domain wall has a lower energy than uniformly ferromagnetic state, resulting in domain formation from single domain state. The result is more rigorously confirmed by calculating spin wave gap[19]. Detailed phase diagram is shown in Fig.3. This domain formation by current would be useful in device applications.

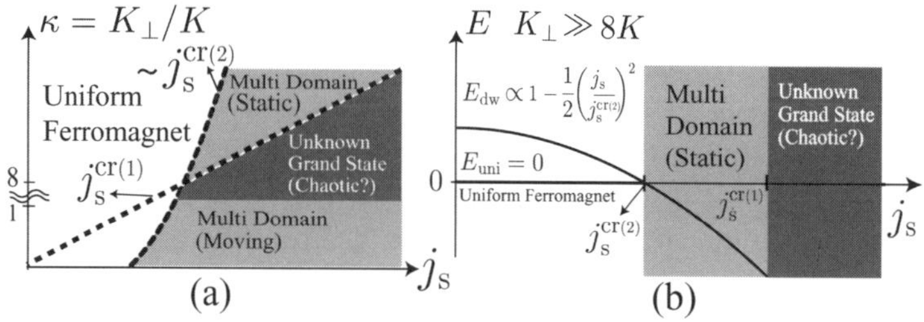

Figure 3. (a)Schematic phase diagram under spin current j_s in the absence of pinning potential. (b)Energy of the single-wall state ($E_{\rm dw}$) for $K_\perp \gg 8K$ is compared with that of the uniformly magnetized state, $E_{\rm uni} = 0$. For $j_s > j_s^{\rm cr(1)}$, the domain wall starts to flow but is unstable, suggesting a new ground state, which is still unknown.

5. More on spin torque

5.1. *Spin torque and Berry phase*

Let us here look more into the mechanism of spin torque from the fundamental aspect. The role of spin torque was represented by the Hamiltonian, (9). It is seen that this term is a spatial version of "the spin Berry phase",

$$L_{\rm B} = \int \frac{d^3x}{a^3} \hbar S \dot{\phi}(\cos\theta - 1), \tag{11}$$

and thus the role of spin current is to induce spatial spin Berry phase along the current. The spin torque has a particularly interesting consequence when the sample geometry is topologically non-trivial as in a ring. In fact, if we apply a spin current in a ferromagnetic ring, an umbrella-like structure of magnetization is favored rather than uniformly ferromagnetized state. The angle of tilt and resultant Berry phase is determined by the spin current.

5.2. *Spin Josephson effect*

Here we can pose an opposite question. Does a spontaneous current arise if we prepare a ring with umbrella structure? The answer is yes, as shown in the adiabatic limit by Loss et al[20]. Physics here would be more easily grasped if we think perturbatively[21,22,23]. When a conduction electron in a conductor is scattered by local spin $\boldsymbol{S}$, the electron wave function is multiplied by an amplitude $A(\boldsymbol{S}) = J_{sd}\boldsymbol{S}\cdot\boldsymbol{\sigma}$, which is generally spin-dependent and is represented by Pauli matrices. (Here J_{sd} is the coupling constant. We consider in this paper only classical, static scattering objects, and assume $\boldsymbol{S}$'s are constant vectors.)

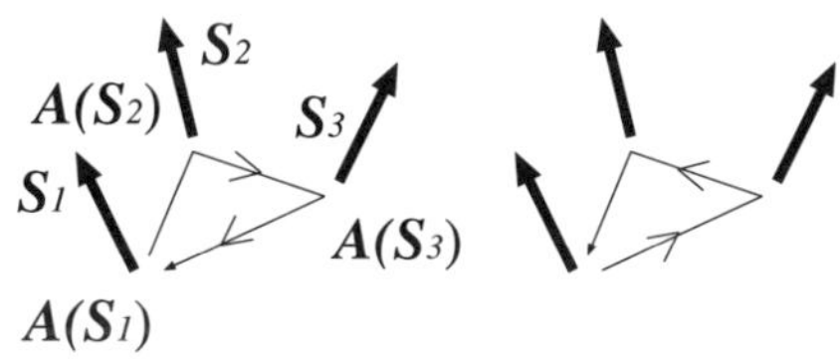

Figure 4. A closed path contributing to the amplitude of the electron propagation from x to x. At X_i, the electron experiences a scattering represented by an SU(2) amplitude, $A(\boldsymbol{S}_i)$. The contribution from one path (left) and the reversed one (right) are different in general due to the non-commutativity of $A(\boldsymbol{S}_i)$'s.

Let us consider two successive scattering events represented by $A(\boldsymbol{S}_1)$ and $A(\boldsymbol{S}_2)$ [Fig.4]. Due to the non-commutativity of σ_i, the amplitude depends on the order of the scattering event; $A(\boldsymbol{S}_1)A(\boldsymbol{S}_2) \neq A(\boldsymbol{S}_2)A(\boldsymbol{S}_1)$ in general. Various features in spin transport, which is under intensive pursuit recently, arise from this non-commutativity. It, however, does not affect the charge transport, since the charge is given as a sum of the two spin components (denoted by tr), and $\mathrm{tr}[A(\boldsymbol{S}_1)A(\boldsymbol{S}_2)] - \mathrm{tr}[A(\boldsymbol{S}_2)A(\boldsymbol{S}_1)] = 0$. Anomaly in the charge transport arises at the third order. We have

$$\mathrm{tr}[A(\boldsymbol{S}_1)A(\boldsymbol{S}_2)A(\boldsymbol{S}_3)] - \mathrm{tr}[A(\boldsymbol{S}_3)A(\boldsymbol{S}_2)A(\boldsymbol{S}_1)] = 4(J_{sd})^3\boldsymbol{S}_1\cdot(\boldsymbol{S}_2\times\boldsymbol{S}_3), \tag{12}$$

where the cross denotes the vector product, *i.e.*, $\boldsymbol{S}_1\cdot(\boldsymbol{S}_2\times\boldsymbol{S}_3) = \sum_{ijk}\epsilon_{ijk}S_1^iS_2^jS_3^k$. (The close relationship between spin chirality and charge current was also discussed in the context of chiral spin liquid in high T_c superconductivity by Ref. [24].) Eq. (12) indicates that in the presence of fixed $\boldsymbol{S}_i$'s with finite spin chirality, $\boldsymbol{S}_1\cdot(\boldsymbol{S}_2\times\boldsymbol{S}_3)$, the contribution from one path, $x\to X_1\to X_2\to X_3\to x$ [Fig.4, left], and its (time-) reversed one, $x\to X_3\to X_2\to X_1\to x$ [Fig.4, right], are not equal. This difference results in a spontaneous electron motion in a direction specified by the sign of spin chirality, namely, a permanent current if the electron is coherent. This spontaneous motion also results in Hall effect in frustrated magnets[25]. This is the meaning of spin torque Hamiltonian, Eq. (9). We thus see that the essence of spin torque is the non-commutativity of the SU(2) algebra, purely quantum mechanical.

A similar Josephson effect is expected also in p-wave superconductors as described by Y. Asano in this volume[26].

Acknowledgments

We thank T. Ono, H. Ohno, Y. Otani, H. Miyajima, H. Fukuyama, S. Maekawa, J. Ferré, J.A.C. Bland, S. Parkin for valuable discussion.

References

1. J.C. Slonczewski, J. Magn. Magn. Mater. **159**, L1 (1996).
2. J. Grollier et. al., J. Appl. Phys. **92**, 4825 (2002); Appl. Phys. Lett. **83**, 509 (2003); N. Vernier et al., Europhys. Lett. **65**, 526 (2004).
3. M. Tsoi, R. E. Fontana, and S. S. P. Parkin Appl. Phys. Lett. **83**, 2617 (2003).
4. M. Kläui et al., Appl. Phys. Lett. **83**, 105 (2003).
5. A. Yamaguchi et al., Phys. Rev. Lett. **92** 077205 (2004).
6. E. Saitoh, H. Miyajima, T. Yamaoka and G. Tatara, Nature **432**, 203 (2004).
7. L. Berger, J. Appl. Phys. **49**, 2156-2161 (1978).
8. L. Berger, J. Appl. Phys. **55**, 1954 (1984).
9. L. Berger, J. Appl. Phys. **71**, 2721 (1992); E. Salhi and L. Berger, *ibid.* **73**, 6405 (1993).
10. G. Tatara and H. Kohno, Phys. Rev. Lett. **92**, 086601 (2004).
11. G. Tatara and H. Kohno, J. Electron Microscopy, **54** (suppl 1), i69 (2005).
12. H. Kohno and G. Tatara, Proceedings of the 7th Oxford-Kobe materials seminar, to appear (2005).
13. H-B. Braun and D. Loss, Phys. Rev. **B53**, 3237 (1996).
14. S. Takagi and G. Tatara, Phys. Rev. **B54**, 9920 (1996).
15. G. Tatara and H. Fukuyama, Phys. Rev. Lett. **78**, 3773 (1997).
16. T. Ono, private communication; S. Parkin, private communication.
17. Ya. B. Bazaliy, B. A. Jones, and Shou-Cheng Zhang, Phys. Rev. B **57**, R3213 (1998).
18. J. Fernández-Rossier, M. Braun, A.S. Núñez and A. H. MacDonald, Phys. Rev. B **69**, 174412 (2004).
19. J. Shibata, G. Tatara and H. Kohno, Phys. Rev. Lett. **94**, 076601 (2005).
20. D. Loss, P. Goldbart and A. V. Balatsky, Phys. Rev. Lett. **65**, 1655 (1990).
21. G. Tatara and H. Kohno, Phys. Rev. **B67**, 113316 (2003).
22. G. Tatara and N. Garcia, Phys. Rev. Lett. **91**, 076806 (2003).
23. G. Tatara, phys. sta. sol. (b) **241**, 1174 (2004).
24. X. G. Wen, F. Wilczek, and A. Zee, Phys. Rev. **B39**, 11413 (1989).
25. G. Tatara and H. Kawamura, J. Phys. Soc. Jpn., **71**, 2613 (2002).
26. Y. Asano, cond-mat/0504419 and in this proceeding.

SQUID OF A RUTHENATE SUPERCONDUCTOR

YASUHIRO ASANO
Applied Physics, Hokkaido University, Sapporo 060-8628, Japan
E-mail: asano@eng.hokudai.ac.jp

YUKIO TANAKA
Department of Applied Physics, Nagoya University, Nagoya 464-8603, Japan

SATOSHI KASHIWAYA
National Institute of Advanced Industrial Science and Technology, Tsukuba 305-8568, Japan

The Josephson effect is theoretically studied in two types of SQUID consisting of s wave superconductor and Sr_2RuO_4. Results show various response of the critical Josephson current to applied magnetic fields depending on the types of SQUID and on the pairing symmetries. In the case of a $p_x + ip_y$ wave symmetry, the critical current in a corner SQUID becomes an asymmetric function of magnetic fields near the critical temperatures. Our results well explain a recent experimental finding [Nelson et. al, Science **306**,1151 (2004)].

1. Introduction

Since the discovery of high-T_c superconductivity [1], a considerable number of scientists had tried to specify the symmetry of Cooper pairs. It is now known that Cooper pairs are characterized by the spin-singlet $d_{x^2-y^2}$ wave symmetry. The multicrystal grain boundary junctions [2] was a good symmetry prove of high-T_c cuprates. In addition, there is no doubt that an experiment on the Josephson current in a corner SQUID consisting of conventional s wave superconductor and high-T_c cuprates also played an crucial role [3,4]. The success is a consequence of a fact that the electric transport in superconducting junctions is essentially phase sensitive [5,6,7]. Thus transport measurements in junctions having a ring topology, such as SQUID, enable us to know a global view of the pair potential on the Fermi surface.

Sr_2RuO_4 (SRO) discovered in 1994 [8] is a candidate of spin-triplet chiral

p wave superconductor. To date, the fundamental symmetry has been implied by $p_x + ip_y$ wave symmetry [9]. The zero-bias anomaly in tunneling spectra [10] is, at least, consistent with the $p_x + ip_y$ wave symmetry [11]. We, however, have never found convincing information on the orbital part of the Cooper pairs from a series of experiments yet. To address the pairing symmetry clearly, the clear-cut distinction is desired.

In this paper, we theoretically analyze recent experimental results of the Josephson effect in two types of SQUID consisting of s wave superconductor and SRO as shown in Fig. 1(a) and (b) [13]. The Josephson critical currents in the two SQUID show different responses to applied magnetic fields (Φ) because of close relations between junction symmetries in real space and pairing symmetries in momentum space. The critical current in the corner SQUID becomes an asymmetric function of Φ [13]. This important feature will be naturally explained in the present paper when the $p_x + ip_y$ wave symmetry is considered in SRO.

2. Josephson current

Before discussing the Josephson effect in the SQUID, the Josephson current between s wave superconductor and SRO in Fig. 1(c) should be summarized. The pair potential in SRO is described by [9]

$$\hat{\Delta}_{\boldsymbol{p}} = i\Delta(\bar{p}_x \pm i\bar{p}_y)\boldsymbol{d}\cdot\hat{\boldsymbol{\sigma}}\hat{\sigma}_2, \tag{1}$$

where $\hat{\sigma}_j$ with $j = 1, 2$ and 3 are the Pauli matrices and the $\boldsymbol{d}$ is a unit vector in the spin space parallel to the z direction (c axis). In the pair potential, $\bar{p}_x = p_x/p_F$ ($\bar{p}_y = p_y/p_F$) is the normalized wave number on the Fermi surface in the x (y) direction with p_F being the Fermi wave number in SRO. The pair potential in the s wave superconductor is described by $\hat{\Delta}_{\boldsymbol{k}} = i\Delta\hat{\sigma}_2$. On the basis of the Josephson current formula [14], the Josephson current just below T_c is derived from a microscopic calculation [12],

$$J_{p_x \pm ip_y} = \pm J_1 \cos\varphi - J_2 \sin(2\varphi), \tag{2}$$

$$J_1 = T_B\, \delta\mu\, \alpha_S \left(\frac{\Delta}{2T}\right)\frac{\Delta}{\Delta_0}, \tag{3}$$

$$J_2 = T_B^2 \left(\frac{\Delta}{2T}\right)^3 \frac{\Delta}{\Delta_0}, \tag{4}$$

where $\alpha_S = (\lambda_e k_F)^2$, $\delta\mu = (\mu_F - \mu_p)/\mu_F$, λ_e is the Compton wave length, k_F and μ_F are the Fermi wave number and the Fermi energy in s wave superconductors, respectively. The Fermi energy in SRO is defined by μ_p.

The transmission probability of the potential barrier is denoted by T_B. The Josephson current is measured in units of $e\Delta_0 N_c/\hbar$, where Δ_0 is the amplitude of the pair potential at $T = 0$ and N_c is the number of propagating channels on the Fermi surface. Constant coefficients of the order of unity have been omitted in Eqs. (3) and (4). We note that the spin-orbit scattering at the potential barrier gives rise to the Josephson current proportional to $\cos\varphi$, which implies the breakdown of the time-reversal symmetry in SRO.

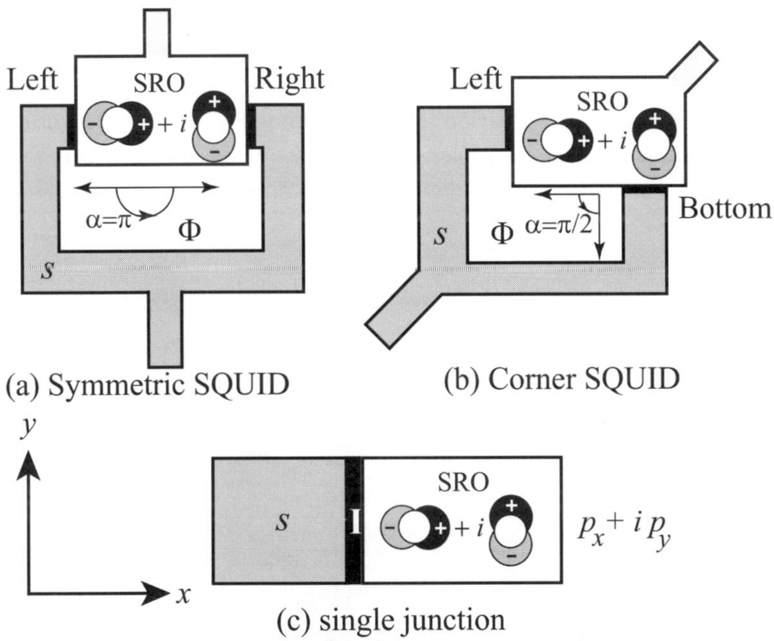

Figure 1. Schematic pictures of two types Josephson junction are shown in (a) and (b). The singly connected Josephson junction is given in (c).

In this paper, we consider two types of SQUID as shown in Fig. 1(a) and (b), where Φ is the magnetic flux passing through the SQUID and α denotes a relative angle between the two junctions. In the symmetric SQUID ($\alpha = \pi$), the Josephson current of the left junctions and that of the right one are given by

$$J_L(\varphi) = J_1 \cos\varphi - J_2 \sin 2\varphi, \tag{5}$$

$$J_R(\varphi) = -J_1 \cos\varphi - J_2 \sin 2\varphi. \tag{6}$$

The current-phase relation in J_R is derived from Eq. (2) with changing $\varphi \to \varphi + \pi$. This derivation is always valid for odd parity superconductors [15]. The Josephson current in the symmetric SQUID is then expressed by [16]

$$J_S(\varphi, \Phi) = J_L(\varphi + \phi_B) + J_R(\varphi - \phi_B), \tag{7}$$
$$= -2J_1 \sin\varphi \sin\phi_B - 2J_2 \sin 2\varphi \cos 2\phi_B, \tag{8}$$

where $\phi_B = \pi\Phi/\Phi_0$ and $\Phi_0 = 2\pi\hbar c/e$. In the corner SQUID ($\alpha = \pi/2$), the Josephson current in the bottom junctions is given by

$$J_B(\varphi) = -J_1 \sin\varphi + J_2 \sin 2\varphi, \tag{9}$$

where the current-phase relation is derived from Eq. (2) with $\varphi \to \varphi + \pi/2$. The derivation is similar to that discussed in an isotropic chiral f wave symmetry [17]. The Josephson current in the corner SQUID is expressed in the same way,

$$J_C(\varphi, \Phi) = J_1 \cos(\varphi + \phi_B) - J_1 \sin(\varphi - \phi_B) - 2J_2 \cos 2\varphi \sin 2\phi_B. \tag{10}$$

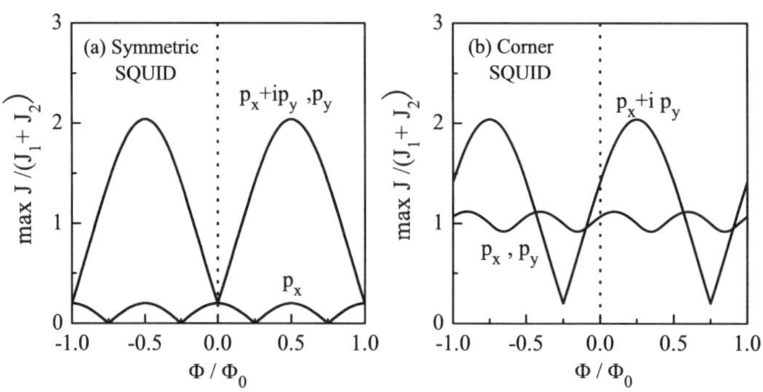

Figure 2. The Josephson critical current is plotted as a function of Φ in the symmetric SQUID in (a) with $J_1/J_2 = 10$. The calculated results for the corner SQUID are shown in (b).

In Fig. 2, the critical Josephson current in the two types of SQUID are shown as a function of Φ. A parameter $J_1/J_2 = 10$ is realized in high temperatures near T_c. Since $J_1 \gg J_2$, we find,

$$\max|J_{\rm S}(\Phi)| \simeq 2J_1 \left|\sin\left(\pi\frac{\Phi}{\Phi_0}\right)\right|, \tag{11}$$

$$\max|J_{\rm C}(\Phi)| \simeq 2J_1 \left|\sin\left(\pi\frac{\Phi}{\Phi_0}+\frac{\pi}{4}\right)\right|. \tag{12}$$

The odd parity symmetry immediately results in the minima of the critical current at $\Phi = 0$ in the symmetric SQUID [15]. In the corner SQUID, the critical current is no longer a symmetric function of Φ. The geometry change from the left junction to bottom one can be described by $\pi/2$ phase shift in the $p_x + ip_y$ wave pair potential. As a consequence, the $\cos\varphi$ component in the left junction becomes the $\sin\varphi$ in the bottom junction as shown in Eq. 9. For comparison, we also show the results for the p_x and p_y symmetries in Fig. 2. The Josephson currents in the single junctions are given by [12]

$$J_{p_x}(\varphi) = -\,J_2 \sin 2\varphi, \tag{13}$$

$$J_{p_y}(\varphi) = J_1 \sin\varphi - J_2 \sin 2\varphi. \tag{14}$$

Effects of the spin-orbit scattering is negligible in the p_x symmetry, whereas the spin-orbit scattering gives rise to the Josephson current proportional to $\sin\varphi$ in the p_y symmetry. In the symmetric SQUID, the critical current for the p_x symmetry has a period of $\Phi_0/2$ as shown in Fig. 2(a) because of $J_1 = 0$. The critical current for the p_y symmetry is identical to that for the $p_x + ip_y$ symmetry. In the corner SQUID, the critical current for the p_x symmetry has a period of $\Phi_0/2$. At the same time, the amplitude of the oscillations is much smaller than the mean value of the critical current. The mean value is proportional to J_1 and the oscillating amplitude is given by J_2. We note in the corner SQUID that the Josephson effect for the p_y symmetry is identical to that for the p_x symmetry.

3. Conclusion

We have studied the critical Josephson current in two types of SQUID consisting of s wave superconductor and Sr_2RuO_4. In the $p_x + ip_y$ symmetry, the critical Josephson current in the corner SQUID becomes the asymmetric function of Φ because the current-phase relation in the two junctions relate to each other by $\pi/2$ phase shift in the pair potential. Our results well explain the recent experimental findings [13].

In real junctions, SRO may have chiral domain structures. Effects of the chiral domains on the interference pattern will be discussed elsewhere [18].

Acknowledgments

The authors thank Y. Maeno and Y. Liu for useful discussion. This work has been partially supported by Grant-in-Aid for the 21st Century COE program on "Topological Science and Technology" from the Ministry of Education, Culture, Sport, Science and Technology of Japan.

References

1. J. G. Bednorz and K. A. Müller, Z Phys. B **64**, 189 (1986).
2. C. C. Tsuei, J. R. Kirtley, C. C. Chi, L. S. Yu-Jahnes, A. Guputa, T. Shaw, J. Z. Sun, and M. B. Ketchen, Phys. Rev. Lett. **73** 593 (1994); C. C. Tsuei and J. R. Kirtley, Rev. Mod. Phys. **72**, 969 (2000).
3. M. Sigrist and T. M. Rice, J. Phys. Soc. Jpn. **61**, 4283 (1992); Rev. Mod. Phys. **67**, 503 (1995).
4. D. A. Wollman, D. J. Van Harlingen, W. C. Lee, D. M. Ginsberg, and A. J. Leggett, Phys. Rev. Lett. **71**, 2134 (1993).
5. Y. Tanaka and S. Kashiwaya, Phys. Rev. Lett. **74**, 3451 (1995).
6. Y. Asano, Y. Tanaka and S. Kashiwaya, Phys. Rev. B **69**, 134501 (2004).
7. S. Kashiwaya and Y. Tanaka, Rep. Prog. Phys. **63**, 1641 (2000).
8. Y. Maeno, H. Hashimoto, K. Yoshida, S. Nishizaki, T. Fujita, J. G. Bednorz, and F. Lichtenberg, Nature **372**, 532 (1994).
9. T. M. Rice and M. Sigrist, J. Phys. Condens. Matter **7**, L643 (1995).
10. Z. Q. Mao, K. D. Nelson, R. Lin, Y. Liu, and Y. Maeno, Phys. Rev. Lett. **87**, 037003 (2001).
11. M. Yamashiro, Y. Tanaka, N. Yoshida, and S. Kashiwaya, J. Phys. Soc. Jpn. **68**, 2019 (1999).
12. Y. Asano, Y. Tanaka, M. Sigrist, and S. Kashiwaya, Phys. Rev. B **67**, 184505 (2003).
13. K. D. Nelson, Z. Q. Mao, Y. Maeno, and Y. Liu, Science **306**, 1151 (2004).
14. Y. Asano, Phys. Rev. B **64**, 224515 (2001).
15. V. B. Geshkenbein, A. I. Larkin, and A. Barone, Phys. Rev. B **36**, 235 (1987).
16. A. Barone and G. Paterno, *Physiscs and Applications of the Josephson Effect* (Wiley, New York, 1982).
17. J. A. Sauls, Adv. Phys. **43**, 113 (1994).
18. Y. Asano, Y. Tanaka, M. Sigrist, and S. Kashiwaya, Phys. Rev. B **71**, 214501 (2005).

PATH INTEGRAL FORMALISM FOR QUANTUM TUNNELING OF RELATIVISTIC FLUXON

K. KONNO[1]* T. FUJII[1] AND N. HATAKENAKA[1,2]

[1] *Graduate School of Advanced Sciences of Matter, Hiroshima University, Higashi-Hiroshima 739–8530, Japan.*

[2] *Faculty of Integrated Arts and Sciences, Hiroshima University, Higashi-Hiroshima 739–8521, Japan.*

Quantum tunneling of a relativistic fluxon through a rectangular potential is discussed. In the relativistic cases in the absence of friction, differences from non-relativistic cases are revealed. In order to extend the study to relativistic quantum tunneling with friction, path integral approach is also developed.

1. Introduction

Recently information-processing technology based on quantum mechanics has been developed by many researchers with intense vigor. Such information-processing technology is quite different from conventional information-processing technology. For the purpose of realizing the quantum information processing, various physical systems have been investigated so far. The most significant problem in the physical systems for the quantum information processing is the effect of environment on the systems. Inevitable interaction between the environment and the systems destroys coherent states or entanglement states that the systems last initially. These physical processes are called *decoherence.*

We focus on a physical system utilizing an elementary excitation with the quantum unit of magnetic flux, i.e., fluxon which propagates along a long Josephson junction with a relativistic speed. The fluxon corresponds to a *topological* soliton obeying the sine-Gordon equation and behaves as a relativistic spin-1/2 particle. In fact, the Lorentz contraction of fluxon was observed in an annular Josephson system.[1] Hence, this phys-

*Present address: Department of Applied Physics, Hokkaido University, Sapporo 060-8628, Japan. E-mail: konno@topology.coe.hokudai.ac.jp

ical system is appropriate not only for processing and storage of information, but also for transmission. However, preceding studies on fluxon are mainly concerned with the classical dynamics. Quantum-mechanical features of fluxon[2] remain unresolved, especially for relativistic cases. Revealing quantum-mechanical aspects of fluxon is then prerequisite for constructing quantum bits in such physical system. Our aims of this study are to clarify quantum-mechanical nature and decoherence processes of fluxon.

The organization of this paper is as follows. First we discuss quantum tunneling of a relativistic fluxon based on the Dirac equation without friction in Sec. 2. To investigate frictional cases, path integral approach is also discussed in Sec. 3. Finally we give a brief summary in Sec. 4.

2. Quantum tunneling of relativistic fluxon in the absence of friction

First we formulate quantum tunneling of a relativistic fluxon without friction. Since the Josephson fluxon has a polarity in phase rotation similar to a spin 1/2 particle, we adopt the Dirac equation as the basic equation governing the dynamics of a fluxon. We now focus on the case that there exists one-dimensional rectangular potential barrier V(x) (see Fig. 1) as

$$V(x) = \begin{cases} V_0 & (a < x < b)\,, \\ 0 & (x < a, b < x)\,. \end{cases} \tag{1}$$

The Dirac equation is reduced to

$$i\hbar\frac{\partial}{\partial t}\Psi(t,x) = \left[-ic\hbar\alpha_1\frac{\partial}{\partial x} + \beta\left(m_0c^2 + V(x)\right)\right]\Psi(t,x), \tag{2}$$

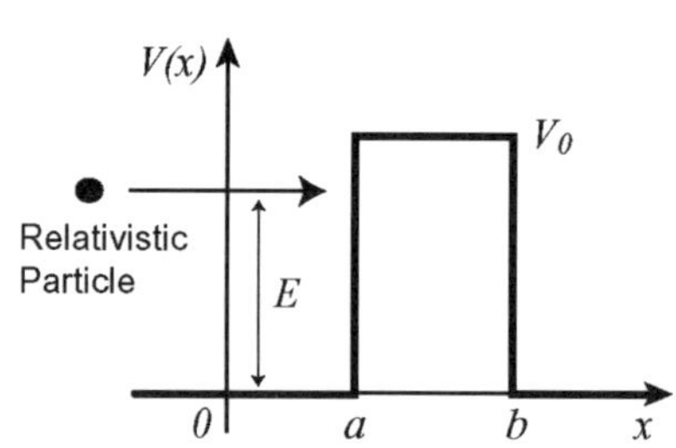

Figure 1. Situation for quantum tunneling of a fluxon through a rectangular potential barrier.

where Ψ denotes the four-component spinor, and α_1 and β are given by

$$\alpha_1 = \begin{pmatrix} 0 & \sigma_1 \\ \sigma_1 & 0 \end{pmatrix}, \quad \beta = \begin{pmatrix} I & 0 \\ 0 & -I \end{pmatrix}, \tag{3}$$

using the Pauli spin matrices σ_i

Let us consider stationary states, i.e., $\Psi(t,x) = e^{-iEt/\hbar}\psi(x)$, where E is the energy. Furthermore, let us discuss the case in which incident wave comes from the left-hand side ($x < a$). Under these conditions, we solved the Dirac equation with the boundary condition $\psi^{(1)}(a) = \psi^{(2)}(a)$ and $\psi^{(2)}(b) = \psi^{(3)}(b)$, where the superscript denotes each region. Here we assumed region (1), region (2) and region (3) as $x < a$, $a < x < b$ and $b < x$, respectively. From the solution,[3] we obtain the transmission probability

$$T = \frac{4f_0^2 f_1^2}{\left(f_0^2 + f_1^2\right)^2 \sinh^2 \dfrac{2(b-a)k_1}{\hbar} + 4f_0^2 f_1^2}, \tag{4}$$

where $f_0 = ck_0/(m_0c^2 + E)$, $f_1 = ck_1/(m_0c^2 + V_0 + E)$, $k_0 = \sqrt{E^2 - (m_0c^2)^2}$, and $k_1 = \sqrt{(m_0c^2 + V_0)^2 - E^2}/c$. Figure 2 shows the tunneling probability T as a function of velocity. For comparison, the result of the non-relativistic case is also shown. Here we specified just two dimensionless parameters $G = (b-a)/(\hbar/m_0c)$ and $\overline{V} = V_0/m_0c^2$, which correspond to the length of the potential barrier normalized by the Compton wavelength and the height normalized by the rest mass energy, respectively. From Fig. 2, we find the following points. In high energy regime, the relativistic tunneling rate becomes larger than that of the non-relativistic case. This is due to significant contribution of higher order power of momentum to the energy via the Lorentz factor. On the other hand, in low energy regime, the relativistic tunneling rate is smaller than that of the non-relativistic case, since the tunneling rate is suppressed by a potential-barrier enhancement due to the rest mass energy m_0c^2. Furthermore, we find that oscillation of tunneling probability of the relativistic case in high energy regime becomes faster with velocity increase. Such feature can be distinguished from that of non-relativistic case.

3. Path integral formalism for incorporating friction

The whole features of quantum tunneling of a relativistic fluxon can be derived by solving the Dirac equation as in the last section if there are no decoherence processes. However, path integral description of the quantum tunneling is also useful. In particular, it is necessary for incorporating the

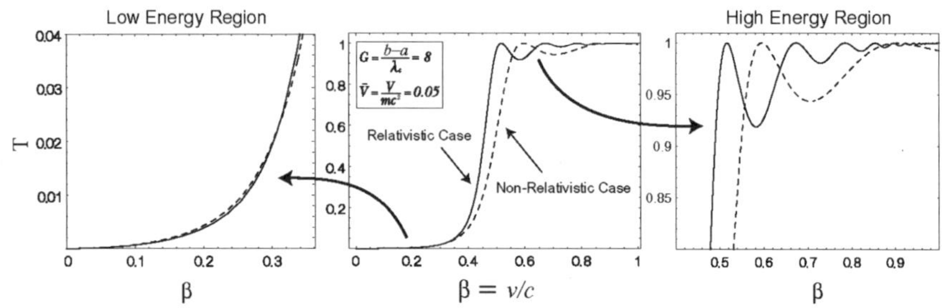

Figure 2. Comparison of the tunneling probability T between the relativistic case and the non-relativistic case.

effects of environment. As discussed by Feynman & Vernon[4] and Caldeira & Leggett,[5] path-integral formalism is straightforward for taking into account any degree of freedom which expresses the environment. For this reason, we also develop path-integral formalism for describing quantum tunneling of a relativistic fluxon, following the path decomposition method developed by Fertig.[6]

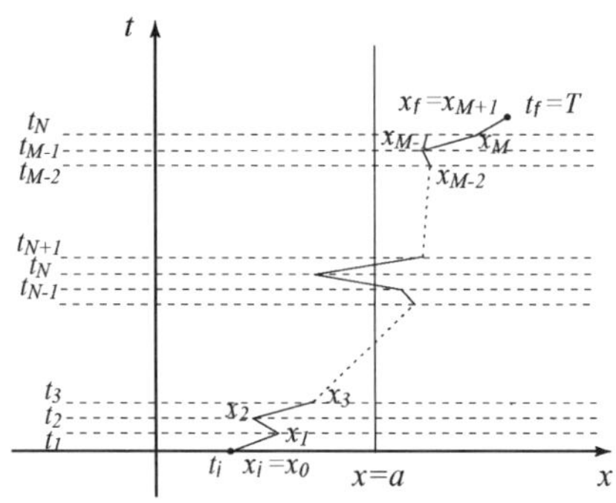

Figure 3. Schematic illustration of path integral.

Here, we temporarily return to non-relativistic cases. The propagator of a non-relativistic particle from $(t, x) = (0, x_0)$ to $(t, x) = (T, x_T)$ is written as

$$K(x_T, x_0; T) = \int_{x(0)=x_0}^{x(T)=x_T} D[x(t)] e^{iS[x(t)]/\hbar}, \tag{5}$$

where $S[x(t)] = \int_0^T dt \left[m \left(dx/dt \right)^2 / 2 - V(x) \right]$. Let us consider a surface of $x = a$ and assume the time at which the path of the particle lastly crosses

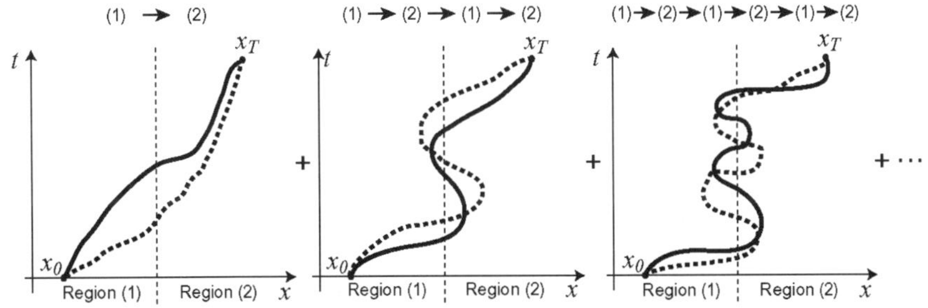

Figure 4. Schematic illustration of path decomposition.

the surface to be $t = t_a$ (see Fig. 3). If we decompose the space into two regions (1) and (2) as $-\infty < x < a$ and $a < x < \infty$, then the propagator can be decomposed as

$$\begin{aligned} iK(x_T, x_0; T) = &\int_0^T dt iK^{(1)}(a, x_0; t)\Sigma_a \left(iK^{(2)}(x_T, a; T-t)\right) \\ &+ \int_0^T dt_1 \int_0^{T-t_1} dt_2 \int_0^{T-t_1-t_2} dt_3 iK^{(1)}(a, x_0; t_1) \\ &\times\Sigma_a \left(iK^{(2)}(a, a; t_2)\right) \Sigma_a \left(iK^{(1)}(a, a; t_3)\right) \\ &\times\Sigma_a \left(iK^{(2)}(x_T, a; T - t_1 - t_2 - t_3)\right) + \cdots , \end{aligned} \tag{6}$$

where $K^{(1)}$ and $K^{(2)}$ denote the propagators defined in the restricted regions (1) and (2), respectively, and $\Sigma_a(\cdots) \equiv \hbar/2m \cdot \partial/\partial x(\cdots)|_{x=a}$. Such path decomposition is schematically illustrated in Fig. 4.

It is useful to find energy states, because the tunneling problems are usually discussed in a constant energy state. For this purpose, we consider the Fourier transformation

$$G(x_T, x_0; E) \equiv i\int_0^\infty dTK(x_T, x_0; T)\, e^{iET/\hbar}. \tag{7}$$

Then, from Eq. (6), we obtain

$$G(x_T, x_0; E) = \frac{G^{(1)}(a, x_0; E)\Sigma_a G^{(2)}(x_T, a; E)}{1 - \Sigma_a G^{(2)}(a, a; E)\Sigma_a G^{(1)}(a, a; E)}, \tag{8}$$

where $G^{(1)}$ and $G^{(2)}$ are the Fourier transform of $K^{(1)}$ and $K^{(2)}$, respectively. Thus, once we know the Green function in each region, we can calculate the total Green function G using Eq. (8).

By applying the above-mentioned path decomposition method to the quantum tunneling through the rectangular potential, in which case there exist three regions, we can obtain the expression

$$T = \frac{4\kappa^2 k^2}{(k^2 - \kappa^2)\sinh^2 \kappa(b-a) + 4\kappa^2 k^2}. \tag{9}$$

In order to find an expression for relativistic cases, we replace k and κ with the relativistic expressions $k = \sqrt{E^2 - m_0^2 c^4}/(\hbar c)$ and $\kappa = \sqrt{(m_0 c^2 + V_0)^2 - E^2}/(\hbar c)$. The derived approximate expression reproduces all the relativistic features. Therefore, the above-mentioned path decomposition method is useful to investigate the effects of friction on quantum tunneling of a relativistic fluxon using the path integral approach[4,5].

4. Summary

We have discussed quantum tunneling of a relativistic fluxon. By solving the Dirac equation, the characteristics of relativistic tunneling probability have been discussed in comparison with the non-relativistic case. Furthermore, path integral approach to relativistic quantum tunneling has been developed for the purpose of incorporating frictional effects. The frictional effects on the tunneling probability for a relativistic fluxon will be revealed by using the path integral approach in our future work.

Acknowledgments

We would like to thank K. Nagai, M. Nishida and S. Matsuo for valuable discussions. This work was supported in part by a research grant from The Mazda Foundation and a Grant-in-Aid for COE Research (No. 13CE2002) from the Ministry of Education, Sports, Science and Technology of Japan.

References

1. A. Laub, T. Doderer, S. G. Lachenmann, R. P. Huebener, and V. A. Oboznov, *Phys. Rev. Lett.* **75**, 1372 (1995)
2. A. Wallraff, A. Lukashenko, J. Lisenfeld, A. Kemp, M. V. Fistul, Y. Koval, and A. V. Ustinov, *Nature (London)* **425**, 155 (2003).
3. K. Konno, F. Kobayashi, S. Matsuo, M. Nishida, and N. Hatakenaka, Proceedings of *International Symposium on Mesoscopic Superconductivity and Spintronics 2004* (in press).
4. R. P. Feynman, and F. L. Vernon, Jr., *Annals of Physics* **24**, 118 (1963).
5. A. O. Caldeira, and A. J. Leggett, *Physica* **121A**, 587 (1983).
6. H. A. Fertig, *Phys. Rev.* **A47**, 1346 (1993).

EXPERIMENTAL STUDY OF TWO AND THREE-DIMENSIONAL SUPERCONDUCTING NETWORKS

S. TSUCHIYA, K. INAGAKI, AND S. TANDA
Department of Applied Physics, Hokkaido University,
Kita 13 , Nishi 8,
Sapporo 060-8628, Japan
E-mail: tuchiya@eng.hokudai.ac.jp

T. KIKUCHI, AND H. TAKAHASHI
Department of Materials Science, Hokkaido University,
Kita 13 , Nishi 8,
Sapporo 060-8628, Japan

We studied two and three-dimensional superconducting networks experimentally. We succeeded in fabrication of three-dimensional structures of nickel and indium by anodizing aluminum, laser irradiation, and electrophoretic deposition. We observed superconducting transition of the indium ring fabricated by the technique. We also fabricated suqure lattice and observed two periods of oscillation of superconducting transition temperature with magnetic field.

1. Introduction

Superconducting networks have been studied in recent years. Since a superconducting network is a frustrated system, network effect is observed as short periods of variation in magnetic field responses. Two-dimensional networks are studied theoretically and experimentally [1,2,3,4,5,6,7,8], for example, simple periodic lattices (square, triangular, honeycomb) and disordered lattices. On the other hand, three-dimensional networks are studied only theoretically [9]. In the case of three-dimensional networks, novel kinds of frustrations are expected to occur because a quantized flux must go inside through one of the faces and go out through another. However, experimental studies of three-dimensional networks have not been reported because it seems to be difficult to fabricate.

We studied two and three-dimensional superconducting networks experimentally. We fabricated three-dimensional networks of indium by anodizing aluminum, laser irradiation, and electrophoretic deposition. We have already demonstrated that three-dimensional structures of nickel and platinum can be fabricated by the technique before [10]. We also fabricated ring of indium and investigated superconducting transition. Because the indium sample containing impurity may not show superconductivity. We prepared a square lattice as a two-dimensional network by standard electron beam lithography and observed variation of transition temperature in magnetic field.

2. Experiment

2.1. *Three-dimensional network*

We considered fabrication of a three-dimensional network of superconductor. We explain the method to fabricate a three-dimensional structure by anodizing aluminum, laser irradiation, and electrophoretic deposition. An aluminum specimen covered with anodic oxide film is irradiated with a pulsed neodymium-doped yttrium aluminum garnet (Nd-YAG) laser in an electro- or electroless plating solution to remove the oxide film by laser ablation of the aluminum substrate. Selective metal deposition is then achieved at the area where film has been removed by electro- or electroless plating. Finally, the specimen is attached to an epoxy resin board before dissolving the aluminum substrate and the oxide film in alkaline solution. Figure 1 (a) shows SEM pictures of three-dimensional networks of nickel. The sample is cylindrical network with 2 mm diameter and 40 μm line width.

We considered fabrication of a three-dimensional structure of superconductor by the technique. In the technique, we need the condition of metal plating and not dissolved in acid or alkaline solution. When we thought that we fabricated three-dimensional structure of superconductor by the technique, we needed to find superconductor satisfied with the condition. Indium was satisfied with the condition. Figure1 (b) shows SEM pictures of three-dimensional structure of indium. Indium deposited at the area where film has been removed by electro- or electroless plating and remained not dissolving in alkaline solution. The three-dimensional network of superconductor can be fabricated by the technique. The sample of indium was broken by itself weight. But the problem will be solved in the near future if the sample size become small.

We should confirm superconductivity of the sample fabricated by the technique because if the other ion in an electro- or electroless plating solu-

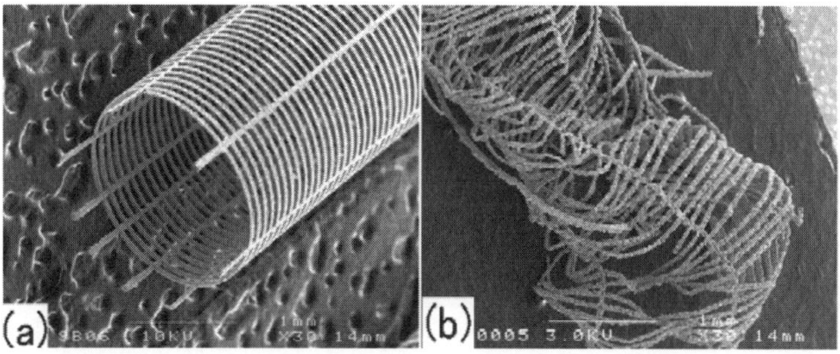

Figure 1. Three-dimensional structure. Both samples are cylindrical network with 2 mm diameter and 40 μm line width. (a) The sample is made of nickel. (b) The sample is made of indium.

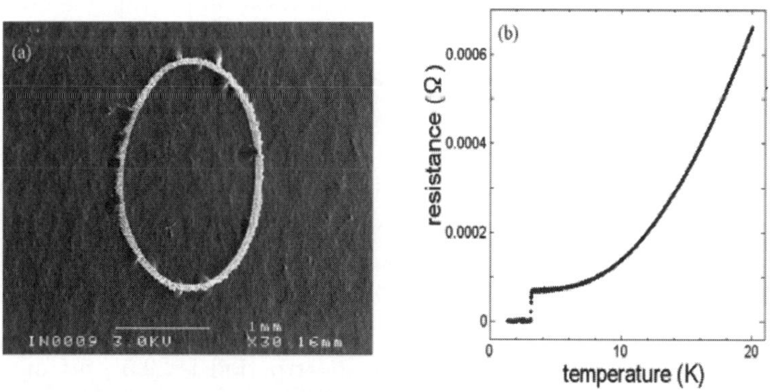

Figure 2. (a) Indium ring. This sample has 2mm diameter and 50 μm line width. (b) Temperature dependence of the samle resistance. We observed superconducting transition temperature at 3.2K.

tion in addition to indium ion deposit, these will be impurity. The indium sample containing impurity may not show superconductivity. So we also fabricated easier structure ring of indium by the same technique to investigate electrical conductivity. Figure 2 (a) shows SEM picture of ring of indium with 2mm diameter and 50 μm line width. Figure 2 (b) shows temperature dependence of the sample resistance. We observed superconducting transition temperature at 3.2K. This value almost agree with transition temperature of indium 3.4K. A resistivity at room temperature is 3×10^{-8} Ωm and a residual resistivity ratio is $\sim 10^2$. We could say that the sample is very pure sample. Accordingly, the technique is valid to fabricate three-dimensional structure of superconductor.

The three-dimensional structure can be made smaller and more precise by the technique. Usage of aluminum rods with smaller diameter enables fabrication of smaller structures, although crack formation during anodizing may confine the thickness of a deposited layer. Hence, three-dimensional structures with 5 μm line width are possible technically, and more precise structure presumably may be fabricated by choosing the optimum laser irradiation condition in the near future.

2.2. *Two-dimensional network*

We also fabricated a square lattice of lead as two-dimensional networks by standard electron beam lithography. Figure 3 (a) shows a picture of the square lattice of lead. This sample has 10 $\times$ 10 cells with a lattice constant of 2 μm , line width of 0.2 μm and thickness of 0.1 μm. We observed Little-Parks oscillation of the square lattice. Little-Parks oscillation [11] is periodic variation of transition temperature (Tc) of superconductivity with magnetic field by the superconducting fluxoid quantization. Experimentally Little-Parks oscillation of Tc can be observed as a periodic variation of resistance with magnetic field at fixed temperature, which was taken near the midpoint of normal-to-superconducting transition. Figure 3 (b) shows magnetic flux dependence of the sample resistance. We observed two periods of oscillation. One is 5.2 $\pm$ 0.5 Gauss, the other 0.70 $\pm$ 0.01 Gauss. The area estimated from the long period is 3.9 μm^2 and correspond to one unit cell. The area estimated from the short period is 29.5 μm^2 and correspond to nine unit cells. After all superconducting current flows through two paths. This is effect of network that has many loops and frustration. It is frustrated system that the two-dimensional superconducting networks with loops like square lattice [2].

3. Summary

In summary, we studied two and three-dimensional superconducting networks experimentally. We fabricated three-dimensional network and ring of indium by anodizing aluminum, laser irradiation, and electrophoretic deposition. We observed superconducting transition temperature of the ring sample at 3.2K. We also fabricated square lattice by standard electron beam lithography and observed two periods of oscillation of superconducting transition temperature with magnetic field. One is 5.2 $\pm$ 0.5 Gauss, the other 0.70 $\pm$ 0.01 Gauss. The area estimated from the short period is 29.5

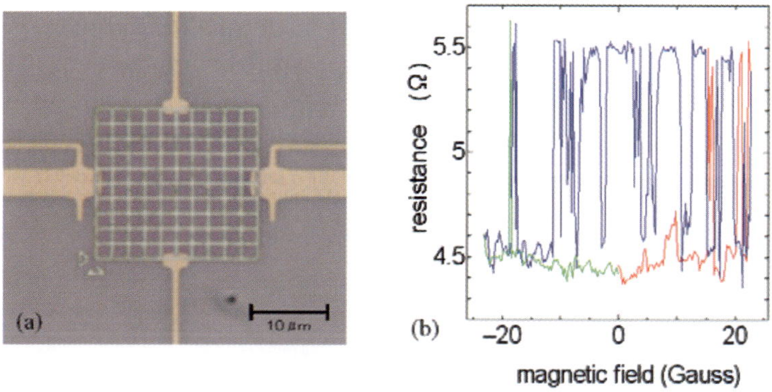

Figure 3. (a) Two-dimensional superconducting network. This sample has 10 × 10 cells with a lattice constant of 2 μm , line width of 0.2 μm and thickness of 0.1 μm. (b) magnetic flux dependence of the sample resistance. We observed two periods of oscillation. One is 5.2 ± 0.5 Gauss, the other 0.70 ± 0.01 Gauss.

μm^2 and correspond to nine unit cells.

Two of the authors (T. K. and K. I.) acknowledge for a partial financial support to Graduate School of Engineering, Hokkaido University.

This work is partly supported by Grant-in-Aid for the 21st Century COE program " Topological Science and Technology ".

References

1. J.Simonin, D.Rodrigues, A. Lopez, *Phys. Rev. Lett.* **49**, 13 (1982)
2. B. Pannetier, J. Chaussy, R. Rammal, *Phys. Rev. Lett.* **53**, 19 (1984)
3. J.M.Gordon, A.M.Goldman, J.Maps, *Phys. Rev. Lett.* **56**, 21 (1986)
4. R.Meyer, J.L.Gavilano, B.Janneret, *Phys. Rev. Lett.* **67**, 21 (1991)
5. H.D.Hallen, R.Seshadri, A.M.chang, *Phys. Rev. Lett.* **71**, 18 (1993)
6. X.S.Ling, H.J.Lezec, M.J.Higgins, *Phys. Rev. Lett.* **76**, 16 (1996)
7. C.C.Abilio, P.Butaud, Th.Fournier, *Phys. Rev. Lett.* **83**, 24 (1999)
8. O.Sato, M. Kato, *Phys. Rev. B.* **68**, 094509 (2003)
9. O.Sato, S.Takamori, M. Kato, *Phys. Rev. B.* **69**, 092505 (2004)
10. T. Kikuchi, M. Sakairi, H. Takahashi, *J.Electrochem. Soc.* **150**, 9 (2003)
11. W. A. Little and R. D. Parks, *Phys. Rev. Lett.* **9**, 9(1968)
12. X.S.Ling, H.J.Lezec, M.J.Higgins, *Phys. Rev. Lett.* **76**, 16 (1996)

AUTHOR INDEX